Springer Series in Wood Science

Editor: T. E. Timell

Springer
Berlin
Heidelberg
New York
Barcelona
Hong Kong
London
Milan
Paris
Tokyo

Springer Series in Wood Science

Editor: T. E. Timell

M. T. Tyree · M. H. Zimmermann †

Xylem Structure and the Ascent of Sap

Second Edition

With 130 Figures

 Springer

Professor Melvin T. Tyree
United States Department of Agriculture
Forest Service
Aiken Forestry Sciences Laboratory
705 Spear Street
South Burlington, VT 05403, USA

Series Editor:
T. E. Timell
State University of New York
College of Environmental Science
and Forestry
Syracuse, NY 13210, USA

Agr
QK
871
.T97
2002

First edition published in 1983 'Xylem Structure and the Ascent of Sap' by M. H. Zimmermann

The author Melvin T. Tyree is an employee of the U.S. government. His constribution remains public domain.

ISSN 1431-8563
ISBN 3-540-43354-6 Springer-Verlag Berlin Heidelberg New York

Library of Congress Cataloging-in-Publication Data
Tyree, Melvin T., Xylem structure and the ascent of sap / M. Tyree, M. H. Zimmermann. p. cm. – (Springer series in wood science.)
Includes bibliographicqal references (p.).
Rev. ed. of: Xylem structure and the ascent of sap / by M. H. Zimmermann. 1983.
    ISBN 3540433546 (alk. paper)
1. Xylem. 2. Sap. 3. Plants, Motion of fluids in I. Zimmermann, Martin Huldrych, 1926- II. Zimmermann, Martin Huldrych, 1926- Xylem structure and the ascent of sap. III. Title. IV. Series.

Springer-Verlag Berlin Heidelberg New York
 a member of BertelsmannSpringer Science+Business Media GmbH
http://www.springer.de
Springer-Verlag Berlin Heidelberg 2002
Printed in Germany

Typesetting: AM-productions, Wiesloch, Germany
Cover design: Design & Production, Heidelberg
Production: PRO EDIT GmbH, 69126 Heidelberg, Germany

Printed on acid-free paper   SPIN: 10860355   31/3130 - 5 4 3 2 1 0

Introduction to Second Edition

Martin Zimmermann wrote the first edition of this book shortly before his death in March 1984. His writing inspired an entire generation of new research and it is with great pleasure that I have accepted the invitation of the publisher to update and expand the original text.

In the first edition of this book, Martin Zimmermann used the old dimensions of atmosphere and bar throughout the text. Even then, most of his colleagues were using the SI unit Pascal, which is the force exerted by 1 kg mass, at an acceleration of 1 m/s^2 (=1 Newton) per m^2 surface. Martin gave a final appeal for retaining the old units, but they are used now only by the 'old-timers', such as I, so in this edition I have adopted SI units.

I have managed to retain a remarkably large proportion of Martin's original writing. Martin slipped into first person narrative occasionally when he wanted to present a personal opinion. In some instances, I thought these to be particularly insightful ideas, some ahead of his time. I have preserved these insights by rewording the sentences to the third person with words such as 'Martin thought', thus leaving the occasional first person wording to express my own opinion in Chapters 3 through 6. In Chapters 7 through 9, Martin was drawn into speculation on a grand scale. So the reader should know that the use of 'I' refers to Martin and that I agree with him in the vast majority of cases.

In a very few instances, Martin presented facts or ideas that I think were mistaken. These, I have taken the liberty of deleting, since I could not ask him what his current thoughts are. The instances I can remember are: (1) On p. 43 of the original edition where Martin suggested that cavitations detected acoustically by Milburn and Johnson (1966) occurred at water potentials of –5 to –10 MPa, which seems unlikely. (2) On p. 64 where it is suggested that dyes or tracers can reveal information on peak velocities in the center of vessels, because it had been shown some decades before (Taylor 1953) that tracers move at the average velocity of laminar flow because of lateral diffusion in small diameter tubes. (3) On pp. 96–97 where it is suggested that tracheids are filled with air that is largely 'water-vapor space' and '...[its] quantitative composition may differ very considerably from atmospheric air'. The wording of these pages suggests that Martin thought tracheids could be permanently filled with water vapor mixed with very little air. Martin did not specify what he thought the composition might be, so I have added words to explain what I think he would accept today. (4) On p. 114, where Martin discusses the difficulty of showing when pathogens have embolized vessels and he writes: 'If we now cut the stem and put the plant into a dye solution, the vapor-blocked vessels will refill as soon as the cut end of the shoot is brought into contact with the dye solution, which is at atmospheric pressure. ... If we remove a piece of stem and make conductance measurements, the same thing happens: in contact with water at atmospheric pressure the xylem refills and conductance is back to normal'. It almost appears from this and his statements on pp. 96–97 that Martin did not think embolized vessels fill with air but instead remain filled only with water vapor and hence were permanently at subatmospheric pressure. However, reading his Chapter 3 you get quite a different impression. Martin was suffering from a fatal brain tumor while writing this book and one wonders if his memory was playing tricks on him when writing the final chapters. I have taken the liberty of rewriting passages to reflect what he might have believed if he had lived to see the literature from 1983 to 2001.

I have never published a paper contributing to our understanding of woody plant anatomy, but I have frequently referred to Martin Zimmermann's writing (and that of others) to help me understand my own research results. Indeed, little of the work I have published in my career would have been possible without a good knowledge of plant structure. It is therefore a disappointment for me to observe, throughout my career, an unending decline in the number of plant anatomists represented in the faculties of universities worldwide. But this is the way 'science' progresses; there are fads. The current 'fad' is molecular biology, which is quickly consuming most of the plant physiologists, too! Consequently, there are few new anatomy papers for me to cite. However, citing old anatomy papers should not be a concern to readers because anatomy, once correctly described, is as true now as it was when published. Martin was very good at seeing the functional significance of structures, hence the strength of the first edition. But I am confident that structure–function anatomy (and whole plant physiology) will have a place in modern biology. Once the molecular biologists get beyond the initial excitement of gene sequencing and on to understanding what the genes actually do, then the elucidation of genes that code for structures and functional physiology will require the assistance of scientists with a knowledge of anatomy and physiology. Anatomy of plants is hardly ever taught today, but when its value is again recognized, perhaps some future molecular biologists will find this little book of some assistance. In the meantime, this book will be of use to physiological ecologists and organismal and evolutionary biologists.

It is particularly disappointing to see how little new work has been published in the plant sciences using the cinematographic technique pioneered by Zimmermann and Tomlinson. However, the technique has been used to great advantage in the medical sciences to elucidate the structure of the brain. A Harvard brain anatomist worked in Martin's laboratory learning the technique and then introduced it into medical science. Shortly before his death, Martin recounted to me the interview he had with the brain specialist he consulted regarding his tumor. His specialist said that tremendous advances had been made in our understanding of brain structure and function because of the cinematographic technique used by his colleagues at Harvard Medical School. Martin told his surgeon, 'Did you know that I pioneered the technique'? His surgeon did not believe him, saying it was the invention of his medical colleagues. Martin decided not to antagonize his medical advisor by insisting he was wrong, 'After all I have to depend on him for my treatment'. However, Martin told me that he was very happy to learn that his ideas had been of such great value to the medical sciences.

The subject matter of this book is drawn from many different fields. I have taken the rather unconventional approach of retaining as much of the text of my late coauthor as possible. I have made the book a little more encyclopedic than the first edition and added a few new topics. But I have retained Martin's original words whenever possible to retain the elements of excitement and exploration conveyed in the first edition. I have attempted to retain the exciting elements of Martin's writing style in the new sections that I have written. I regret that Martin was not around to help me correct errors in sections that I exclusively authored. It is quite obvious that I cannot be expert in all of these. Readers may therefore find errors, inaccuracies, or gaps in areas with which they are particularly familiar. I would greatly appreciate if they would communicate these to me, or let me know any comments they might have.

Burlington, Vermont, USA   MELVIN T. TYREE

Introduction to First Edition

Wood is a marvelous tissue; it never ceases to fascinate me, be it as a construction material for buildings, ships, fine musical instruments, or as the delicate structure one sees in the microscope. Wood is described in standard plant anatomy texts (e.g., Esau 1965; Fahn 1974), and in books that are entirely devoted to it (Harlow 1970; Meylan and Butterfield 1972, 1978a; Bosshard 1974; Core et al. 1979; Panshin and de Zeeuw 1980; Baas 1982 a, and others). Microscopic anatomy is a useful tool for wood identification, it is also of interest to biologists who are concerned with the evolution of plants, because the xylem is easily fossilized.

The study of wood function is certainly not new. Technologists have been investigating the mechanical properties of wood for years, and specific microscopic features have been recognized as serving certain aspects of water conduction. Nevertheless, the xylem as a system of pipelines, serving the plant as a whole, has received relatively little attention since Huber (1956) published his chapter on vascular water transport in the *Encyclopedia of Plant Physiology*. Plant anatomy and physiology have drifted apart. Plant physiology is largely concerned with biochemistry today and pays relatively little attention to the plant as a whole. Reviews of sap ascent have become so mathematical that they are not read by many plant anatomists (Pickard 1981; Hatheway and Winter 1981). Plant anatomy, on the other hand, has become largely preoccupied with evolutionary changes. But we cannot merely look at a structure and speculate what it might be good for. On the other hand, mathematical treatment should have a sound basis in structural understanding. We must study the function of structures experimentally. This is the only valid approach and it is certainly not a new one. When de Vries (1886) investigated the Casparian strip, he did not just speculate about its function, he pressurized the root and tested the effectiveness of the pressure seal! We can learn a great deal from the early botanists in this respect. I was astounded how often my literature search took me back to the last century, often to ideas and approaches, which have been forgotten. The reader will undoubtedly become very much aware of this fact.

The concept of evolutionary adaptation, i.e., the concept that everything about the plant has a specific purpose, can become misleading in those cases where a feature serves more than one purpose but where the investigator is preoccupied with a single one. The xylem provides a good example. In spite of this danger, the purpose of this book has been confined to xylem structure from the point of view of water conduction throughout the plant. Other functions, such as mechanical support, the periodic storage and mobilization of reserve materials in axial and radial xylem parenchyma (Sauter 1966, 1967), nutrient transport in the rays (Höll 1975) and in the axial xylem (Sauter 1976, 1980) are not covered.

I have used the old dimensions of atmosphere and bar throughout the text. The two are, for our purposes at least, practically identical (they differ by only about 1.3%). Most of my colleagues are now using the SI unit Pascal, which is the force exerted by 1 kg mass, at an acceleration of 1 m/s^2 (=1 Newton) per m^2 surface. I find this unit utterly unpractical for our purposes. First, the atmosphere, 1 kg weight cm^{-2}, is so very easy to visualize. Second, it is so convenient to have ambient pressure equal one. The SI pressures are very abstract units. They may be useful for engineers, but certainly not for biologists. In the range we use them, we even have to switch back and forth between MPa and kPa, because of their awkward size. Thus I see nothing but disadvantages in using SI units in dealing with pressures in plants. I do not mind being considered either old-fashioned or reactionary for the advantage of being practical; units are supposed to be our servants, we are not their slaves! As a concession, I am giving the conversion at strategic locations of the text in parentheses. Pressures are always given as absolute values, i.e., ambient pressure is +1 atm (or bar), and vacuum is zero, except where specifically noted.

Writing this little book has been a very rewarding experience. As one writes one has to give the matter a great deal of thought. One often reaches suddenly a deeper level of understanding. It happened to me at least three times during writing that I suddenly thought, „ah, this is why!" First, I discovered the phenomenon of what I have subsequently called „air seeding," a mechanism by which a minute amount of air can enter a xylem duct, thus inducing cavitation (Fig. 3.6). The water-containing compartment embolizes because it has been air-seeded, but it contains only a minute amount of air. This can easily redissolve later, should the xylem pressure rise to atmospheric again. I am sure the German botanist Otto Renner, who worked on such problems in the first third of this century, was fully aware of this situation. But he never spelled it out, and I always assumed naively that air enters the cell until atmospheric pressure was reached. The second discovery was the presence of water storage in the xylem by capillarity (Chap. 3.4). I do not know if this was clear to anyone else before, but it is certainly unavoidable and must be very important. I had always known, and given lip service to the presence of intercellular spaces that are the ducts for air movement in the xylem. But it was not until we made our paint infusion experiments that I stumbled over them, so to speak, and realized that they were very conspicuous. Capillarity is an unavoidable phenomenon, and it became clear to me that one cannot have pressure changes in the xylem without changes in capillary water storage (Chap. 3.4). This led to my third major discovery. I knew that below-atmospheric pressure provides the xylem with a ready sealing mechanism (Chap. 3.5), but I had not been fully aware of the fact that below-atmospheric pressures provide the plant with water-storage space and air ducts in the xylem. What, then, do aquatic angiosperms do in which xylem pressures are always positive in submerged parts? Of course they must separate the two compartments! They have an air-duct system, which must be sealed off from the special water-conducting „positive-pressure xylem" (Chap. 5.4). What a rich field of investigation lay ahead!

Successful competition depends on evolutionary adaptation. A few years ago Carlquist (1975) pointed out some of these problems with a book entitled *Ecological Strategies of Xylem Evolution*. This was a step in the right direction, but it also showed how very little we really know about xylem function in the whole plant.

We all enjoy speculating; the reader will certainly find plenty of speculations in this volume. But what we really need is experimental evidence, and it seems to me that we have only started to scratch the surface in this respect. This book cannot be, at this stage of our ignorance, a book of encyclopedic nature. It is an „idea" book that tries to build a bridge between the study area of structure and that of function. Functional xylem anatomy may well develop into an exciting new field of endeavor. If this is accomplished, the book will have served its purpose well.

The subject matter of this book is drawn from many different fields. It is quite obvious that I cannot be expert in all of these. Readers may therefore find errors, inaccuracies, or gaps in areas with which they are particularly familiar. I would greatly appreciate if they would communicate these to me, or let me know any comments they might have.

MARTIN H. ZIMMERMANN

Contents

1 Conducting Units: Tracheids and Vessels

1.1 Evolutionary Specialization

The development of upright land plants depended on the development of a water-conducting system. Many of the earliest land plants, e.g., species of *Rhynia* and *Cooksonia*, had little woody stem, depending mostly on turgor pressure of soft tissues for support (Niklas 1993a,b). As plants evolved to taller sizes, water conduction and mechanical support were more closely linked; in fact, this still is the case in many present-day plants that have no vessels, like the conifers. Both water conduction and rigidity depend largely upon cell-wall lignification, and it is thought that it was the evolution of the biochemical synthesis of lignin that made upright land plants possible (Barghoorn 1964).

The first more-or-less continuous record of upright land plants dates back to the upper Silurian era, about 400 million years ago (Andrews 1961; Banks 1964), although there are sporadic earlier records. Tracheids appear to have been the only highly specialized water-conducting elements in existence for some 300 million years. At the end of this time, when the flowering plants arose, xylem became more specialized by a separation of water conduction from mechanical support. This can perhaps best be shown by the imaginative illustration of Bailey and Tupper (1918; Fig. 1.1). This figure suggests how, during evolution, cells became specialized as support cells on the one hand (the fibers) and water-conducting cells on the other (the vessel elements). In 1953, Bailey published a paper in which he discussed various aspects of the evolution of tracheary tissue.

Figure 1.1 also illustrates the significance of cell length. Tracheids in vessel-less woods are generally not only much longer than vessel elements in vessel-containing woods, but also longer than fibers. The fibers serve a primarily mechanical function; this tells us that the great length attained by some tracheids before the evolution of vessels served a hydraulic rather than a mechanical purpose. In the flowering plants, where long-distance water conduction is mostly via vessels, the bulk of the water flows through cell series, namely, the vessels, rather than through individual cells.

Sediments of the lower Cretaceous (ca. 125 million years ago) contain fossil wood that looks quite like modern dicotyledonous wood (Andrews 1961, p. 181; Kramer 1974). How long such vessel-containing wood had been in existence prior to that time is by no means certain.

The construction of a vessel is shown in Fig. 7.5 (left), which illustrates three successive vessel elements of which the middle one is entire and the two outer ones are cut open. These vessel elements (of red maple) have simple perforation plates, i.e., their end walls are completely dissolved. Figure 7.5 (lower right) shows

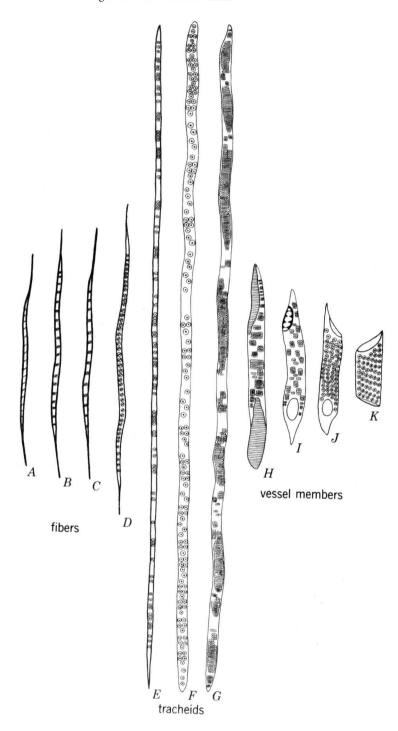

fibers

tracheids

vessel members

an example of a scalariform perforation plate in birch wood. End walls are not completely hydrolyzed during the final stages of development in this species. The stages in vessel development, e.g., the degradation of the perforation plate (end wall pairs), have been reviewed by Butterfield and Meylan (1982). The cytochemical studies of Benayoun et al. (1981) are also of interest.

Vessels show a great variety of structural features. Some are wide, others narrow, their perforation plates are of many forms, some occur in clusters, others more or less solitary, etc. This is not the place to describe this diversity in detail; there are books that illustrate it beautifully (e.g., Meylan and Butterfield 1972, 1978a). However, during the course of discussion of hydraulic properties we shall have to explore the possible functional significance of some of these features in more detail.

1.2 Vessel-Length Distributions

The overall dimensions of tracheids, their length and width, can be grasped relatively easily by looking at macerated xylem. Some tracheids are quite long, but those of most of our present-day conifers can at least still be shown on a single page without distortion of the proportion, by diagrammatically "folding" them (e.g., Fig. 11.6 in Esau 1965). However, others are rather too long for convenient illustration. In *Agathis cunninghamii*, lengths up to 10.9 mm have been reported; a 6-mm length is even exceeded occasionally in several pine species (Bailey and Tupper 1918). In a (carboniferous) *Sphenophyllum* species, lengths range up to 3 cm (Cichan and Taylor 1982)! It would be rather difficult to illustrate these to scale.

Vessels are almost impossible to illustrate on a printed page; in fact, their extent and shape was poorly known until recently. Inside vessel diameter is hydraulically an extremely important parameter; this will be discussed in the next section. Vessel diameter has been measured many times in the past. It is somewhat variable and depends, for example, upon age of tree and location within the tree (leaves, branches, trunk, etc.), a feature that will be discussed in Chapter 7.2.

It is rarely possible to see vessels throughout their entire length because they consist of small cells which need a microscope for observation, and at the same time they are so long that a microscope is far too myopic to grasp their extent. They can never be seen on single sections or even on short series of transverse or longitudinal sections. To observe them in their entirety, we need to apply the technique of cinematographic analysis: we look at long series of wood transverse sec-

Fig. 1.1. Diagrammatic illustration of average size and structure of tracheary elements in the mature wood of some conifers and dicotyledons. **E–G** Long tracheids from primitive woods (**G** showing *Trochodendron* or *Dioon*, axially foreshortened). **D–A** Evolution of fibers showing decrease in length and reduction in size of pit borders. **H–K** Evolution of vessel elements, decrease in length, reduction in inclination of end walls, change from scalariform to simple perforation plates, and from scalariform to alternate vessel-to-vessel pits. (Bailey and Tupper 1918)

tions, recorded on film, with a special movie projector, a so-called analyzer. By running the film forward or backward, we can move up or down the stem in axial direction at any speed, go frame by frame, or stand still at any one point. The intractable axial dimension, otherwise inaccessible to the microscope, is thus translated into time, and we can move from one end of a vessel to the other end by observing a "moving" transverse stem section on the projection screen. Vessel ends thus become visible, the course of vessels can be plotted and the vessel network can be reconstructed. Cinematographic analysis will be discussed briefly in the next chapter in connection with vessel network reconstructions. Let us now look at vessel length in another way.

Vessels consist of series of individual cells, the vessel elements, whose end walls are partly or completely dissolved during late stages of cell maturation, thus forming together long capillaries. The ends usually taper out; it is very important for the understanding of water conduction to realize that the water does not leave a vessel in axial direction through the very end, but *laterally* along a relatively long stretch where the two vessels, the ending and the continuing one, run side by side. The principle is shown in Fig. 1.2. A vessel consisting of nine elements is shown on the left in its entirety. Parts of three others are shown. The illustration is rather diagrammatic, because vessels consist in reality of a far greater number of elements and lengths of overlap can be very much longer. The overlap area between two vessels is of a peculiar structure, shown on the right of Fig. 1.2; this will be discussed later (Sect. 1.5).

The older literature contains only information on maximum vessel length (e.g., Greenidge 1952) and "average" vessel length (Scholander 1958). The concept of vessel-length distribution has been introduced by Skene and Balodis (1968). The only accurate way to obtain information on vessel-length distribution is to use the cinematographic technique, which is far too labor intensive to be practical for routine studies. One way to obtain a qualitative impression of vessel-length distribution is to perfuse cut stems with a dilute latex paint. Suitable paint mixes have pigment particles that are small enough to pass through all vessel lumina and large enough to be stopped by the pit membranes. Ideal particle sizes are between 0.2 and 2 µm in diameter. At the cut surface of the stem, all vessels are cut open so paint particles can freely enter when perfused under pressure. Particles presumably travel the full length of the vessels and accumulate along the walls of the vessel wherever water flows through a pit to an adjacent vessel. The stem is then cut into segments of equal length (e.g., 2-cm length) and the number of paint-filled vessels observed under a microscope in each 2-cm segment. A plot of the number of paint-filled vessels versus distance from the point of infusion gives a length distribution for the length of vessels from the infusion surface to the end of the vessel. Without further analysis, these data cannot yield information on vessel-length distribution because it cannot be determined whether the infusion surface was near the distal, basal or median portion of any particular vessel.

Quantitative information on the probable vessel-length distribution can be derived if we assume that vessel ends occur randomly over the length of a stem segment. Vessel ends are randomly distributed if successive stem segments of the same length, dL, are likely to contain the same number of vessel ends. Skene and Balodis (1968) provide a rigorous statistical analysis to deduce vessel-length distributions from a paint-infusion experiment by use of a double-difference (DD)

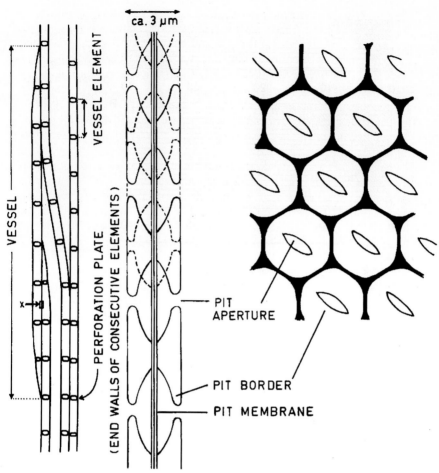

Fig. 1.2. *Left* Diagrammatic representation of a vessel network. Vessels are of finite length, their ends are overlapping. Water moves from one vessel to the next laterally through bordered pits. *Center* Diagrammatic section through a bordered pit field as we would see it at higher magnification in the *boxed area at X* of the left-hand drawing. The secondary walls arch over the primary wall pair, the pit membranes, providing mechanical strength with minimal obstruction of the membrane area. *Right* Vessel-to-vessel pit membrane in surface view. *Black hexagonal pattern* is the area where the secondary wall is attached to the primary wall, thus leaving much of the membrane accessible to water flow. (Zimmermann and McDonough 1978)

algorithm. The DD algorithm is used to convert the counts of paint-filled vessels into a frequency distribution of the number or percentage of vessels in size classes of length L. To illustrate the DD algorithm, let us consider a stem containing vessels of equal length but with randomly located vessel ends. Figure 1.3 A is a two-dimensional representation of a stem containing 10 vessels in a cross section. The stem is 40 cm long and contains 50 open or closed vessels. (An open vessel is a ves-

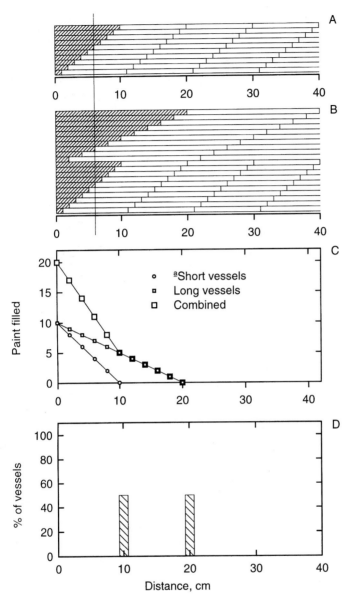

Fig. 1.3. A, B Theoretical stem segments containing randomly distributed vessels (as defined in the text). *Vertical bars* indicate vessel ends. *Hatched area* indicates extent of paint-filled vessels infused from distance 0. **C** Count of paint-filled vessels versus distance for short vessels (*open circles*) and long vessels (*closed circles*). **D** Output of DD algorithm for theoretical stem **B** above

sel cut open at one end and a closed vessel is an intact vessel.) The vertical lines represent the vessel ends; the vessels have been arranged in Fig. 1.3A so that the longest open vessel on the left is at the top, the shortest is at the bottom, and all others are ranked in between. This is done for clarity, but it does not limit the generality of the argument that follows since the assumption of random distribution of vessel ends, along the axis of a stem, places no restriction on the location of the vessels in a cross section.

The hatched area in the ten open vessels on the left represents the distance that paint will travel when perfused from the left (basal) surface. The stem is cut into 2-cm segments and the paint-filled vessels are counted on the surfaces proximal to the infusion surface. There would be 10, 8, 6, 4, 2, and 0 paint-filled vessels at distances of 0 2, 4, 6, 8, and 10 cm from the left, as shown in Fig. 1.3C. In the DD algorithm, the first difference (Table 1.1) gives the number of vessel ends between the distances where the raw counts were made. In our example, this difference is 2 for every count <10 cm from the left side. The second difference represents the rate of decrease in vessels of a size class. In our example, all vessels are 10 cm long and there is only one size class, so there is only one second difference >0. The second difference multiplied by the number of increments (steps to zero) gives the number of vessels in the size class. The number can be expressed as a percentage of the number of paint-filled vessels in the size class. The frequency distribution is computed correctly by the DD algorithm in Table 1.1.

The same result would be obtained if the stem segment in Fig. 1.3A had been cut to the right of the surface shown. For example, if the stem were cut at 6 cm from where the thin vertical line is drawn, then the top four vessels would be paint-filled to the right of the line. The bottom six vessels would be cut open and filled with paint to the right of the line up to their vessel ends. The distribution of the paint-filled vessels in the cross section would change, but the count of the paint-filled vessels at 2-cm intervals to the right of the line would be unchanged.

Table 1.1. Example of calculation of vessel-length distribution from data in Fig. 1.3A using the DD algorithm

Distance (cm)	Count of paint filled	First difference	Second difference	Steps to zero	Number of vessels	Vessels (%)
0– 2	10	2	0	1	0	0
2– 4	8	2	0	2	0	0
4– 6	6	2	0	3	0	0
6– 8	4	2	0	4	0	0
8–10	2	2	2	5	10	100
10–12	0	0	0	6	0	0

Note: The first difference in any row of column 3 is the difference between the count in column 2 of the same row minus the count in the row below. The second difference in any row of column 4 is the difference between the number in column 3 of the same row and the number in the row below. The number of vessels in any row of column 6 is the product of the numbers in the same row of columns 4 and 5

A more complicated example is shown in Fig. 1.3B, where a stem contains 20 vessels in a cross section with 10 vessels 10-cm long and 10 of length 20 cm. If this stem is cut into 2-cm-long segments, the paint counts for the long and short vessels would be given by the solid and open circles, respectively, in Fig. 1.3C, and the total count would be shown by the squares. The reader can verify that the DD algorithm yields the expected result, i.e., 50% of the vessels are 10 cm long and 50% are 20 cm long (Fig. 1.3D).

This theoretical consideration has been widely used to estimate vessel-length distribution in many species using the paint-infusion method. The stem must be absolutely fresh, i.e., vessels must still be water-filled when the experiment is started. This is an almost impossible requirement if the plant is under stress during a summer day, because a stressed plant will suck air into the stem as soon as it is cut. However, there are various ways of overcoming the problem. One can cut a piece of stem that is much longer than needed. Air will be drawn into the vessels at both ends. Successive disks are then cut off both ends and the tension is thereby released. Another method is to vacuum-infiltrate the cut end with distilled water; alternatively, stems can be flushed with degassed water under pressure to dissolve embolisms prior to paint perfusion. A suspension of very small paint particles such as very dilute latex paint can then be infused for several days. Particles must be small enough to pass through perforation plates (Fig. 1.4A) but large enough to be stopped by the minute pores of the pit membranes (Fig. 1.4B).

Unfortunately, latex paint particles tend to clump and hence block vessels before they reach vessel ends. Better results are obtained using paint pigment which consists of fine suspensions of mineral crystals because crystals tend to clump less. Alternative methods have also been used that take advantage of the fact that air can pass through a wet pit membrane only if a very large pressure difference exists between the air and water interface at the pit (see Chap. 4). One can cut a stem segment longer than the longest vessel and push air through the segment at 100 kPa of air pressure; this pressure is enough to displace water from the open vessels, but not enough to push the air through pits. If no air comes out, the stem segment is long enough. Then the distal end of the stem is cut back in equal increments of length. As vessels are cut open at the distal end air will flow through. The rate of air flow will be approximately proportional to the number of vessels cut open. The air flow rate through large-diameter vessels will be more than through narrow vessels. However, if diameter distributions do not change along the length of the segment and if there is no correlation between vessel diameter and vessel length, the approximation between open vessel count and flow rate will be within a satisfactory error tolerance.

A more serious error will arise if the vessel ends are not randomly distributed along the axis of a stem. For example, it is often presumed that vessels end more frequently near the nodes of stems than in the internodes. The nonrandom distribution of vessel ends in stems deserves more study by independent methods such as the cinematographic method. There is reason to believe there are stem regions where vessel ends are clustered together because the DD algorithm frequently results in negative second differences. A negative second difference makes no sense since this implies a negative number of vessel ends in the corresponding stem segment where the negative 2nd difference occurs. Counting errors could account for negative second differences, but negative differences are very com-

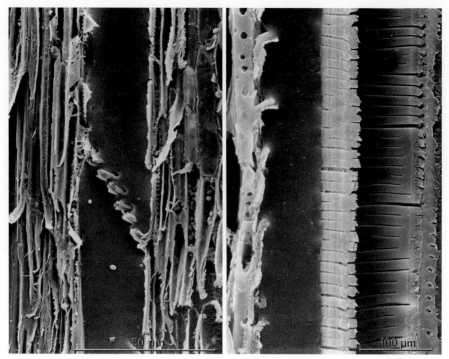

Fig. 1.4. Longitudinal sections through metaxylem vessels of the stem of the small palm *Rhapis excelsa*, infused with a latex paint suspension. *Left* Paint particles are small enough to penetrate scalariform perforation plates. *Right* The particles are too large to penetrate the pores of the vessel-to-vessel pit membranes from the paint-containing vessel on the left of the one on the right

mon in correctly counted data sets and are a theoretical consequence on nonrandom vessel ends, as shown below.

Figure 1.5A is an example of a stem with a mixture of random and nonrandom vessel ends. In Fig. 1.5A, at any given cross section, there are 10 vessels 10 cm long with randomly distributed vessel ends, and 10 vessels up to 36 cm long with nonrandom vessel ends. This is a special case of nonrandom ends where all vessel ends are clustered between distances of 26 and 36 cm from the base but randomly distributed within this range. The hatched area shows the distance that paint would perfuse up to the end of each vessel. Again, the stem is cut into 2-cm segments and the count of paint-filled vessels in the long- and short-vessel classes is shown by the solid and open circles in Fig. 1.5B. The total count is shown by the squares. When the DD algorithm is applied to this data set, the uncorrected frequency distribution in Fig. 1.5C is obtained. The short, randomly distributed vessels are correctly computed to have a frequency of 50%, but this is followed by a negative value of –130% and then a positive value of 180%. The negative number occurs at the break in the combined count at 26 cm and is a result of the second difference calculation in the DD algorithm. Zimmermann and Jeje (1981) were

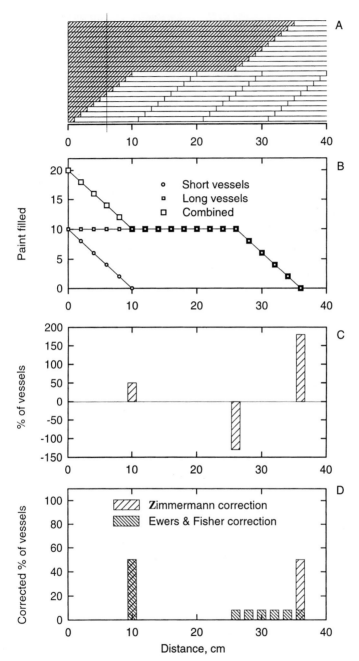

Fig. 1.5. A Theoretical stem segment as in Fig. 1.3A, but this example illustrates a stem with ten vessels with randomly distributed vessel ends and ten vessels with nonrandomly distributed vessel end. **B** Similar to Fig. 1.3C. **C** Output of the DD algorithm. **D** Output of the DD algorithm after using the Zimmermann correction algorithm or the Ewers and Fisher correction algorithm

aware of this theoretical problem and suggested a correction algorithm, without proof of its generality, to deal with the negative numbers.

Zimmermann's correction algorithm consists of removing the negative number (–130% at 26 cm) from its size class and adding it to the next larger size class (28 cm). If this still results in a negative number (which it will because this size class has a frequency of 0), then the process is repeated until a positive number results. So –130% is added to the size classes of 30, 32, 34, and 36 cm until, at 36 cm, the sum of –130+180% yields the positive result of 50% shown in Fig. 1.5D. Zimmermann's correction algorithm produced the anticipated result, i.e., 50% of the vessels are 36 cm long.

Zimmermann and Jeje (1981) assumed that the correction algorithm would work in all cases, but they provided no other examples to show that it always provides the correct result. Examination of Fig. 1.5A shows that there is a problem with nonrandom vessel ends. If the stem had been cut 6 cm to the right of the infusion point (where the fine vertical line is drawn) and then infused, the result would have been different. For the randomly distributed short vessels, the number of vessels filled with paint for long and short distances would be unchanged. But the distance the paint would travel in the long, nonrandom vessels would all be reduced by 6 cm. This would move the 50% bar for the long vessels from 36 cm over to 30 cm in Fig. 1.5D. So, in most cases, the length of nonrandom vessels would be underestimated because, in general, we would not know how much of the nonrandom vessels are cut off basal of the infusion surface.

Ewers and Fisher (1989) suggested an alternative algorithm but without theoretical justification. They suggested that negative numbers should be removed by "grouping [length] categories to arrive at positive values ... To do this, negative numbers were averaged with adjacent positive numbers(s) ... when a choice had to be made between averaging with a length class above or below the length class with the negative number, the adjacent length class with the greater positive vessel number was used." If this algorithm is applied to the example in Fig. 1.5C, the alternative distribution in Fig. 1.5D would result. The –130% at 26 cm is averaged with the zeros in the 28-, 30-, 32-, and 34-cm size classes and with the +180% in the 36-cm size class and divided by the number of size classes (=6) to yield 8.33% in each size class from 26 to 36 cm. A rationale for using this algorithm is never stated, but the rationale appears to be to obtain frequency distributions without zero values. When the Zimmermann correction is used, some zero values always result in the analysis on real sites, but with the Ewers and Fisher correction zero values rarely occur.

It could be argued that the Ewers and Fisher correction algorithm produces a better representation of the distribution of nonrandom vessels. We do not know how far the infusion surface is from the basal end of the nonrandom vessels or whether nonrandom vessels end randomly over a narrow range. Hence it might be better to interpret the frequency distribution of nonrandom vessels as the percentage of vessels that come to an end in a given segment length. The Ewers and Fisher correction algorithm more nearly gives this distribution than does the Zimmermann correction algorithm.

Tyree (1993) has examined sources of error computing vessel-length distributions. The approach has been to apply the DD algorithm and the alternative correction algorithms to known, realistic vessel-length distributions. Figure 1.6A

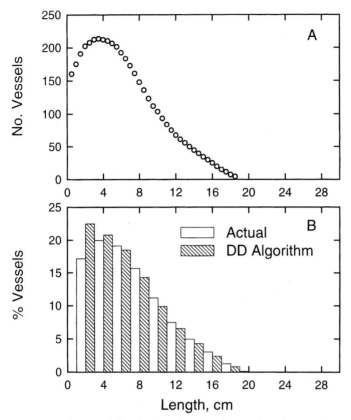

Fig. 1.6. **A** Theoretical distribution of 37 size classes of random vessels used as an input for the DD algorithm. **B** Actual frequency distribution compared with the output of the DD algorithm

shows a hypothetical vessel-length distribution. The length categories are distributed into 0.5-cm length categories and have randomly distributed vessel ends. However, if the stem is cut into 2-cm pieces for the analysis of paint-filled vessel counts, the 0.5-cm length categories will be lumped together. This tends to result in an overestimation of the number of short vessels and an underestimation of the number of long vessels, as shown in Fig. 1.6B. The situation becomes more complex if we overlay some nonrandom vessel (Fig. 1.7). If 80% of the vessels are randomly distributed and 20% are nonrandom, as in Fig. 1.5A, then negative second differences result in the DD algorithm. How this effects the length distribution predicted by the algorithm depends on (1) the correction algorithm used and (2) whether or not the nonrandom vessel ends occur beyond the random distributions >18.5 cm (Fig. 1.7A) or inside the random distributions <18.5 cm (Fig. 1.7B). If the nonrandom vessel ends are >18.5 cm beyond the infusion surface, the random vessel-length distribution is fairly accurate (Fig. 1.6A). If the nonrandom distributions are <18.5 cm, then the correction of the negative second differences

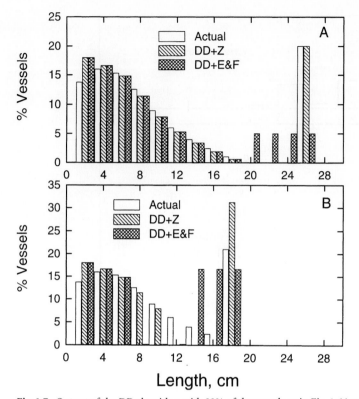

Fig. 1.7. Output of the DD algorithm with 80% of the vessels as in Fig. 1.6A and 20% nonrandom as in Fig. 1.5A. **A** Results when region of nonrandom vessel ends is to the right of the region of paint-filled random vessels. **B** Results when region of nonrandom vessel ends within the region of paint-filled random vessels. DD+Z (output of the DD algorithm with Zimmermann correction algorithm) and DD+E&F (output of DD algorithm with Ewers and Fisher correction algorithm

tends to 'rob' vessel-length classes from the random category and places them into longer nonrandom vessel-length categories.

Tyree (1993) basically concludes that if the DD algorithm is used on a data set and no negative second differences result, then it is safe to assume that vessel ends are randomly distributed and that the vessel-length distributions predicted are fairly accurate. However, if negative second differences result, it is very difficult to know how accurate the vessel-length distributions are. Negative differences in the DD algorithm tend to occur mostly in the longer vessel-length size classes where the number of vessels is small in the sample. Hence some negative numbers might result because the subsample of long vessels is too small to be truly random. Some valuable resolution to this problem awaits further studies comparing vessel-length distributions from cinematographic analysis to paint-infusion experiments on the same stems. For the remainder of this section, we will ignore the uncertainties in order to compare vessel-length distributions in a variety of contrasting species.

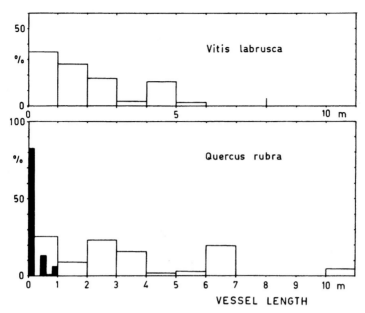

Fig. 1.8. *Top* Vessel-length distribution in the stem of the vine *Vitis labrusca. Small vertical line* at 8 m indicates longest vessel length. *Bottom* Vessel-length distribution in the trunk of red oak (*Quercus rubra*). The longest vessels were in the 10–11 m length class. *Black bars* illustrate length distribution of the narrow latewood vessels. (Redrawn from Zimmermann and Jeje 1981)

It is important that the reader clearly distinguishes between the two kinds of percentages referred to in this chapter, namely (1) the vessel count, i.e., the percentage of vessels that are open at a given distance from a cut across the stem, and (2) the vessel-length distribution, the percentage of vessels in a certain length class that we calculate from the vessel count. We shall have to deal with the first when we are concerned about injuries (Chap. 2.5) and diseases (Chap. 9.2). Assume again the hypothetical stem in Fig. 1.3B. If we make a saw cut into the tree trunk at distance zero, 25% of the vessels will be air-blocked 10 cm away from the saw cut and 75% of the vessels will still be intact (water filled). Vessel-length distributions given in the following illustrations are all percentages of vessels in a given length class. The DD method of calculation of length classes depends on random distributions of vessels within the stem in which it is measured. The vessel count line should always be concave or straight (Fig. 1.3C). If it is convex, some calculated percentages are negative because the second difference is negative and some correction algorithm is needed (Skene and Balodis 1968; Zimmermann and Jeje 1981; Tyree 1993).

Random distribution of vessels is found in trunk wood of diffuse-porous tree species where wood tissue is relatively uniform and vessels are short. This is indicated by the fact that results are repeatable. The method works best here. In ring-porous trees some vessels may be as long as the entire trunk. Random distribution is obviously not possible in this case, and the calculation yields negative numbers;

Fig. 1.9. *Top* Vessel-length distribution in the trunk of *Acer saccharum*. *Small vertical line* at 32 cm indicates longest length. *Bottom* Vessel-length distribution in the stem of *Vaccinium corymbosum*. The longest vessels were in the 1.2–1.3 m length class. (Redrawn from Zimmermann and Jeje 1981)

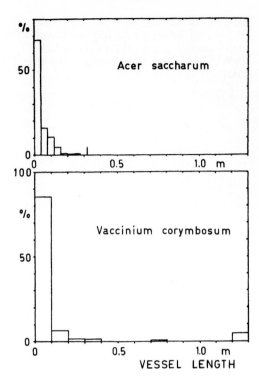

however, these may be compensated for by adjacent positive numbers (Tyree 1993). Other examples of nonrandom distributions are petioles where more, shorter vessels may be present near the stem and near the blade. Branch junctions merit special attention, because, in some cases, vessel ends may be preferentially located at the fork. This will be discussed in more detail in connection with the concept of hydraulic architecture (Chap. 5).

Let us now look at vessel-length distribution in some species. It is quite obvious that vessel length is positively correlated with vessel diameter. This has been indicated by earlier reports on longest vessel lengths (Handley 1936; Greenidge 1952). The north-temperate ring-porous trees, which have large-diameter (ca. 300 μm) earlywood vessels, as well as the grapevine, whose vessel diameters are also wide, have lengths of many meters (Fig. 1.8). The vessels of diffuse-porous trees and shrubs are much narrower in diameter (of the order of 75 μm), they are also much shorter (Fig. 1.9). It should be noted that the horizontal scales of the two illustrations (Figs. 1.8 and 1.9) are quite different, and that the narrow diameter latewood vessels of oak have a length distribution not unlike that of the vessels of diffuse-porous trees. It is also quite interesting to note that the longest vessels of shrubs [several species have been investigated by Zimmermann and Jeje (1981)] are remarkably long, often as long as the shrub is tall. However, the percentages of the longest length classes are often so low that the height of the bars in the diagram does not exceed the thickness of the base line.

Fig. 1.10. Average vessel diameters (with SD) and vessel-length distributions in early- and latewood of the outermost growth ring of a 19-cm-thick trunk of *Acer rubrum*. (Zimmermann and Potter 1982)

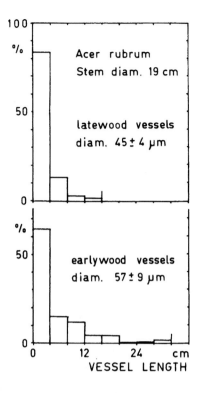

Figure 1.8 shows quite dramatically that the wide earlywood vessels of oak are long and the narrow latewood vessels are much shorter. In a much more subtle way this appears to be the case generally, if we can generalize from the half-dozen species that have been investigated in this respect. In *Acer rubrum*, a diffuse-porous species in which vessel diameters diminish only very slightly from early- to latewood, length also diminishes (Fig. 1.10).

Vessel-length distribution has also been reported for the small palm *Rhapis excelsa*. The vessels are comparable to those of *Acer* in width and length. We shall come back to this in a later section in connection with the vessel network.

The most important finding of this survey was the fact that xylem contains generally more short vessels than long ones. Figures 1.8 and 1.9 show this clearly. An extreme case is represented by the shrub *Ilex verticillata* in which the longest vessels were 130 cm long, but 99.5% of them were shorter than 10 cm. In another test at shorter intervals, 66% were found to be shorter than 2 cm (Zimmermann and Jeje 1981). The functional significance of the greater number of short vessels is one of safety (see Sect. 1.4).

1.3 The Hagen-Poiseuille Equation and Its Implications

The flow of water through vessels can be compared with flow through smooth-walled capillaries of circular cross section. We shall make this assumption, although this is not exactly correct, and discuss the complications later. The flow rate (dV/dt) through a capillary is proportional to the applied pressure gradient (dP/dl) and the hydraulic conductivity ($K_{capillary}$),

$$\frac{dV}{dt} = -K_{capillary}\frac{dP}{dl} \tag{1.1}$$

where the minus sign comes from a sign convention. Most people do not bother writing transport equations like Eq. (1.1) with a minus sign, and I have frequently been guilty of that omission too. However, failing to adhere to a strict sign convention can lead to mistakes in interpreting data, an important example of which will appear in Chapter 3. The sign convention is that positive flow is from left to right on a graph. Since water flow is from high pressure (on the left) to low pressure (on the right), dP/dl, which is the slope of P versus l, is negative. So to make flow come out positive, we need a minus sign. Most people get around this by just reporting dP/dl as an absolute quantity without a sign, but this practice can lead to errors on rare occasions.

We should dwell a moment on the meaning of conductance versus conductivity. The value of $K_{capillary}$ is a conductivity which means flow rate (say in m^3 s^{-1} or sometimes kg s^{-1}) per unit pressure gradient (Pa m^{-1}). So, during a measurement of conductivity of a tube of length L, the length if factored into the pressure drop, ΔP, so $dP/dl=\Delta P/L$. Conductance, on the other hand, does not take into account the length and sometimes the length is not even known and is defined as flow rate per unit pressure drop. In this volume, conductivity will be given an upper case K and conductance a lower case k. Resistivity and resistance are the inverse of conductivity and conductance, respectively. We will switch between resistance and conductance or resistivity and conductivity as appropriate. When we talk about two or more pipes in parallel it is more convenient to talk about conductance, because conductances in parallel are additive, i.e., the conductance of all pipes in parallel is the sum of their individual conductances. When we talk about two or more pipes in series it is most convenient to talk about resistance, because resistances in series are additive, i.e., the total resistance of several pipes in series equals the sum of their individual resistances. In a complex network of pipes that might be found in the crown of a tree where stems are both in series and in parallel, there is no strong advantage of giving values in resistance or conductance units.

What interests us most here is the nature of the term that describes conductivity in Eq. (1.1). Nonturbulent capillary flow was investigated independently by Hagen and by Poiseuille, who published in 1839 and 1840, respectively. Their experiments are described in some detail by Reiner (1960). Essentially, they discovered that a liquid is stationary on the capillary wall, and that its velocity increases toward the center of the tube. Visualize the following. At time zero we mark all water molecules that are located in a transverse-sectional plane in a capillary. We then let them flow and stop them again after time t. We would then find them neatly lined up on the surface of a paraboloid (Fig. 3.15). The ones touching the capillary walls have not moved at all, the ones in the center of the capillary have moved farthest. Hagen and Poiseuille found empirically that

$$K_{capillary} = \frac{r^4 \pi}{8\eta} \tag{1.2}$$

Equation (1.2) applies to laminar flow; when flow rates get too high the flow is turbulent, but turbulent flow is unlikely for flow rates in capillaries the size of vessels. Laminar flow causes the flow rate to be proportional to the fourth power of the capillary radius (r in m; Reiner 1960). This is the relationship that interests us most. The other terms are constants, except for η, the dynamic viscosity of the liquid (Pa s). Viscosity depends on solute content (e.g., a concentrated sugar solution is quite viscous and slows down the flow considerably). The solute concentration of xylem sap is negligible and does not measurably influence viscosity. Viscosity is also temperature-dependent, decreasing about 2.4% per °C (see Chap. 6.5).

It is important to note that flow rate is proportional to the fourth power of the radius of the capillary. This means that whenever evolution brought about a slight increase in tracheary diameter, it caused a considerable increase in conductivity and dramatically changed the amount of wood needed to conduct water at a given rate. Figure 1.11 shows three blocks of wood in transverse section with closely packed vessels. Their relative diameters are 1, 2, and 4 (e.g., 10, 20, and 40 μm). The relative transverse-sectional area of individual vessels are 1, 4, and 16, and relative flow rates under comparable conditions, 1, 16, and 256! Each 'block of wood' in Fig. 1.11B,C has the same hydraulic conductivity as the single vessel in Fig. 1.11A. In a real block of wood, the spacing between vessels will be wider to accommodate differences in cell-wall thickness and the space occupied by other types of cells in wood (ray cell, wood fiber cells, etc.). This tells us that if we want to compare conductivities of different woods, we should *not* compare their respective transverse-sectional vessel area, vessel density or any such measure. We must compare the sums of the fourth powers of their inside vessel diameters (or radii). If we look at vessels in xylem transverse sections, we must realize that small vessels next to large ones carry an insignificant amount of water. If three vessels of 10, 20, and 40 μm diameter are adjacent to each other in a vascular bundle, the smallest one would carry 0.4%, the middle one 5.9%, and the large one 93.8% of the water.

If such a slight evolutionary (and developmental) change as the widening of vessels makes them so very much more efficient, why do not all modern plants have wide vessels? The answer to this question is that efficiency comes only at a price, and in this case the price is safety. We shall look into this problem in the next section. However, before we do so, we should briefly consider the question of how vessels differ from ideal capillaries.

We have seen that vessels are of finite length and that some of them are quite short. Water must move many times from one vessel to the next laterally through the membrane of the bordered pit areas. Furthermore, water must flow through perforation plates. This surely represents a considerable flow resistance if they are scalariform. Vessel walls are often not very smooth, but may contain warts, ridges

Fig. 1.11. Hagen-Poiseuille law prediction of relationship between conduit diameter and conductance. The vessel blocks in **A**, **B**, and **C** have equal conductance, i.e., one vessel 40 μm diameter (**A**) is as conductive as 16 vessels 20 μm diameter (**B**), or as conductive as 256 vessels 10 μm diameter (**C**). Note that more wood cross section is needed to contain many small vessels having a combined conductance of a few large vessels

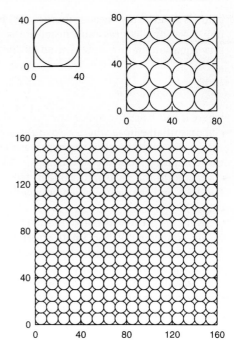

and other irregularities, especially due to the presence of pits. Real vessels are rarely circular in cross section but are more nearly described by rectangles or ellipses. The flow pattern is therefore not ideally paraboloid, but a good deal more complex (Jeje and Zimmermann 1979). How does all this modify flow rate at any given pressure gradient?

Many years ago, Ewart (1905), Berger (1931), Riedl (1937), and Münch (1943) compared the flow rates that they had experimentally measured with the flow rates that they calculated, assuming that the vessels are ideal capillaries. In some cases, average velocities and not flow rates were measured by a heat-pulse method. The results were expressed in percent efficiency (see pp. 198 and 199 in Zimmermann and Brown 1971). While vines appeared to be 100% efficient (i.e., they behaved like ideal capillaries), other dicotyledonous xylem was found to be about 40% efficient (they carried about 40% as much water as ideal capillaries of the same diameter). The result obtained with conifers was surprising. Ewart (1905) reported an efficiency of 43% for *Taxus* and Münch (1943) 26–43% for *Abies alba*. This is quite remarkable, considering that conifers contain only tracheids. The pit membranes of fern tracheids double their resistivity, i.e., the conductance doubles when the pits are digested with cellulase (Calkin et al. 1986). Tyree and Zimmermann (1971) who reported 33–67% for red maple (*Acer rubrum*) and Petty (1978, 1981) who reported 34–38% for birch (*Betula pubescens*), which has scalariform perforation plates, and 38% for sycamore (*Acer pseudoplatanus*). One of the most recent studies compares the theoretical to measured vessel conductance in a grass leaf, *Festuca arundinacea* (Martre et al. 2000). This study is noteworthy because there is a wide range of vessel diameters in the developing leaf of this grass and because the authors used a corresponding laminar flow equation for an ellipse. For an ellipse the conductance is:

$$K_{ellipse} = \frac{\pi}{64\eta} \frac{a^3 b^3}{a^2 + b^2} \tag{1.3}$$

where a and b are the major and minor diameters of the ellipse. A circle is a special case of an ellipse where $a=b$ and Eq. (1.3) reduces to Eq. (1.2) for this special case. Every vessel in a leaf segment was measured and the theoretical conductivity was computed from the sum of $K_{ellipse}$ for all vessels in each segment ($=K_t$) and compared to the measured value in the segment ($=K_h$). The values are potted in Fig. 1.12, and the slope, which equals the relative efficiency, is 0.36 (or 36% efficiency).

By far the greatest problem of such measurements is the difficulty in measuring inside diameter reliably. Small errors in measurement of vessel diameter yield large errors of calculated flow rates. For example, a 10% error in the diameter measurement increases the calculated flow rate by 46% ($1.1^4=1.4641$) or decreases it by 34% ($0.9^4=0.6561$). By the same token, we can say that xylem that has vessels that are 50% as efficient as ideal capillaries are equal in efficiency to ideal capillaries, which are only 84% as wide. In other words, it is not unreasonable, in some cases, to simply ignore the problem.

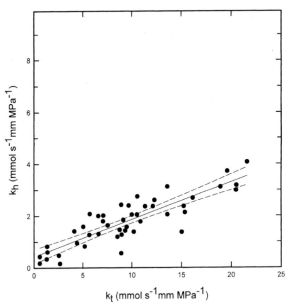

Fig. 1.12. Relationship between theoretical (K_t) axial hydraulic conductivity and measured hydraulic conductivity (K_h) of 30-mm sections of tall fescue elongating leaf blades. Each point represents a single section taken from 30 to 170 mm from the base. K_t was computed from the middle of the section where K_h was previously measured. Straight line, linear regression (y=0.36 x+0.04, r=0.86, P<0.05); *dashed lines* are 95% confidence intervals. (Martre et al. 2000)

1.4 Efficiency Versus Safety

Wide vessels are very much more efficient water conductors than narrow vessels; it is therefore not surprising that evolution in many cases was toward wider vessels. However, there seems to be an upper limit of the useful vessel diameter at approximately 0.5 mm. This limit seems to have been reached many times during evolution, because there are trees and vines with wide vessels in many diverse plant families. On the other hand, the single genus *Quercus* has representatives with wide vessels (the ring-porous species of the north temperate areas) and others with narrow vessels (the evergreen species of dry or subtropical areas).

Increased vessel diameter increases efficiency of water conduction dramatically, but at the same time it decreases safety. There are various ways of looking at this. Let us compare a narrow-vessel tree like maple with a wide-vessel tree like oak, and assume that their respective vessel diameters are 75 and 300 μm. Let us assume that both trees are of equal size, transpire the same amount of water and have similarly functioning root systems. In order to carry equal amounts of water, maple must have 256 times as many vessels as oak per transverse section of the trunk. The transverse-sectional conducting area is then 16 times greater. But the vessels of oak are also much longer than those of maple. The difference can be seen in Figs. 1.8 and 1.9. As a rough estimate, we may consider the average vessel length of the wide earlywood vessels of oak to be 3 m, those of maple 0.1 m. The narrow latewood vessels of oak can be ignored, because they are hydraulically insignificant as long as there are conducting earlywood vessels (as we shall see later, Chap. 6. 1, this may not be the case in early spring). Taking length into consideration, the trunk of maple must contain about 30×256=7680 as many functioning vessels as that of oak. Consider now an injury, perhaps inflicted by a foraging bark beetle damaging the vessel wall at one point. Air is drawn into the injured vessel, thus blocking it permanently. The damage done in oak is 7680 times more serious than in maple. One could argue, of course, that if the damage is limited to a specific height in the stem, maple is only 256 times as safe as oak. But in addition to this we have to consider the fact that the functioning vessels in ring-porous species are all located very near the cambium, i.e., in a superficial and therefore vulnerable position. We shall have to come back to this in the last chapter when we discuss vascular wilt diseases.

We can look at accidental damage in another way. If cavitation of the water columns were a random event, it would have to happen once in a given volume within a given time span. Maple has 16 times as much vessel volume as oak. A random event will therefore cavitate a maple vessel 16 times more often, but because of the 7680 times greater number of vessels in maple, the damage in oak is still 480 times more serious. These arguments provide only a first approximation to the issue of safety versus efficiency. A full consideration will have to wait a more detailed discussion of the mechanisms of vessel dysfunction.

Other risks that become more serious in wide-vessel woods will be discussed later. They include the possible negative correlation of tensile strength of water with compartment size (Chap. 4.5), ice formation in the winter (Chap. 8.2), and prevention of the coalescence of bubbles upon thawing of frozen xylem water (Chap. 7.3).

1.5 Conduit-to-Conduit Pits

Bordered intervessel pits are fascinating structures, which we shall have to discuss repeatedly in relation to different functions. Vessels with scalariform pits (shown in Fig. 1.4, left and Fig. 7.5, right) are assumed to be primitive; they are developmentally similar and related to protoxylem with annular or helical wall thickening. In many organs such as in the leaves of palms, one can find a continuous transition from annular or helical protoxylem tracheids to metaxylem vessels with scalariform intervessel pitting (Zimmermann and Sperry 1983).

Intervessel pits were briefly introduced when Fig. 1.2 was discussed; they are invariably found where vessels run parallel and in contact. Scanning electron micrographs show this quite dramatically (Fig. 1.13). Regions of intervessel pitting are the locations where water moves from one vessel to the next on its way up the stem. Their characteristic bordered structure insures mechanical strength; the overarching secondary wall exposes a large membrane area (the primary wall) for water movement, but still provides support.

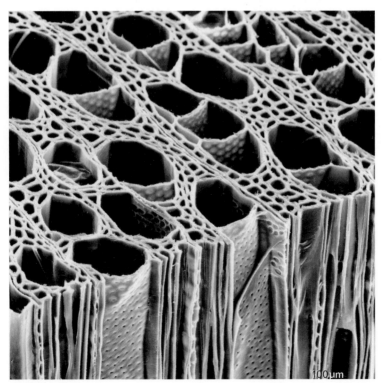

Fig. 1.13. Block of wood of *Populus grandidentata* showing intervessel pitting on the walls between two vessels wherever they run parallel (scanning electron micrograph courtesy of W.A. Côté)

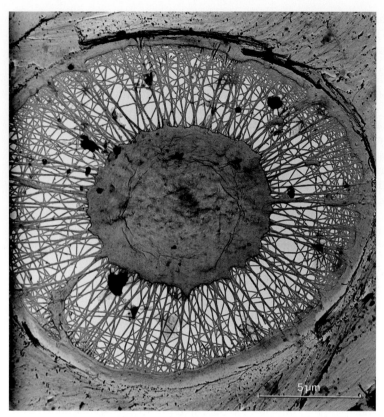

Fig. 1.14. Bordered pit of eastern hemlock (*Tsuga canadensis*), solvent-dried from green condition. The pit membrane consists of the net-like margo and the central torus (transmission electron micrograph courtesy of W.A. Côté)

During vessel differentiation, intervessel pit membranes appear to swell, lose their encrusting substances and, when viewed under a transmission electron microscope, become very transparent to electrons. Pit membranes between vessels and parenchyma cells show a similar but asymmetrical development (Schmid and Machado 1968). A number of papers on this topic have appeared during the years since Schmidt and Machado's original description; a review by Butterfield and Meylan (1982) is informative. Interpretation of the chemical events is not without controversy. Most authors believe that membrane differentiation involves hydrolysis of matrix material of the original cell wall (e.g., O'Brien and Thimann 1967), whereby a cellulosic microfibrillar web remains. However, using various histochemical techniques, a French group interprets events differently, they say that cellulose and certain hemicelluloses ("vic-glycol polysaccharides") disappear, while methylated pectins remain unchanged (Catesson et al. 1979). However, the details of this process will be resolved, the important result of differentiation is increased permeability of intervessel pit membranes to water movement.

It is perhaps appropriate to mention the differentiation of coniferous, bordered pits here. This has been reviewed by Butterfield and Meylan (1982). There are of course many types, but it is sufficient if we just look at one example here (Fig. 1.14). While the torus of the pit membrane remains more or less unchanged, the margo loses the wall matrix substances. A cellulosic network of fibril bundles then remains which appears quite permeable. The gaps between these fibril bundles will be of particular interest to us later, because they are so wide that they not only facilitate water movement, but may also let an air-water interface pass. To prevent this, the pit can act like a valve (Chap. 4.4, Fig. 4.9).

Pit areas must be relatively large in order to permit water to pass relatively freely from one vessel to the next. Their importance will be discussed in connection with the vessel network. Some tree species have what we call xylem with solitary vessels. On a transverse section, less than one out of 50 vessels can be seen to form a pair. In other words, most vessels are single and not associated with other vessels. How does the water get from one vessel to the next in this case? Are the vessels very long? Are vessel overlap areas very short, thus increasing resistance to flow considerably? We do not yet know, but techniques are now available to elucidate this question.

In some species the pit borders of intervessel pits have peculiar outgrowths known as vestures. Vestured pits were described by Bailey (1933) in remarkable detail, considering that he had nothing more than a light microscope at his disposal. Some of Bailey's drawings are reproduced in Fig. 1.15. In every modern textbook on wood, vestured pits are illustrated with scanning electron micrographs. Detailed surveys of vestured pits were published by Ohtani and Ishida (1976) and van Vliet (1978). They show a great variety of forms and discuss their diagnostic and systematic value. Sometimes in the preparation of the specimens, the pit field was split longitudinally and both split surfaces are shown, illustrating the structure dramatically (Fig. 1.16).

Intervessel pit membranes of functional vessels are not under much stress, because the pressure drop across them, even during peak flow rates, must be rather minute. However, the pressure drop across them can become very considerable as soon as one of the vessels has admitted air by injury, or has been "air-seeded" and is thus vapor-blocked (see Chap. 4.3). We then have a situation in which the pressure in the vapor-blocked vessel lumen is either atmospheric (+101.3 kPa) or "vacuum", i.e., water vapor pressure (+2.4 kPa), but in the neighboring functioning vessel is quite negative, perhaps as low as −10,000 or −20,000 kPa. This enormous pressure drop across the pit membranes exerts considerable stress on the pit membrane.

Zweypfenning (1978) suggested that the vestures are structures which support the pit membrane when they are subjected to such one-sided stress, and prevent them from tearing. Wood anatomists interested in functional adaptation might be interested to see whether vestured pits occur preferentially in trees of xeric habitat where unilateral stresses on the pit membrane can be most severe. The problem is not quite as simple, because there are at least three additional factors that affect the danger of tearing. Firstly, membranes of different species may differ in their mechanical properties. A membrane that is elastically or plastically deformable may be able to lodge against the pit border without tearing. Secondly, the pit cavity may vary in depth; the shallower it is, the sooner the membrane will come to rest on it. Thirdly, intervessel pits vary in size considerably among species. The smaller the individual surface area, the smaller the force across it at a given pressure drop (because pressure is a force per unit area). It is thus not a simple matter to set up correlations between pit structure and habitat (Zweypfenning 1978).

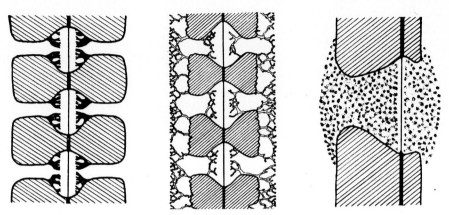

Fig. 1.15. Three drawings of sectional views of vestured pit pairs taken from Bailey (1933). **A** Coralloid outgrowth from the pit border into the pit cavity in *Combretum* sp. **B** Branched and anastomosing projections from the pit border into the pit cavity and from the inside vessel wall into the vessel lumen (*Vochysia hondurensis*). **C** Bordered pit pair between a vessel (*left*) and a tracheid (*right*), mats of fine texture fill the entire pit cavity (*Parashorea plicata*)

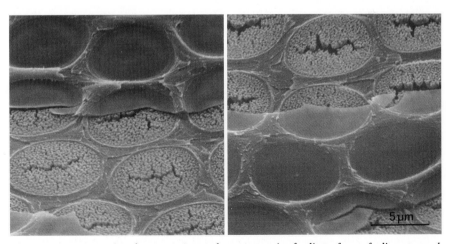

Fig. 1.16. *Lagerstroemia subcostataq.* A complementary pair of split surfaces of adjacent vessel elements showing vestured pit pairs. (Ohtani and Ishida 1976)

2 The Vessel Network in the Stem

2.1 Dicotyledons

The vessels of a dicotyledonous tree stem do not all run neatly parallel; they deviate more or less from their axial path. This deviation differs in successive layers of the xylem; the vessels form a network. This has been known to foresters on a macroscopic scale as "cross grain". Cross grain was considered an exceptional characteristic of certain species. We know today that virtually all xylem is microscopically "cross grained" in respect to the path of their vessels. This was described first by Braun (1959). He meticulously reconstructed the vessel network of *Populus* from a series of transverse sections and showed the structure in a block of the dimensions 0.3×1×2.5 mm. He also described the nature of vessel ends. Vessel ends are not easy to recognize, particularly not in single sections. They have occasionally been "seen" where India ink infusions terminated. However, this is by no means reliable, because India ink particles are quite large and ragged; lateral water loss from the vessels concentrates them, and they may clog the vessels somewhere midway. Anyone who has tried to fill a fountain pen with India ink will heartily agree! Vessel ends have occasionally been reported as terminal elements in macerations (e.g., elements that have a perforation only at one end; Handley 1936; Bierhorst and Zamora 1965). This is one of the most reliable methods, but the vessel end is then seen only in isolation.

Another, even more meticulous, analysis of the vessel network was reported by Burggraaf (1972). It concerns a 1.5×2.0×10 mm block of European ash *(Fraxinus excelsior)*. It is somewhat ironic that such an enormous effort was made at a time when a very much easier method was already available.

The introduction of cinematographic methods by Zimmermann and Tomlinson (1965, 1966) very considerably simplified the recognition of the three-dimensional vessel structure and the localization of vessel ends. The principle is the following. A 16-mm motion picture film is assembled, frame-by-frame, from either serial sections in the microscope, or by photographing, with a close-up arrangement, the surface of a specimen in the microtome. The various methods are described in detail in Zimmermann (1976). Interestingly enough, the idea is about 75 years old (Reicher 1907). The important recent innovation which made the methods useful was (1) a continuous advance microtome clamp which permits the direct advance of the specimen (rather than part of the clamp), and (2) a method for aligning successive microtome sections optically in the microscope. This can be done either with an auxiliary drawing, or more elegantly with the shuttle microscope in which two adjacent sections are viewed in rapid succession. These methods transform the axial dimension of the specimen into time. With a

Fig. 2.1. Diagrammatic illustration
of the process of reconstruction
of a vessel network from 16-mm film

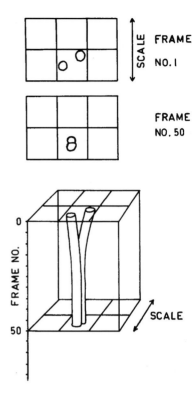

motion picture analyzer, one can then study the three-dimensional vessel structure on a "moving" transverse section, which is projected onto a table in front of
the investigator. One can move up or down in the stem by running the film forward or backward. Needless to say, such films are also very useful for teaching
purposes. A film showing the three-dimensional structure of dicotyledonous
wood has been available since the early 1970s (Zimmermann 1971).

The principle of vessel-network reconstruction is shown in Fig. 2.1. The film is
projected onto a grid of squares; individual vessels are drawn, and the frame number is recorded. The film is then slowly moved, a few frames at a time, until vessels
have been displaced from their former positions, but their identity is still recognized. They are then drawn again, either in another color on the same sheet of
paper, or on a new sheet. This is repeated along the desired axis length. A simple
example is shown in Fig. 2.1 with film frames no. 1 and no. 50, and a grid of six
squares. In a second process, the vessel transverse sections are redrawn in perspective along a scale of frame numbers, which represents the stem axis. The vessel network in a block of wood is thus easily reconstructed and shown in a perspective drawing. The tangential and radial scales of the block are provided on the
first frame of the film by a photographed micrometer scale. The longitudinal
(axial) scale is taken from the frame number, whereby each frame corresponds to
the distance between individual microtome slides. In dicotyledonous wood, the
axial scale is foreshortened about ten times, so that the reader has to visualize that

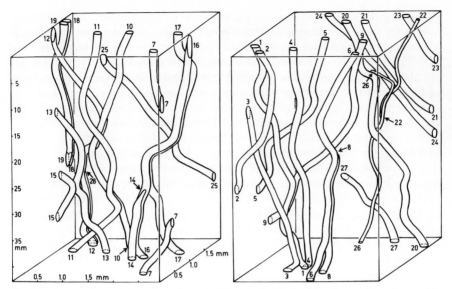

Fig. 2.2. Course of vessels in a piece of wood of *Cedrela fissilis*. Individual vessels have been arbitrarily separated into two blocks for clarity. Vessels are numbered where they exit from the block. *Arrows* show vessel ends. Note that the axial scale is foreshortened ten times. (Zimmermann and Brown 1971)

the diagram shown should in reality be stretched about ten times to regain its natural proportions.

Figure 2.2 shows the reconstructed vessel network of cigarbox wood, *Cedrela fissilis*. Here, individual vessels have been separated arbitrarily into two separate blocks so that they do not "hide" each other. In reality, all vessels are in the same block, and the block should be visualized ten times longer than it appears on paper. The vessels are numbered where they exit from the block. Numbers with arrows are vessel ends. It can be seen that all ends are in contact with other, continuing vessels. Where neighboring vessels run parallel (i.e., where they touch), we have to visualize intervessel pit areas. For example, in the center of the right-hand drawing, we see the end of vessel no. 8, which runs along no. 6 for a distance of about 1.7 cm. Vessels are cut off on all faces of the block.

The cinematographic method is very labor intensive and this may explain, in part, why it has been used very little in recent years. More recently, an easier method, the microcasting technique (André 1993; Fujii 1993; Kitin et al. 2001), offers another easier solution of the problem of observing the three-dimensional structure of vessels. Briefly, the vessel lumina of a wood sample are visualized in the form of silicon filament casts made by perfusing the vessels with a polymer before it polymerizes. This resulting cast retains the vessel structure with a resolution better than 1 μm. The lumina casts keep the original form and position of each element of the vascular system in three dimensions after the cellulose is enzymatically dissolved. Numerous data obtained by André with this method, from more than 50 plant species of angiosperms, reveals vessel structures more complex than the "classical" vessel pattern with its linear biperforated elements having the same secondary wall ornamentation, with uniperforation at each end. Vessel shapes, which have been rarely or never mentioned in the literature, have been discovered. These nonclassical vessels, which seem to be very frequent, are circular vessels, zigzag vessels, branched vessels and heterogeneous vessels. Some of them seem to characterize particular plant families (ramified shapes in Poaceae), or position in the plant: zigzag, circular and heterogeneous shapes, in the connecting zones between organs (André 1998, 2000, 2002; André et al. 1999).

The existence of circular vessels has been recognized for about a century but hypotheses on their formation are recent. They are described as closed endless ring-shaped vessels. They are found in large quantities between the wood of two axes, e.g., between a stem and a branch, and, especially, in the wood formed during healing around a branch stump. They also have been found in crown galls induced by *Agrobacterium tumefaciens* (Aloni et al. 1995). In young woody stems, the secondary vessels commonly present short segments, with more-or-less complicated "zigzag" shapes (Fig. 2.3), which are located in the nodes, on each side of the bulging insertion zone of petioles and axillary branches. Branched vessels with triperforated elements are scarce in the wood of dicotyledons but present in almost every monocotyledon species studied.

The term "heterogeneous" vessel denotes a vessel type in which a short sequence of about 5–20 members with different ornamentation of the secondary walls is inserted between two files of pitted members. They seem to be always present in the zone connecting two organs or two organ parts, or within an abscission zone. The pitted parts of their length are distributed on each side of this zone. They have been found in the foliar traces of numerous dicotyledons (e.g., maple, oak, plane tree), in the rachis of compound leaves of walnut, and in foliar and fruit pedicels.

As one watches a vessel pair in a motion-picture film, one usually can observe them drifting apart tangentially within the growth ring. This occurs even in "straight-grained" trees like ash *(Fraxinus)*. The physiological significance of this network is the tangential spreading of the axial path of water transport. This has been shown by MacDougal et al. (1929, Fig. 15), and subsequently by many others. If a dye solution is injected into a radial hole bored into a tree trunk, and the dye pattern analyzed from above, one can see that it has spread tangentially within the growth ring (Fig. 2.4). This shows the axial vessel path on a much larger scale and

Fig. 2.3. *Top* Picture of segments of zigzag vessels in the earlywood (A–E) and in the latewood (F, G) of a twig of pedonculate oak (*Quercus pedonculata*). *Bottom* Figure showing the localization and morphology of zigzag secondary vessels of the above picture. **A, B** Localization in the node, on each side of the petiole insertion (*p*) and of the axillary branch insertion (*ax*). *Arrows* indicate the supposed direction of the tangential pressure exerted on the axis cambium by the lateral organ growth. **C** Zigzag vessel segment (cast) in spring wood of *Quercus pedonculata*. **D, E** Compact form (in situ) and extended form (picture) of the vessel cast. **F** Diagram of the zigzag segment, split in short linear parts, each of them hypothetically deriving from a different cambial cell file. *Arrows* indicate the supposed shearing strength as resulting from the lateral organ growth. (André 2001)

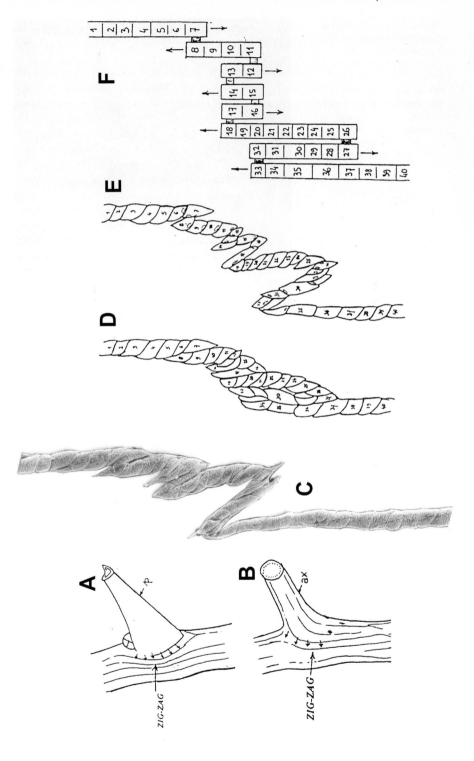

Fig. 2.4. Dye ascent from a radial hole in the trunk of *Populus grandidentata.* The dye pattern is shown in disks cut from 0, 50, and 100 cm above the point of injection. The steep pressure gradient spreads the dye tangentially even at the injection level by moving it up and down in successive vessels

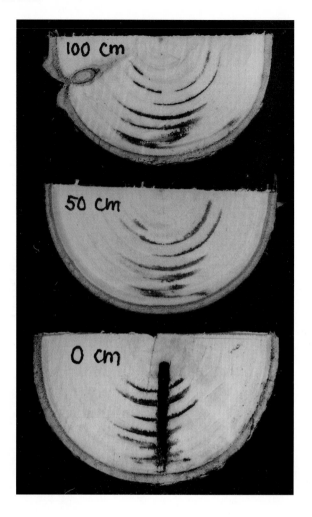

we can see, in addition to the spreading, that, within each growth ring, the vessels follow a helical path up the stem. The helices usually differ from one ring to the next. This helical path is rather slight in the case shown in Fig. 2.4. The spread within individual growth rings at the application level might appear puzzling. However, we must realize that, by applying the dye at ambient pressure, we introduce a very steep pressure gradient not only axially, but also laterally. Dye therefore moves up and down from vessel to vessel, penetrating the growth ring horizontally. Whenever we make a dye ascent, we alter pressure gradients and hence flow directions. For example, we always also get downward movement. Dye tracks must therefore always be interpreted with caution.

The physiological significance of this complicated layout of vessels is quite obvious: from any one root, water ascends in the trunk and spreads out within the growth rings, thus reaching not only a single branch, but also a large part of the crown. Looking at it another way, each branch of the crown obtains its water from many different roots. It is clear that this is a considerable safety feature. Loss of one or more roots does not impede the growth of individual branches, it merely diminishes the water supply to the crown slightly. It also shows that the cambium has a built-in flexibility: vessel differentiation is reoriented relatively easily and injuries can therefore be bypassed.

Spiral and wavy grains of wood have been known for a long time. However, it has only relatively recently become known that the pitch of the helical path of vessels can vary within a single growth ring. The split-disk technique shows this macroscopically. When a thin, transverse disk of a tree stem is broken along a straight line that has been scored radially on one of the transverse surfaces, the break becomes slightly wavy on the other side, because it follows the "grain" of the wood. Mariaux used this technique to recognize growth rings in tropical trees, because a very thin layer of wood at the growth ring border often follows a different helical pitch (Bormann and Berlyn 1981). This can be seen very clearly when one watches a film of *Cedrela* stem transverse sections (Zimmermann 1971). Morphogeneticists have tried to unravel the hormonal regulation of axial direction of vascular differentiation in the past (see the discussion in Burggraaf 1972). The matter has received renewed attention with the work of Polish scientists who are studying cambial domains and hormone wave patterns (Hejnowicz and Romberger 1973; Zajaczkowski et al. 1983).

The exact layout of vessels in even apparently simple herbaceous dicotyledons is amazingly complex and poorly understood. An entire army of plant anatomists could spend years mapping vessels, even though we now have a relatively simple method to do it. Such work would be so endless that it would have to be restricted to specific purposes such as comparative studies or for the localization of pathways for water, nutrients, or hormonal signals.

2.2 Monocotyledons

It may seem like a paradox, given the apparent complexity of the primary vasculature of monocotyledons, but the network of vessels of monocotyledons is not more difficult to study than that of dicotyledons, simply because the vessels are contained within easily identified vascular bundles. The pattern of vasculature has been analyzed in a number of arborescent monocotyledons, e.g., some palms (Zimmermann and Tomlinson 1965, 1974), *Prionium* (Juncaceae; Zimmermann and Tomlinson 1968), *Dracaena* (Agavaceae; Zimmermann and Tomlinson 1969, 1970), members of the Pandanaceae (Zimmermann et al. 1974), *Alpinia* (Zingiberaceae; Bell 1980), and members of the Araceae (French and Tomlinson 1981a-d). In spite of this relative wealth of knowledge of the vascular structure of monocotyledonous stems, the vessel network within the vascular bundles is well known only for the small palm *Rhapis excelsa* (Zimmermann et al. 1982). However, in analyzing films of other monocotyledons, we have seen enough to recognize that *Rhapis* is representative in principle. But before we discuss the vessel network, let us first briefly review the vascular bundle network.

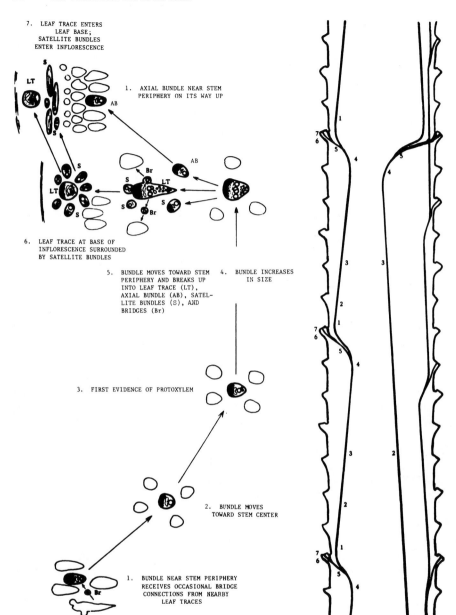

Fig. 2.5. Course of vascular bundles in the stem of the palm *Rhapis excelsa*. In the diagram on the *right* several bundles are shown in radial coordinate projection. The diagram is foreshortened about four times in relation to the stem diameter. (Modified from Zimmermann and Tomlinson 1965)

The network of vascular bundles in the stem of *Rhapis* is less complex than in many other arborescent monocotyledons and can be regarded as the fundamental monocotyledonous pattern. However, no evolutionary meaning should be attached to the word "fundamental" here; the expression should be taken didactically. Once the vascular structure of *Rhapis* is understood, the more complex, larger palms and other monocotyledons are more easily comprehended. The following brief description is based on the original publication of Zimmermann and Tomlinson (1965) and the more recent demonstration films (Zimmermann and Mattmuller 1982a,b).

Axial bundles run along the length of the stem. In the central part they follow a helical path, which we shall ignore for the sake of clarity. The easiest way to illustrate the path of the bundles is in radial coordinate projection: we show them on a radial plane, after having "untwisted" the central helical part (Fig. 2.5, right). If we follow a single axial bundle up in the stem, we can see that it gradually approaches the stem center. It then quite suddenly turns outward, splits into a number of branches of which one, the leaf trace proper, enters the leaf base, another one, the continuing axial bundle, continues up the stem, and several others, the bridges, fuse with neighboring axial bundles. Additional branches connect the stem with the axillary inflorescence, but these need not concern us further here. Figure 2.5 (right) shows bundles of different lengths. There are major (axial) bundles that reach the stem center; in these the distance from one leaf contact to the next is quite long. There are minor bundles that do not reach the stem center, their leaf-contact distance is shorter. If we inspect the bundles under the microscope (as illustrated in Fig. 2.6), we see that in their lower parts they usually contain a single, relatively wide metaxylem vessel. This is shown diagrammatically in Fig. 2.5 (left) and Fig. 2.6 as positions 1 and 2. As we follow the bundle up, a point is reached where it contains a few narrow-diameter protoxylem elements (position 3). This point is about 10 cm below the leaf contact. As we continue to follow the bundle up, the number of protoxylem elements and metaxylem vessels increase (position 4). The bundle now turns sharply toward the stem periphery and we refer to it here as a leaf trace (this is somewhat arbitrary and is done merely for convenience). On its way toward the stem periphery, the bundle begins to break up into branches (position 5). One (or more, or none) of these branches becomes an independent axial bundle and repeats the cycle (position 1). Other branches are short bridges, which fuse with neighboring axial bundles; still another branch, the leaf trace proper, enters the leaf base (position 6).

What interests us particularly here is the fact that all wide metaxylem vessels remain in the stem, and that the entire complement of narrow protoxylem elements enters the leaf base. This can be seen in Fig. 2.6, and even more dramatically in a film assembled from serial sections. We can again plot individual vessels as we did with the dicotyledonous wood. Such an analysis, covering an entire leaf contact distance, is illustrated in Fig. 2.7. On the right side, the radial coordinate projection is shown, on the left, the vascular bundle is projected radially onto a tangential plane. Each solid line represents a single metaxylem vessel. The hatching between individual lines indicates that the vessels run next to each other and are connected by intervessel pit areas. The dashed line indicates protoxylem, regardless of how many elements there are. The horizontal scale is very greatly expanded. If we showed the horizontal scale at the same magnification as the axial

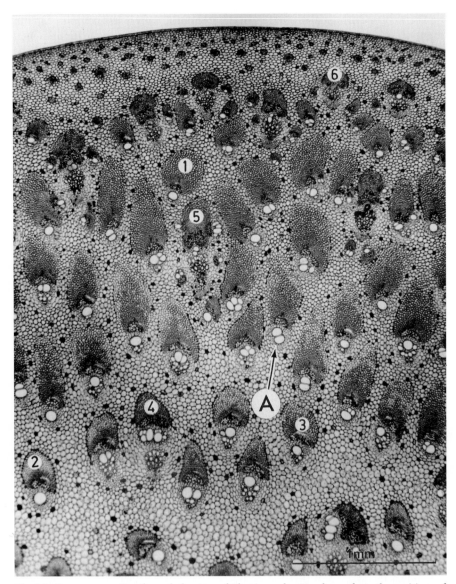

Fig. 2.6. Transverse section through the stem of *Rhapis excelsa*. *Numbers* refer to the positions of bundles indicated in Fig. 2.5. A vessel overlap is shown in **A**. (Modified from Zimmermann et al. 1982

Fig. 2.7. Vessels of a single vascular bundle of the stem of *Rhapis excelsa*, shown over a complete leaf-contact distance of 40 cm, projected onto a tangential plane and horizontally expanded. *Solid lines* are metaxylem vessels. *Hatching* between *solid lines* represents intervessel pit areas. *Dashed lines* indicate that the bundle contains protoxylem, regardless of quantity. *AB* Axial bundle; *SAT* satellite bundle (leading into axillary inflorescence). *Arrows* indicate bridges (*Br*) always pointing away from the leaf. (Zimmermann et al. 1982)

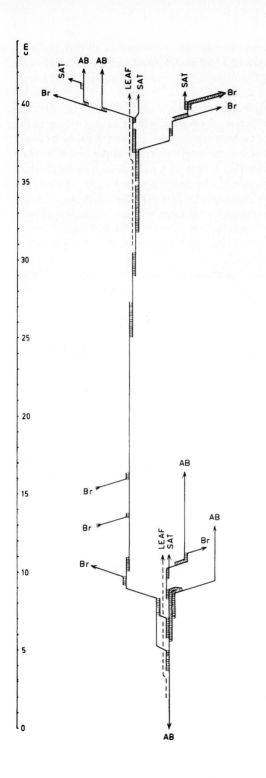

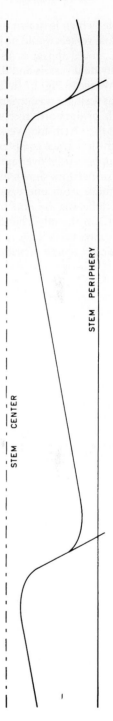

scale (which is drawn on the left and extends over a distance of 44 cm), the individual vessels would be just a fraction of a millimeter apart and the entire illustration would appear as a single line. The expanded horizontal scale lets us illustrate individual vessels and their relative location.

We see in Fig. 2.7 (left), for example, that the lower part of the axial bundle contains a single metaxylem vessel, which is 17 cm long. About 13 and 16 cm of the scale bridges are "received" from neighboring departing leaf traces. At its lower end the 17-cm-long vessel overlaps for 1 cm length with the upper end of a vessel of the leaf-trace complex. Such an overlap is seen at A in Figs. 2.6 and 2.9 (left). At its upper end it overlaps about 2.5 cm with the continuing vessel. As we study the vessel network shown in Fig. 2.7, we realize that water that ascends in the stem has to move from one vessel to the next quite frequently. Clear vessel lengths are greater along the axial bundle, and considerably shorter in the leaf-trace complex area. On the other hand, the leaf-trace complex region provides many alternate pathways via bridges to other axial bundles.

Vessel-length distribution measured with the latex-paint method in the *Rhapis* stem show the metaxylem vessels of different parts of the vascular bundle quite nicely (Fig. 2.8). Vessels of the 0–5 cm length class are quite obviously the vessels of the leaf-trace complex area (bridges, etc.), while the longer vessels are primarily those of the axial bundles in positions 1–4.

Let us now look at the vessel-to-vessel contact in quantitative terms. Figure 2.9 (left) shows a metaxylem vessel pair in transverse section. Intervessel pits are scalariform. If we take the vessel to be circular in transverse section and its diameter to be 60 μm (Fig. 2.9, left), the transverse sectional area of the vessel lumen is 2.8×10^{-3} mm^2. The width of the scalariform pit area is ca. 35 μm. This gives us a vessel-to-vessel contact area of 0.35 mm^2/cm of overlap. However, not all this area is available to water conduction. The cross section across the intervessel pit area shown in Fig. 2.9 (right) shows that ca. 40% of the contact area is pit membrane and 60% is occupied by the secondary wall. We can therefore say that 0.14 mm^2 of pit membrane area is available per centimeter of vessel overlap length for the flow of water from one vessel to the next. For a 2-cm vessel overlap the cross-sectional area through which the water must flow from one vessel to the next is ca. 100 times larger than the transverse-sectional area of a single vessel. We do not have, at this time, any information on how much this increases the resistance to flow.

So far, we have been concerned entirely with the wide metaxylem vessels and have disregarded the narrow protoxylem elements that are the pathway from stem to leaf. This is a very important point, but it will have to be considered later (Chap. 5.7).

To summarize water movement in the palm stem, we can say that it is almost exclusively via wide metaxylem vessels. In *Rhapis* and many other small palms, there is basically a single metaxylem vessel per axial bundle; such palms are called single-vessel palms. In other species, especially the larger palms, there are often two vessels ("two-vessel palms"), but the two vessels are not connected by pit membrane areas, because there is a layer of parenchyma between them. We have not yet investigated the vessel network of any two-vessel palm (nor of any other monocotyledon!). Nevertheless, we know that water can move most freely through the long vessels of the lower part of the axial bundle. In the leaf-trace complex area vessels are shorter, but there are several per bundle. The leaf-trace complex

Fig. 2.8. Vessel-length distribution in the stem of the small palm *Rhapis excelsa*. (Zimmermann et al. 1982)

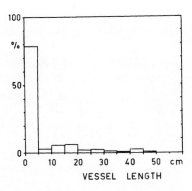

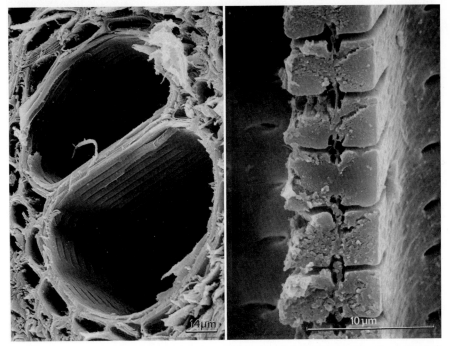

Fig. 2.9. Scanning electron micrographs of metaxylem vessels in the stem of *Rhapis excelsa*. *Left* Transversely cut stem showing intervessel pit area. *Right* Intervessel pits in longitudinal section (cutting across the wall). (Zimmermann et al. 1982)

area also provides direct connections, via bridges, to other bundles, and thus lateral transport paths. These are very important alternate pathways, providing the necessary by-pass pathways around injuries. A dye ascent illustrates the path of water in the stem visually. Figure 2.10 shows a disk cut from a palm stem 25 cm above a dye injection into a radial hole, similar to that illustrated in Fig. 2.4 for

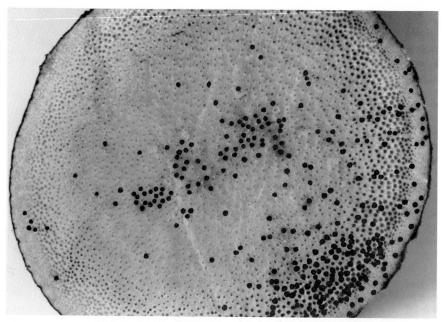

Fig. 2.10. Dye ascent from a radial hole in the stem of the palm *Chrysalidocarpus lutescens*. The disk shown was taken 25 cm above the point of injection. Stained vascular bundles are marked with a *black spot*. (Zimmermann and Brown 1971)

poplar. The dye has spread out into most of the stem area by lateral movement across bridges and by the helically twisted bundle path of the stem center.

Let us now consider what appears to be a new paradigm found in rattan palms, genus *Calamus* (Arecaceae-Calamoideae-Calaminae). The genus *Calamus* is the most species-rich of all palms, with an estimated 370 species (~15% of all palms). They are often vine-like and reach lengths of well over 100 m. The stem and leaf anatomy of species of *Calamus* has been reported by Gudrun Weiner (Weiner 1992; Weiner and Liese 1992, 1993) and the cinematographic technique has been applied to *Calamus longipinna* (Tomlinson et al. 2001). The unusual aspect of this species is the discontinuity in metaxylem between the base of the stem and the leaves. *Calamus* is a single-vessel palm having one large metaxylem vessel per vascular bundle (average diameter 210±13 µm) and typically five protoxylem vessels (average diameter 40±10 µm). However, the metaxylem vessels are absent in the leaf insertions, where the vascular bundles contain 10–20 protoxylem vessels. The metaxylem vessels run for several meters but start and end in blind ends. The metaxylem elements of adjacent bundles are occasionally interconnected by transverse connections (transverse commissures; Tomlinson et al. 2001), but there is no direct connection between metaxylem and protoxylem (Fig. 2.11). Hence the only direct connection between stem and leaf is via the protoxylem and the question arises regarding the relative hydraulic contributions of the protoxylem versus the metaxylem and where it exists.

We can use the Hagen-Poiseuille equation to compute the relative conductivity of five protoxylem vessels of 40-μm diameter to one metaxylem of 210-μm diameter; this ratio will be given by $5 (40/210)^4 \sim 0.007$, i.e., somewhat less than 1%. However, we cannot conclude that 99% of the water in *Calamus* stems is carried by metaxylem vessels, because we must add to hydraulic resistance (=inverse conductance) of metaxylem vessels the resistance for water to enter and exit the vessels. Metaxylem vessels have no direct connection with protoxylem vessels. The metaxylem is completely surrounded by living parenchyma cells and it is clear that water must pass from protoxylem to metaxylem via several layers of living cells (Fig. 2.11B).

There are two parallel pathways for water movement through the four layers of cells between the protoxylem and metaxylem, i.e., transcellular and apoplastic. The transcellular pathway would require water to pass through eight membranes (two per cell). The apoplastic pathway occupies about 5% of the tissue cross section between cells and is via cell walls. From Fig. 2.11B it appears that water would enter the metaxylem over about one fourth of the circumference. Let us also assume that water passes between protoxylem and metaxylem over a distance of 1 m (about one half of the length vessels of the plants up to 10 m long, studied by Tomlinson et al. 2001). The surface area of one fourth of the tissue annulus would be $1/4\pi DL$, where D is the diameter of the tissue sleeve around the metaxylem and $L = 1$ m. The known range of membrane permeability (conductance) of plant cell membranes is $k_m = 1 \times 10^{-13}$ to 20×10^{-13} m s^{-1} Pa^{-1} (Boyer 1985; Smith and Nobel 1986; Tomos 1988). So the conductance of eight membranes in series comprising the transcellular pathway is $k_{trans} = (1/4\pi DL) k_m/8 = 2.5 \times 10^{-16}$ m^3 s^{-1} Pa^{-1}, when k_m is 20×10^{-13}. The hydraulic conductivity of a cube of primary cell wall (K_{cw}, measured in *Nitella*; Tyree 1969) is 1.4×10^{-16} m^2 s^{-1} Pa^{-1} so the apoplastic conductance of the sleeve is given by $k_{apo} = K_{cw}\alpha(1/4\pi DL)/\delta$, where α is the fractional area of the tissue that is cell wall (=0.05), and δ is the thickness of the sleeve (=50 μm). So k_{apo} equals 1.3×10^{-16} m^3 s^{-1} Pa^{-1}; however, any lignification of *Calamus* cell walls in stems would likely reduce K_{cw} and hence would reduce k_{apo}. Since the total conductance of the sleeve $k_{sleeve} = k_{trans} + k_{apo}$, it seems unlikely that k_{sleeve} could exceed 4×10^{-16} m^3 s^{-1} Pa^{-1}. This figure should be compared to the axial conductance of a 2-m length of metaxylem. From the Hagen-Poiseuille equation, the lumen conductance is 2.5×10^{-13} m^3 s^{-1} Pa^{-1} for 2 m.

The conclusion is that the conductance of the lumen is much more than the sleeve, and since water must enter and exit the metaxylem by two sleeves (one near the base and one near the apex), it seems unlikely that the total conductance of the metaxylem (including the two sleeves) could be much more than 2×10^{-16} m^3 s^{-1} Pa^{-1}. Here, we have used the Ohm's law principle that conductances in series add inversely, and we must take this as the maximum likely value since we have used the maximum values of membrane and cell wall conductance reported so far in the literature. This leads to a rather surprising conclusion that the total metaxylem conductance is likely to be less than that of the five protoxylem vessels in parallel=1.8×10^{-15} m^3 s^{-1} Pa^{-1} (for a 2-m length of protoxylem parallel to the metaxylem). There appears to be little adaptive advantage to having metaxylem in *Calamus*. The situation would be improved if metaxylem vessel lengths scale up in proportion to plant length for long plants, because the sleeve conductance will increase with surface area on longer vessels whereas protoxylem

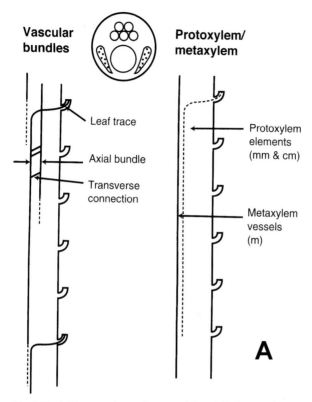

Fig. 2.11. **A** Diagram shows the essential and distinctive features of the vascular pattern in the stems of *Calamus*. *Left* Course of individual axial bundles. These end blindly in a basipetal direction, but are connected distally by narrow transverse connections. *Right* Distribution of protoxylem and metaxylem within a single axial strand and leaf trace. Protoxylem (*dotted*) is continuous distally into a leaf, but is never in contact proximally with the metaxylem of an axial bundle (*solid line*). Protoxylem elements (vessels and tracheids) are mm and cm long, metaxylem vessels are m long. The *insert cartoon* shows the appearance, within a transverse section of a single central vascular bundle, of protoxylem (*small circles*), metaxylem (*large circle*), and lateral metaphloem (*dotted*) (Tomlinson and Fisher 2000).

conductance would decrease inversely with distance. So if a 100-m cane had 20-m-long metaxylem vessels then the conductance of two sleeves each 10 m long would be 2×10^{-15} compared to the conductance of five protoxylem vessels of 1.8×10^{-16} (20 m long). So the importance of metaxylem versus protoxylem will increase as metaxylem vessel lengths increase. The break-even point where conductances are about equal is approximately 5–6 m. However, a few experiments are superior to days of theoretical computations, so I hope someone will study the hydraulic architecture of rattans to elucidate this fascinating anatomy.

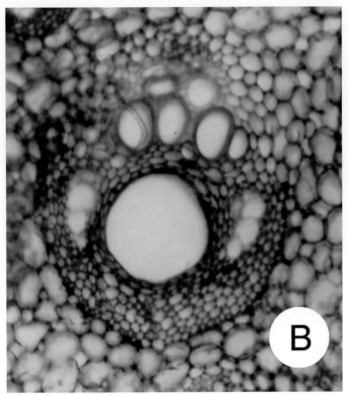

Fig. 2.11. B Micrograph shows more than four layers of living cells between the protoxylem and the metaxylem. (Tomlinson et al. 2001)

2.3 A Comparison with Conifers

Coniferous wood consists almost entirely of tracheids. In most cases, the tracheids are connected on their radial walls by bordered pit pairs. This means that water can spread easily within a growth ring in a tangential direction. The axial path of the water can be traced with a dye ascent from a radial hole (Vité and Rudinsky 1959). The flow patterns made visible in this way are not too different from those of dicotyledonous trees: the dye spreads out within the growth ring and the fan-shaped dye track twists around the stem on its way up. Vité and Rudinsky (1959) found some quite distinct patterns in different groups of conifers. In the soft pine group (Haploxylon) they found what they called "left-turning spiral ascent". With this they mean that, although the dye spreads within each growth ring, each successive growth ring has a more twisted spiral grain so that disks taken above the point of injection show a spiral track, turning clockwise on the disk from the center out. In the hard pine group (Diploxylon), larch, fir, and spruce, they found "right-turning spiral ascent" in which the spiral turns the other way. They also distinguish "interlocked", "sectorial straight", and "sectorial winding" ascent in certain other conifers. The pattern shown in Fig. 2.4 fits the last of these groups best. These classifications need not concern us further here, although it is interesting that the cambium succeeds in "tilting" successive tracheids one way or the other. The important point for us is that these patterns have the same function as the vessel network in angiosperm wood, namely that individual roots contribute to a very large part of the crown rather than to a small section of it. This is, of course, a marvelous safety feature of the xylem.

An interesting exception to the relatively free water movement in tracheids comes from a study of the pathway of water movement in *Thuja* trees growing on the rock face of cliffs in Ontario. These trees are dramatically sectored hydraulically such that one root is linked directly with one branch (Larson 1994).

While the path of water up the tree stem is quite comparable in conifers and dicotyledons, and functionally equivalent in the monocotyledons, the lengths of the compartments, as discussed in Chapter 1, are quite different. If narrow- and short-vessel trees are less efficient but safer water conductors, i.e., more redundant, than the wide- and long-vessel trees, conifers are even more conservative. The mere fact that among the world's tallest trees there are conifers as well as dicotyledons shows that both designs have been successful.

2.4 Some Quantitative Considerations

It should be clear from Chapter 1.3 that a quantitative description of the xylem is only useful if the sum of the fourth powers of individual inside vessel radii (or diameters) is known. Transverse-sectional vessel area or vessel densities, etc., are of very limited value for describing water conduction capacity, though they may be useful for other purposes. In this chapter we are mainly concerned with the three-dimensional aspects of structure. There are various ways in which these can be quantified.

Bosshard and Kučera (1973) give a quantitative description of the vessel network in a block 0.5×1.0×10 mm of *Fagus sylvatica* taken near the cambium from a mature tree. They followed the path of 35 vessels and found a total radial spread of 0.504 mm and a total tangential spread of 0.314 mm. The first item need not concern us further at this moment because water normally remains confined within the growth ring, at least along the trunk of the tree. Movement of water across the growth ring border will be discussed later. The tangential spread of vessels of 0.314 mm over a length of 10 mm corresponds to an angular spread of ca. 1.8°. We can extrapolate and say that if the spread of 0.314 mm cm^{-1} axial length is maintained along a 10-m-tall trunk, the spread would be ca. 30 cm along the stem circumference. From dye ascents like the one illustrated in Fig. 2.4 we can also estimate the angle of tangential spread of the axial water path. It is usually of the order of 1°. It is likely that this varies quite a bit from one species to another, but it has not been systematically investigated. The angle of tangential spread of phloem transport in *Fraxinus americana* is of a similar magnitude, i.e., somewhat less than 1° (Zimmermann 1960).

Anatomical measurements of the possible spread of water within a growth ring predict a fan-shaped path that, along a tall stem, would eventually enclose the entire circumference somewhere higher up in the tree. This rarely happens with a dye ascent from a point source. In Fig. 2.4 there is very little, if any, spread from 50–100 cm height. Indeed if we crudely analyze the shape of the dye path in a stem, we find the pattern to be more diamond- than fan-shaped. Although this has never been investigated in detail, we can speculate that the problem is one of pressure gradients. Pressure gradients will be discussed in more detail in Chapters 3.7 and 3.8, but we can at least say at this point that the pressures in the trunk of a transpiring tree are usually quite negative (something like –1 MPa), while the dye solution we inject is at ambient pressure, i.e. +0.1 MPa absolute. Pressure gradients in the immediate neighborhood of the injection point are therefore very steep. This means that the dye moves down as well as up, and indeed tangentially out by moving up and down repeatedly via individual vessels, thus getting away from the hole along a zigzag path. The steep pressure gradient then causes dye initially to fan out in the greatest possible angle. Far away from the injection hole, pressure gradients are not as steep and the dye stream "competes" with the water stream that comes from the roots. Very little research has been done in this direction, but it is worth giving these matters some thought, because tree injection is of considerable practical importance. We will return to it later (Chap. 9.6).

We now have to consider the degree of mutual vessel contact: the more vessels are connected with each other, the more alternate paths for water flow are available along the stem: the longer the overlaps, the less the resistance to flow from one vessel to the next. Extreme cases are dicotyledonous wood types that show, on the one hand, many vessel clusters on a single transverse section, and, on the other, a so-called solitary vessel arrangement. Truly solitary vessel structure is physiologically perhaps not impossible, but would certainly be very peculiar. Research on vessel-length distribution and the three-dimensional structure of xylem with "solitary" vessels has never been completed (Butterfield, pers. comm.). It will be very interesting to see how water moves in stems of such trees.

Braun (1959), as well as Bosshard and Kučera (1973), made quite complex calculations to describe the degree of vessel-to-vessel contact. Although this has given us a more precise description of wood structure than was available before, it is difficult to extract from these computations the extent of possible lateral water movement. This depends not only on the number of lateral contacts, but also on contact lengths, which Bosshard and Kučera (1973) found to range up to ca. 2 mm in beech. A long contact length improves the vessel-to-vessel movement (provided there is a pressure gradient across the pit area). Where vessels merely touch, water movement is not possible, at least not in quantity. One essential item that is missing from these analyses is vessel length. A given contact length represents a greater resistance to flow if it has to be negotiated frequently because the vessels are very short. It will be necessary to come to grips with the functional aspects of alternate pathways and resistance to flow due to vessel-to-vessel movement if we want to compare different wood species quantitatively. Undoubtedly, anatomical measurements alone will not be sufficient, some experimental approaches will be necessary.

The degree to which vessels are in contact with rays has been described for beech by Kučera (1975). Vessel-to-ray contacts are the link between xylem and ray transport of water and solutes. It is interesting that Bosshard et al. (1978) found that occasional vessels ended in contact with rays rather than with other vessels. However, these vessel ends were of 10–15 µm diameter, which is rather narrow for oak, even for seedlings. Such diameters are insignificant in terms of axial flow. Nevertheless, it is developmentally interesting.

Comstock and Sperry (2000) have provided theoretical insight into the optimal conduit length for water transport in vascular plants. Their theoretical considerations combine the concepts we have reviewed so far concerning safety versus efficiency of complex vessel networks. Vascular plants have shown a strong evolutionary trend towards increasing length in xylem conduits. Increasing conduit length affects water transport in conflicting ways hence they argue evolution should reach a compromise that ultimately defines an optimal conduit length. Increasing length reduces the number of high-resistance sequential transfers of water from conduits over the distance that they are in contact; therefore, the overall hydraulic resistance is reduced. The optimal conduit length distribution will have the shortest conduits where the safety margin from cavitation is least. For plants with the same vulnerability curve (see Chap. 4) throughout, this translates to shortest vessels in the leaves, longest in the roots, which seems to be the case. There are a number of advantages of reduced hydraulic resistance in woody stems that will be discussed in Chapters 5 and 6. There are, however, disadvantages to long conduits related to the impact of cavitation on loss of conductance in trees with long conduits (Chaps. 3 and 4). The theoretical considerations of Comstock and Sperry (2000) provide a useful basis for future studies that might explain evolutionary tradeoffs in xylem structure and function.

2.5 Implications: the Remarkable Safety Design

We have discussed safety of water conduction before and shall have to come back to it repeatedly. At this point we are specifically concerned with the fact that mutual vessel contact contributes very significantly to the safety of water conduction by providing alternate pathways. This is perhaps most dramatically illustrated by the so-called double saw-cut experiment. If saw cuts are made along points of the stem, halfway across the tree trunk and from opposite sides, one often finds that water transport is not interrupted and that the tree survives. The concept of this experiment is very old and was tested at least as long ago as 1806 by Cotta (cited in Hartig 1878). It has been repeated many times more recently (e.g., Preston 1952; Greenidge 1958) and is often regarded as rather mysterious and inexplicable. There is nothing mysterious about the experiment; the path of water around such injuries should by now be quite clear to the reader (see also Mackay and Weatherley 1973).

Let us first consider the problem with the sugar maple tree whose vessel-length distribution is illustrated in Fig. 1.9. If we want to know how far from a saw-cut vessels are interrupted, we should not consult the vessel distribution percentages,

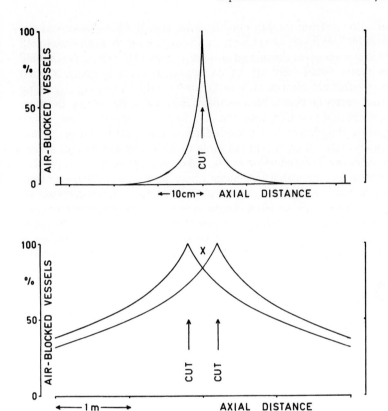

Fig. 2.12. *Top* Cut into the stem of a mature sugar maple (*Acer saccharum*) opens all vessels at that point. The graph shows the decreasing number of opened (injured) vessels at distances away from the cut. This number (the count of opened vessels) is the same as derived from paint infusions (Chap. 1.2). There are no insured vessels beyond 32 cm in maple. *Bottom* Two saw cuts made 40 cm apart into the oak (*Quercus rubra*) stem whose vessel length is illustrated in Fig. 1.8 (*bottom*). Note that the axial-distance scales of the two graphs are very different

but the vessel counts. The original vessel counts from which the distribution in Fig. 1.7 was computed were as follows:

Number of vessels injected at 0 cm; number of paint-containing vessels downstream

0 cm	10,786 or	100.0%	20 cm	33	0.3%
4 cm	2,147	19.9%	24 cm	12	0.1%
8 cm	762	7.1%	28 cm	2	0.02%
12 cm	229	2.1%	32 cm		0%
16 cm	69	0.6%			

In other words, 20 cm from the saw cuts there are only 0.3% of the vessels air-blocked and 99.7% intact (see also Fig. 2.12). Consider now that the vessel path fans out from any point up or down, and also twists around the stem. This should make it clear that a double saw cut can be bypassed relatively easily. If worst comes to worst, water can always move tangentially in the tree by zigzagging up and down from vessel to vessel. We can make this visible by cutting the stem across through a radial injection hole (Fig. 2.4). Sperry et al. (1993) have shown that one pair of overlapping cuts 1 cm apart did not significantly change the soil-to-leaf hydraulic conductance of 5-m-tall birch trees. It took 4 to 12 cuts spaced 1 cm apart to achieve a 47% reduction in whole tree conductance.

To be fair, it should be stated that some of the above-cited double saw-cut experiments were made with elm, a ring-porous tree whose earlywood vessels are very long. So far, we have only one (unpublished) vessel-length-distribution measurement of a large elm tree; length distributions are comparable to those of oak or ash. Let us therefore look at the earlywood vessel counts for the oak shown in Fig. 1.8. The number of vessels injected was 286. At a distance of 0.5 m, 200 (69.9%) contained paint and, at a distance of 1 m, 158 (55.2%). Greenidge (1958) made his saw cuts 30–40 cm apart; the interruption is rather more drastic in this case. If we assume the cuts to be 40 cm apart and plot the vessel count graphically, we can see that only about 10% of the vessels are intact 10–30 cm away from the cut (Fig. 2.12). The experiment would probably fail with oak, but elm has an extremely wavy grain, i.e., as one follows single vessels along the stem one can see that they weave back and forth tangentially at a wavelength of some 20–50 cm. A film of transverse sections shows this very dramatically. Anyone who has tried to split elm wood knows this! The forester calls it cross-grained. This means that if two saw cuts completely interrupt the trunk in the axial direction, they do not interrupt all vessels because they twist around on both sides of the cuts.

3 The Cohesion-Tension Theory of Sap Ascent

3.1 The Motive Force Behind Water Movement in Plants

The cohesion-tension theory is also sometimes called the cohesion theory. It is a rather old theory dating back to the late nineteenth century, which attempts to explain the principal forces that govern the 'ascent of sap' from the soil to the leaves in plants. It was a very controversial theory when first proposed (Böhm 1893; Dixon 1914, pp. 142–154) and it has been the focus of controversy about every 20 years since it was postulated. At the time of writing, there has been more controversy over the cohesion theory than at any other time, excluding the initial controversy when it was originally proposed. The theory, at the very least, is incomplete in the sense that it describes a 'universal' motive force for water movement in plants. This should not surprise us because nineteenth century science had an incomplete understanding of what makes water move in plants because the science of thermodynamics was not mature enough to provide plant physiologists with the necessary insights. While the cohesion-tension theory is not perfect and has a few unsolved problems, I personally feel much of the current debate has proved to have little merit, as will become evident in this and subsequent chapters.

The principal motive force driving long-distance sap ascent in plants is water pressure and pressure plays a major role in the cohesion theory. The controversial aspect of the cohesion-tension theory is the 'improbable' concept of water under tension, which is equivalent to negative pressure in a fluid. Tension is easy to understand in a solid object, e.g., a rope used to lift a weight. While a weight is lifted with the aid of a rope, the rope is under tension; the fibers in the rope are being pulled apart by two opposing forces – the force of gravity acting on the weight and the counterforce at the other end of the rope where the person or machine is pulling on it. The cohesion theory proposes that water can be under tension in the special circumstances that exist in plants. The tension is supposed to be generated in the leaves at the liquid-air interface where water evaporates from leaves. However, before we get into the question of how the tension is generated, how much tension is needed to move water up a large tree, and how water can be stable under tension, we must review the nuances of the motive force for water movements in plants. This will allow us to view the cohesion theory in a more complete and modern context.

Most people are willing to accept the hypothesis that water moves in plants passively, which simply means that plants do not have metabolic water pumps. The equivalent organ in animals for a metabolic pump would be a heart. When water and solutes can move equally freely, e.g., in a xylem vessel, then passive movement is driven mostly by pressure gradient. When solute movement is restricted much more than water, e.g., water flow across a membrane, passive water movement is driven by the chemical potential of water, μ, which gives the energy state of water in J mol^{-1}. Water moves passively, i.e., spontaneously, from a place 'A' where the chemical potential is μ_A to place 'B' where the chemical potential is lower, μ_B, i.e., water moves passively from high μ_A to lower μ_B. Plant physiologists prefer to measure the chemical potential of water in pressure units, Pa=N m^{-2}, and give it the name 'water potential', Ψ. Plant physiologists also define pure water at atmospheric pressure as having a value of Ψ=0. The conversion from J mol^{-1} requires dividing μ by the partial molar volume of water, V_w, which has units of m^3 mol^{-1}. If μ_0 is the chemical potential of pure water at atmospheric pressure, then:

$$\Psi = \frac{\mu - \mu_o}{V_w} \tag{3.1}$$

It is critical to know that Ψ is an intensive variable, which means that Ψ is defined by a value at a 'point', e.g., a small region of space much larger than the size of a water molecule. You could view a point as being as small as a single water molecule if you are willing to define Ψ as a statistical time-averaged value, because the energy of a single molecule will change radically with time as it bounces around and interacts with other molecules and force fields.

Ψ is usually written as the sum of just two components, $P+\pi$, pressure plus osmotic potential, for reasons explained below. In the vapor phase, Ψ is approximately a log function of relative humidity (see Eq. 3.12, Sect. 3.3). In reality, the value of Ψ is the sum of several component values all of which are also intensive in the same sense as temperature, pressure and concentration are defined as values measured at a point in space.

The biggest factor that determines the value of Ψ is the product of Kelvin temperature and molar entropy, T times $\bar{S}$ (since Ψ is the difference in water potential between water in our system and pure water in the standard state, the same 'difference' meaning applies to $\bar{S}$.

Two drops of water that differ in temperature by 1 K differ in Ψ by about 7.7 MPa! In practice, biologists generally ignore $T\bar{S}$ because it is almost impossible to measure it with the same precision as the other components, i.e., ±0.01 MPa. In practice it is very dufficult to measure Kelvin temperature to ±0.1 °C and heat fluxes have to be measured to an equally high tolerance to obtain values of $\bar{S}$ or even changes in $\bar{S}$. Ignoring $T\bar{S}$ implies that we have to compare Ψ between water samples at the same temperature. This may seem like an impractical restriction given that the temperature of water can differ by several degrees K between water in the soil and leaf. However, temperature gradients have little effect on liquid water transport through vessels except for the effect of temperature on viscosity (Chap. 6.5). Temperature gradients will impact vapor phase transport of water and hence must be measured carefully when measuring water potential with psychrometers (Sect. 3.3).

The second largest component in Ψ is pressure, P. Unlike entropy, the effect of pressure on water movement is an everyday experience. It is common knowledge that the rate of water flow, F, though a pipe is proportional to the difference in pressure, ΔP, between the inlet and outlet ends of the pipe (see also the Hagen-Poiseuille equation, Chap. 1.2). This can be generalized to any water-filled porous medium, such as a collection of vessels in wood:

$$F=k\,\Delta P \tag{3.2}$$

where k is the hydraulic conductance of the porous medium. If the flow is in kg s^{-1} and ΔP in MPa, then k would be in units of kg s^{-1} MPa^{-1}. On the other hand, if we require F in terms of pressure gradient, as in the Hagen-Poiseuille equation, the proportionality constant would be called the hydraulic conductivity, K_h.

$$F=-K_h\frac{dP}{dl} \tag{3.3}$$

In Eq. (3.2) we adhere to the sign convention of positive flow from left to right, but, since the high pressure is on the left and the low on the right, ΔP is positive so there is no minus sign. However, when we deal with pressure gradients as in Eq. (3.3), dP/dl is negative, hence we need the minus sign. So far, thermodynamics makes perfect sense because it agrees with common experience and the Hagen-Poiseuille equation.

Another most important component of Ψ is solute potential, π, which measures how much the Ψ of water is lowered by the solutes being in solution with the water. There are instruments available to measure π more-or-less directly, but a simple computational formula, which is usually accurate to within 5 or 10%, is:

$$\pi =-RTC \tag{3.4}$$

where RT is the gas constant times Kelvin temperature (=2.48 MPa kg/mol at 25 °C) and C is the osmolal concentration in mol/kg of water. In an osmolal concentration you have to count each solute species separately, hence a 1 molal NaCl solution would be 2 osmolal because Na$^+$ and Cl$^-$ are both in solution. Similarly, a solution containing 1 molal NaCl plus 1 molal sucrose would have a 3 osmolal concentration. However, the influence of π on water flow in plants is more complex than it is for pressure. A $\Delta\pi$ difference or a gradient, $d\pi/dl$, has virtually no influence on the rate of water flow in a pipe, except insofar as the viscosity of a solution is increased by the presence of solute (see Eq. 1.2). However, $\Delta\pi$ does have an impact at the membrane level, depending on the solute. For solutes that permeate membranes very slowly compared to water, the flow rate across the membrane is $F=k_m\,\Delta\pi$, where k_m is the hydraulic conductance of the membrane, but, for solutes that are more permeable, we have to write:

$$F=k_m\,\sigma\Delta\pi \tag{3.5}$$

The new quantity, σ, is the reflection coefficient of the solute. The concept behind σ can be derived from basic principles of irreversible thermodynamics. However, we need not get into that, all we need to know is that σ is usually a number between 0 and 1. The more leaky the membrane to the solute, the lower is σ for the solute. We can now see why $\sigma\Delta\pi$ has no influence on water flow through the lumen of a vessel; there is no semipermeable membrane in a vessel, so $\sigma=0$. However, π can have some influence on long-distance transport in plants because π can influence the rate of transport through roots, specifically when water passes from the outside surface of fine roots to the vessels. Water passes through a membrane on its path into a root. Also, root cells can actively accumulate salts in the xylem by an active transport mechanism located in the membranes. Hence if J_s is the rate of solute uptake by a root in osmol s^{-1} per m^2 of root surface area, and if the rate of water flow into the root is J_w kg s^{-1} m^{-2}, the resulting osmolal concentration in the xylem will be $C_x=J_s/J_w$. The difference in π between the soil solution at the surface of the root will be given by $\Delta\pi=\pi_s-\pi_x=-RT(C_s-C_x)$. If we factor in the influence of pressure on water flow into roots, we have:

$$J_w=k_r\,[\Delta P+\sigma\Delta\pi]=k_r\,[(P_s-P_x)-\sigma RT(C_s-C_x)] \tag{3.6}$$

where k_r is the root hydraulic conductance for the nonvascular pathway, i.e., from the root surface to the first vessels. [Sometimes Eq. (3.6) is written with two $\Delta\pi$ terms, one with $\sigma=1$ and the other with $\sigma<1$. The first term is for the presumed, impermeable solutes and the other for the permeable solute under study in an experiment. The derivation of Eq. (3.6) is beyond the scope of this book but interested readers can consult Nobel (1983) for a simple introduction and further references.] When describing water flow into roots, water can enter from both the left and the right (if we draw the root vertically), so our sign convention is that positive flow is from outside to inside and the 'forces' are computed as outside minus inside values; the same sign convention holds for cells where water can enter from any direction. Equation (3.6) tells us that the driving force on water flow into a root will change with the rate of water flow because $C_x=J_s/J_w$ hence, at high flow rates, C_x becomes quite small and C_s is small in most soils except salt marshes and some saline desert soils. When typical values are put into the parameters in Eq. (3.6) it turns out that $\Delta\pi$ is much smaller than ΔP at midday when J_w is large but $\Delta\pi$ can be larger than ΔP overnight or on rainy days when J_w is small.

The discussion leading up to Eq. (3.6) allows us to conclude that pressure is likely to be the most important driving force for water transport in plants, where long-distance transport is through xylem conduits (vessels or tracheids). Pressure is also the main driving force for sap movement in the sieve tubes of the phloem, where water movement is against the water potential gradient because σ is zero in the lumen of sieve tubes. Hence we can ignore osmotic effects in our consideration of the cohesion theory. But first, let us resume the question of when you can and when you cannot ignore temperature effects on Ψ and on water transport.

In water-filled pipes, temperature gradients do not induce water flow. However, temperature increase does increase the rate of water flow induced by pressure gradients because the viscosity of water decreases about 2.4% for every °C rise in temperature. So the Hagen-Poiseuille law would predict a 2.4% rate of increase in K_h per °C in a stem or any water-filled porous medium. Also, as the temperature rises above 4 °C, water or entrapped air bubbles in vessels do expand a fraction of a % per °C causing transitory water displacement (we will come back to this topic when we discuss spring sap flow in sugar maples).

Temperature is predominant in vapor-phase transport, e.g., when water evaporates from a warm drop of water and diffuses to a cooler drop of water. This process is called vapor distillation and is an important mechanism of water transport in fairly dry soils. The other place where vapor-phase diffusion is important is when water evaporates from mesophyll air spaces inside leaves and diffuses to the outside air via stomates. In this case, the temperature differences between the leaf and air have a major influence on the driving force. We could write an equation for vapor-phase transport through a porous medium such as stomates, which is analogous to the equation for liquid transport. So if E is the vapor transport flux density from a leaf we could write:

$$E = k_{gas} \Delta \Psi \tag{3.7}$$

where k_{gas} is the vapor transport conductance of gaseous pathways through the leaf. This equation is rarely useful because k_{gas} is not constant; in theory, k_{gas} is strongly dependent on Ψ. This type of equation is useful only when trying to compare the relative conductances or resistances in the continuous pathway from soils, through plants to the atmosphere, i.e., in the so-called soil-plant-atmosphere continuum. So we will have to return to this equation in a later section. For tunately, for people studying gas exchange in leaves, there is a more useful equation:

$$E = g_L \Delta X \tag{3.8}$$

The leaf vapor conductance, g_L, is independent of the driving force ΔX, where ΔX can be the difference between the leaf and air of the vapor concentration, vapor pressure, or mole fraction of water. Equation (3.8) can be derived rigorously from Eq. (3.7) using thermodynamic principles, but this derivation is beyond the scope of this volume.

The mechanical properties of cell walls and water transport in cell walls are the defining characteristic most central to the water physiology of plants. The evolution of cell walls allowed the plant kingdom to solve the problem of osmoregulation in freshwater environments; confining protoplasm inside a rigid exoskeleton prevented cell rupture as a result of osmotic inflow of water. To illustrate this important difference in osmoregulation between plant and animal cells, let us imagine that two identical cells start out with equal water and osmotic potentials, say $\Psi = \pi = -2$ MPa. The plant cell differs only in that it is surrounded by an elastic cell wall. Both cells have membranes with identical hydraulic conductivities, i.e., they both restrict the rate of water flow identically. At time zero they are both placed in pure water which has $\Psi = 0$ by definition. Both cells will begin to swell in volume, as indicated in Fig. 3.1. As water enters, the concentration of solutes, C, inside is diluted making π less negative (remember that $\pi = -RTC$). In the case of the animal cell, this dilution will continue without limit because the membrane is too weak to resist the swelling of the cell. The water potential of the animal cell, $\Psi_{animal\ cell} = \pi$ can reach zero only at infinite dilution which means infinite volume. Obviously, this never happens so at some time, marked by '*', the cell bursts. In the case of the plant cell, the increase in cell volume stretches the elastic cell wall, which starts to apply an opposing pressure which plant physiologists call turgor pressure, P_t, i.e., the hydrostatic pressure in the cell. This turgor pressure causes a rapid rise in the plant cell's water potential since $\Psi_{plant\ cell} = P_t + \pi$. Although the plant cell solute content is diluted as the cell swells, at some point P_t rises to be equal to $-\pi$ at which point $\Psi_{plant\ cell} = 0$ and is in equilibrium with the pure water. The effect of the cell wall on P_t is quantified by the bulk modulus of elasticity, ε, which measures the rate of change of pressure per unit relative volume change of the cell, i.e., $\varepsilon = \Delta P_t / (\Delta V / V)$ and it can be shown that the time constant for the approach to water potential equilibriums is given by $t = V/(A\ L_m (\varepsilon + \pi))$, where L_m is the membrane hydraulic conductance and A is the area of the membrane. When a plant cell is exposed to pure water it typically swells 2–20% in volume depending on π and how rigid the cell wall is. Readers interested in more details should consult Tyree and Jarvis (1982) and Nobel (1983).

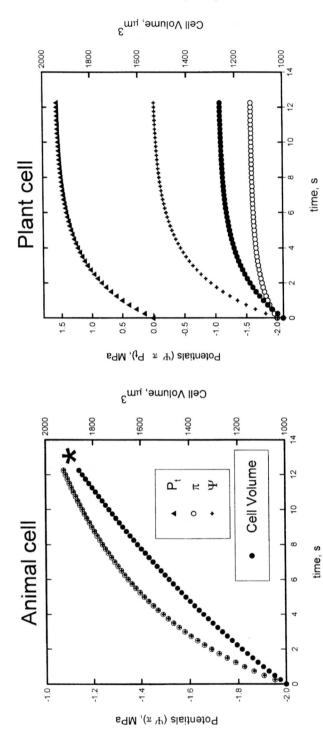

Fig. 3.1. Differences in osmoregulation of animal versus plant cell. These plots show the volume changes and water potential changes of two identical cells starting out at $\Psi=\pi=-2$ MPa at time zero when they are placed in pure water. The plant cell is surrounded by an elastic cell wall outside the plasma membrane, the animal cell has only the plasma membrane. Water flows into the animal cell uncontrollably until it bursts at *. In the plant cell, the cell wall resists swelling because of the elastic properties of the cell wall. The pressure inside the plant cell (P_t) increases until equilibrium is reached with the water.

The cost of cell walls was a loss of motility. Another cost is the inability to be holotrophic or phagotrophic; hence all 'foods' entering plant cells must be small enough to pass through the pores of cell walls. In contrast, in the animal kingdom, osmoregulation involved the evolution of a vascular system that bathed most cells in iso-osmotic blood plasma; this avoided rigid walls and permitted cell and organismal motility. Cell walls also placed constraints on the evolution of long-distance transport systems. Tissues were too rigid to evolve a heart pump mechanism. Instead plants evolved two novel transport systems.

One transport mechanism is a positive pressure system that moves concentrated, sugar-rich sap in the phloem from leaves to growing meristems or storage areas. Phloem transport uses a standing-gradient osmotic flow mechanism similar to that found in some animal excretory organs, but it is unique in that it occurs at very high pressure (up to 3 MPa) and requires two standing-gradient systems in tandem. One loads sugar and pushes phloem sap and the other unloads sugar and pulls phloem sap (Van Bel 1993). The other long-distance transport system controls sap ascent from soils to leaves and it is proposed that negative pressure is responsible for sap ascent.

3.2 Cohesion-Tension Theory: the Basic Elements

The cohesion-tension (C-T) theory was proposed over 100 years ago by Dixon and Joly (1894). Van den Honert (1948) introduced the Ohm's law analog of sap flow in the soil–plant–atmosphere continuum.

According to the C-T theory, water ascends plants in a metastable state under tension, i.e., with xylem pressure (P_x) more negative than that of a perfect vacuum. The driving force is generated by surface tension at the evaporating surfaces of the leaf and the tension is transmitted through a continuous water column from the leaves to the root apices and throughout all parts of the apoplast in every organ of the plant. Evaporation occurs predominately from the cell walls of the substomatal chambers due to the much lower water potential of the water vapor in air. The evaporation creates a curvature in the water menisci of apoplastic water within the cellulosic microfibril pores of cell walls; the curvature of the menisci lowers P_x (as will be explained in more detail in Sect. 3.7). Surface tension forces consequently lower P_x in the liquid directly behind the menisci (the air/water interfaces). This creates a lower water potential, Ψ, in adjacent regions including adjoining cell walls and cell protoplasts. The lowering of Ψ is a direct consequence of P_x, being one of the two major components of water potential in plants, the other component being the solute potential, π:

$$\Psi = P_x + \pi \tag{3.9}$$

The energy for the evaporation process ultimately comes from the sun, which provides the energy to overcome the latent heat of evaporation of the water molecules; i.e., the energy to break hydrogen bonds at the menisci.

Water in xylem conduits is said to be in a metastable condition when P_x is below the pressure of a perfect vacuum, because the continuity of the water column, once broken, will not rejoin until P_x rises to values above that of a vacuum. Metastable conditions are maintained by the cohesion of water to water and by adhesion of water to walls of xylem conduits. Both cohesion and adhesion of water are manifestations of hydrogen bonding. Even though air/water interfaces can exist anywhere along the path of water movement, the small diameter of pores in cell walls and the capillary forces produced by surface tension within such pores prevent the passage of air into conduits under normal circumstances. When P_x becomes negative enough, however, then air bubbles can be sucked into xylem conduits through porous walls (see Sect. 3.7).

This tension (negative P_x) is ultimately transferred to the roots where it lowers the Ψ of the roots to below the Ψ of the soil water. This causes water uptake from the soil to the roots and from the roots to the leaves to replace water evaporated at the surface of the leaves.

Van den Honert (1948) quantified the C-T theory in a classical paper in which he viewed the flow of water in a plant as a catenary process, where each catena element is viewed as a hydraulic conductance (analogous to an electrical conductance) across which water (analogous to electric current) flows. Thus, van den Honert proposed an Ohm's law analogue for water flow in plants. The Ohm's analogue leads to the following predictions: (1) the driving force of sap ascent is a continuous decrease in P_x in the direction of sap flow, and (2) evaporative flux density from leaves (E) is proportional to the negative of the pressure gradient $\left(-\dfrac{dP_x}{dx}\right)$ at any given 'point' (cross section) along the transpiration stream. Thus, at any given point of a root, stem or leaf vein, we have:

$$-\frac{dP_x}{dx} = \frac{F}{K_h} + \rho g \frac{dh}{dx} \tag{3.10}$$

where F is the rate of the water flow in a stem segment with hydraulic conductivity K_h and $\rho g\, dh/dx$ is the gravitational potential gradient where ρ is the density of water, g is the acceleration due to gravity and dh/dx the height gained, dh, per unit distance, dx, traveled by water in the stem segment. F is positive for flow from base to apex and is often estimated from AE, i.e., (the leaf area distal of the stem segment)×(average evaporative flux density from the distal leaves). Since branches can be pointing (from base to apex) in any direction (up, down, or horizontal) we must allow dh/dx to adopt any value from +1 to –1 for completely vertical to inverted, respectively.

In the context of stem segments of length L with finite pressure drops across ends of the segment, we have:

$$\Delta P_x = LF/K_h + \rho g \Delta h \tag{3.11}$$

In Chapter 4 we will discuss how Eq. (3.11) and a 'hydraulic map' of a tree can be used to reconstruct pressure-gradient profiles in large trees.

3.3 Negative Pressures (Water Tension)

Pressures in the xylem of plants are usually negative, although, in rare cases, positive values can be expected. This happens in certain species, such as *Acer*, *Betula*, *Vitis*, etc., during late winter when roots are active but leaves have not yet unfolded. In *Acer* and a few other species positive stem pressure can develop in response to freeze-thaw cycles even though the roots are inactive (see Sect. 3.9). If at that time the xylem is injured, it bleeds and this is a sure sign of positive pressure above atmospheric pressure. In certain herbs, xylem pressures become positive at night when transpiration is suppressed. Leaves then guttate from hydathodes, which are normally located at the leaf margins. Under the humid conditions of the tropical rain forest, positive pressures have been recorded in the xylem of several species even in the presence of leaves. Guttation from leaves was then observed (e.g., von Faber 1915); walking in the forest under such conditions reportedly gave the impression of walking in a drizzle of rain.

During the growing season water is lifted up into the leaves by pressures that are less than atmospheric created in the leaf xylem by transpiration. In wide-vessel trees such as oaks one can hear, upon injury of the xylem by a knife or axe, the hissing sound of air being drawn into the vessels. There are many other indications that water is pulled up. Kramer (1937) showed that absorption of water by the roots lags considerably behind transpiration from leaves. Huber and Schmidt (1936) showed with their thermoelectric method that water begins to move in the most distal parts of the tree after sunrise, and later in the basal part of the trunk. In the evening when transpiration declines, movement again slows down in the branches first. When velocities in the top are plotted vs. velocities at the base of the tree, a hysteresis curve is obtained (Fig. 3.2). This happens because the tree trunk is slightly elastic and can contract, and because of the xylem's storage capacity that is provided by capillarity (Chap. 4.8). Transpiration can thus begin without instantaneous water uptake from the soil. This phenomenon was also shown by Huber and Schmidt (1936), who mounted dendrographs at two different heights on tree stems, found shrinkage first in the upper and a little later in the lower part when transpiration began in the morning. Dobbs and Scott (1971) made similar measurements with the same results.

When conducting experimental work we very often depend on the fact that cut branches take up water when put into a container (or more poetically, when flowers are put into a vase). Moreover, dye ascents almost invariably depend on uptake of the dye solution by suction. However, this does not necessarily indicate negative pressures. Indeed, we use positive absolute pressures when we pull water through a stem with a vacuum pump or when we drink a beverage through a straw. The working pressures are then between +101 kPa (ambient) and near zero (vacuum), i.e., they are positive, absolute. Most pressure sensors in common use measure pressure relative to current atmospheric pressure, which is 101.3 kPa at sea level ± a few kPa depending on weather and less as elevation increases. All pressures in this book will be relative to atmospheric unless otherwise specified as absolute.

A 100 years ago, when the cohesion theory was first introduced, there was no direct way to measure negative pressure (absolute). More or less indirect methods were used to show that xylem pressures are actually negative, i.e., below zero abso-

Fig. 3.2. Velocities in the upper and lower part of the tree in *Fraxinus* (a wide-vessel tree) and *Betula* (a narrow-vessel tree) over the course of 24 h. The points on the curves are 06:00, 12:00, and 18:00 h. Absolute velocities are much higher in the wide-vessel species, and relative velocities are larger at the base than at the top. The reverse is true in *Betula*. However, in both species, the water begins to move first at the top, and later at the bottom. (Redrawn from Huber and Schmidt 1936)

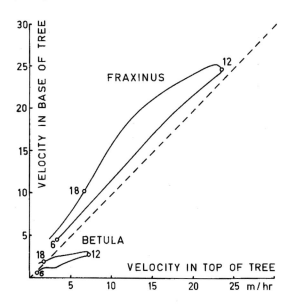

lute. In one experiment, probably first performed by Renner (1911), transpiring leaves were found to be able to pull considerably more water through an artificially constricted stem than a vacuum pump could (Fig. 3.3). Jost (1916) showed that *Sanchezia* shoots could pull a great deal more water from their root system than a vacuum pump. It is also possible to have transpiring shoots take up water from a sealed container in which the air space above the water is evacuated (Ursprung 1913). This works only with cut shoots of species whose stem contains sufficient tracheids and short vessels to provide a path from the water-contact area to the nearest intact vessels. It usually fails with branches of ring-porous species, which have wide and long vessels. Dye ascents performed under such conditions are crucial for the documentation of embolization. We will discuss this in more detail in the last chapter of this book.

The C-T theory of sap ascent is usually ascribed to H.H. Dixon, who gave a clear and detailed account in his book (Dixon 1914). However, the idea that water is under negative pressures in the xylem was certainly "in the air" around 1900. Böhm (1893) probably first demonstrated water under tension in a *Thuja* twig, that pulled up mercury via a water column beyond atmospheric height. For many decades after these early efforts, the C-T theory remained controversial. Few investigators doubted that negative pressures could occur in the xylem, but many did not believe that the C-T theory provided the full explanation of sap ascent. Two reasons caused such doubt: (1) it was hard to visualize how water could exist in such a metastable state for long periods of time, in many cases for years, and (2) nobody had demonstrated that negative pressures and gradients of negative pressures existed for any length of time. The first question, how water can remain in a metastable state for long periods of time, received little attention once negative pressures were shown to exist. The answer to this question undoubtedly involves

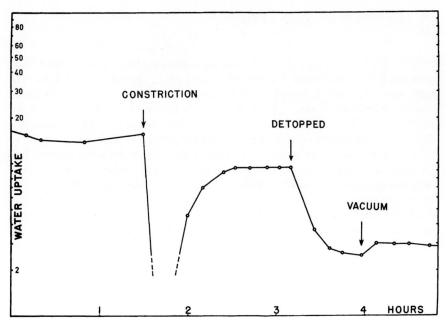

Fig. 3.3. Uptake of water by a branch from a potometer. For 1.5 h the branch transpired freely. A constriction was then made into the axis by saw cuts and clamping. This resulted in an initial reversal of water flow at the basal end. However, transpiration continued and stabilized only at a slightly lower level than before at 3.5–3 h. The leafy part of the branch was cut off and a vacuum pump was allowed to pull in place of it. Water consumption fell to about one third, indicating a pressure in the xylem prior to removal of the leafy part of –200 kPa. In this way, Renner (1911) showed pressures down to –1000 kPa, i.e., –900 kPa absolute. (Zimmermann and Brown 1971)

the intricate anatomical structure of the xylem. This is described in the first two chapters of this book. Water is confined to an extremely intricate network of compartments. The fact that they are so small and so numerous makes the seemingly impossible situation possible. We also have already encountered the limit: no plant has been able to evolve vessels much in excess of 0.5 mm width. If the compartments are too large, any accidental loss becomes too serious. The second question was resolved by Scholander, one of the critics of the cohesion theory, by the introduction of the pressure bomb (Scholander et al. 1965). The pressure bomb was actually a rediscovery. Dixon (1914, pp. 142–154) had experimented with such a device. Unfortunately, as Dixon's container was made of glass and exploded twice when he used higher pressures, he discontinued the experiments. Both questions are reviewed in more detail below.

3.4 The Tensile Strength of Water

It is not easy to visualize water under negative pressure. It is the same problem as visualizing water superheated above the boiling point. There is no liquid water in outer space, because it would all evaporate. When we use a vacuum pump to remove air from a container filled with water, we have no problem until the vapor pressure of water is reached. At room temperature this is 2.3 kPa (absolute) or 17.5 mm Hg on a barometer, which measures absolute pressure. Water begins to boil at that pressure and it is difficult or impossible to lower the pressure much further until all the water has evaporated. How can it then be subjected to zero or even to negative pressures? In an experimental setup it is the bubbles that are trapped on the wall of the container which expand and thus expose the liquid surface from which evaporation begins. This can be observed at home: boiling in a pan on the stove can be seen to begin as steady trains of bubbles emerging always from the same points, the gas nuclei trapped on the wall of the pan. But if the liquid and container walls are entirely bubble-free, the liquid may not evaporate in the presence of negative pressure. If the C-T theory is correct then plants must have accomplished this feat. Xylem walls must be completely bubble-free and water uptake by roots must exclude air bubbles, although not dissolved air. Water must hold together quite solidly by cohesion in the myriads of small xylem compartments.

As the pressure of water drops, a point will be reached where the column finally breaks. It may do so with a click, i.e., an acoustic emission (AE). Acoustic emissions have a broad sound spectrum from >1 MHz to 50 Hz or less, i.e., from ultrasonic to audible. Audible AEs can be amplified by a standard stereo-amplifier and listened to real-time (Milburn and Johnson 1966; Milburn and McLaughlin 1974) and ultrasonic AEs (UAEs) can be amplified and counted electronically or digitized and played back at a 1000 times slower speed to be made audible to the human ear (Tyree et al. 1984; Tyree and Sperry 1989b). The breaking of water columns is also referred to as cavitation. Cavitation is an engineering problem in the design of ship propellers. In order to propel the ship, a propeller has to "grab" water from forward and push it back. If this "grabbing" happens too vigorously, e.g., if the propeller runs too fast, a point may be reached where cavitation takes place and the propeller loses efficiency because it runs more or less freely in a mixture of water and vapor.

How can we show the tensile strength of water? There are indeed many ways. Let us begin with an example of the effect of air pressure. Every plant anatomist has probably had difficulties, at one time or another, separating individual microslides or coverslips. They sometimes stick together rather tenaciously. This is not because they are actually sticky, but because the pressure of the atmosphere pushes them together. The trick then is to pry them apart at one end in order to let air (and air pressure) enter between the slides. If we do not let air in from the side, it is much more difficult to separate two pieces of glass. A normal microslide measures 2.5×7.5 cm, its surface area is therefore 18.75 cm^2. One atmosphere (ambient air pressure) acting upon such a surface is equivalent to about 19 kg weight! In other words, without prying the slides apart at an edge to let the air pressure act between the two slides, we would have considerable difficulty in separating them.

Instead of elastic glass plates, visualize now two round steel plates with a diameter of 11.28 cm so that the surface area is exactly 100 cm^2. The plates are machined very precisely so that they are absolutely flat. We mount one of them on the ceiling of a room, flat side down, and push the other with its smooth side up to it. The plates are thick enough so that they cannot be bent easily and thus let air in from the side. A person weighing less than 100 kg (220 lbs) can hang on the lower plate without it letting go. It is held to the upper plate by air pressure. However, let a second person hang on it, air pressure is now insufficient and the plate will come loose. Let us modify the experiment. We now wring the two plates together with water from which all bubbles have been removed, for example, by boiling. The plates now hold together much more tenaciously by adhesive forces between steel and water, and cohesive forces within the water. In fact, it is likely that an entire class of 20–50 students could be supported. Budgett (1912) performed an experiment similar to this. He found the tensile strength of water under these conditions to be 0.4–6 MPa, which corresponds to 400–6000 kg weight in our 100-cm^2 plate pair described above.

Donny (1846) was able to hang up a column of sulfuric acid in a glass tube without atmospheric support from below. This has been repeated many times with carefully prepared water columns (e.g., Dixon 1914, p. 85).

Berthelot (1850) described how he filled water into glass capillaries and sealed them with only a small vapor bubble left in the liquid. By warming up the capillary, the vapor bubble was dissolved. By cooling again, the bubble reappeared only at a distinctly lower temperature, then often with an audible click. This could be induced by a slight shock to the tube, or by rubbing. The appearing bubble expanded immediately to equilibrium size. During cooling, just before the noisy reappearance of the bubble, the water was under tensile stress. Dixon (1914) spent a considerable amount of time measuring the tensile strength of water with this method; he arrived at values of 4–20 MPa. Dissolved gasses in the water did not change the results; he also performed the experiments with sap that had been extracted from the xylem of plants. Many investigators repeated these experiments, obtaining somewhat different results by assuming slightly different conditions and/or modifying the calculations.

Another way to test the tensile strength is to spin water in glass capillaries in a centrifuge, thus exposing it to stress. Briggs (1950) used a Z-shaped capillary, which was open at both ends. It was spun in its Z plane, which was horizontal and so mounted that its center intersected the projected spin axis. Even at high speeds breakage of the water column could be observed by a marked change in refraction of the tube. The tensile strength thus found ranged from ca. 2 to 28 MPa. The interesting result was the fact that these values were temperature-dependent. Between 0 and 5 °C the tensile strength rose from 2 to about 26 MPa; it reached a peak at ca. 10 °C and then slowly declined again (Fig. 3.4). While the slow decline above 10 °C is explainable, the sharp rise between 0 and 5 °C is rather mysterious.

So far, we have looked at tests of the tensile strength of water that were made in artificial containers. Internal surfaces of artificial containers always have many cracks and crevices which contain gas, hence considerable efforts are required to remove this gas by boiling or by application of high pressures before the tensile strength is tested (Apfel 1972). The stability of water in a tensile state is dependent on the absence of bubbles of a critical size (Oertli 1971). The relative ineffective-

Fig. 3.4. Tensile strength of water measured with a Z-shaped glass capillary of 0.6–0.8 mm diameter spinning at various angular velocities. The *upward pointing arrows* indicate that the water column was not broken. The remarkable result here was the tenfold increase in tensile strength from near 0 to 5 °C. (Redrawn from Briggs 1950)

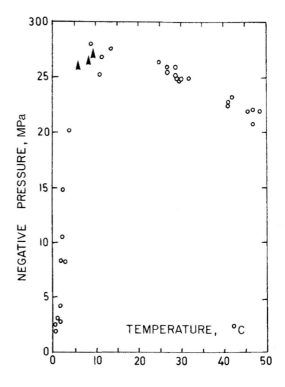

ness of bubble removal is probably responsible for the rather erratic results of tensile strength measurements in artificial containers. We might even go a step further and say that what was measured was not the tensile strength of water, but the size to which bubbles had been reduced prior to the experiment! Calculated values of the tensile strength of water are very high [130–150 MPa (Apfel 1972); 1500 MPa cited by Greenidge (1957)]. Others have attempted measuring tensile strength of water held in glass tubes coated with substances to simulate real vessels, but such exercises are of limited utility. Even the experiments of Briggs (1950) and the calculations of Apfel (1972) tell us only the maximum tensile strength in an artificial experimental situation. The important point is to measure the cohesive strength of water in plant stems. We now have many different ways to induce and measure negative pressures in plant stems, which induce increasing embolisms as pressure declines. We can measure the effect of this embolism on the loss of hydraulic conductivity in the stems. A plot of percent loss of hydraulic conductivity versus negative pressure is called a vulnerability curve and we will review these experiments in Chapter 4.3. Some plants evidently are able to provide much better conditions in the xylem than we can in the laboratory with glass capillaries, steel plates, and the like. The plant can grow xylem cell walls entirely bubble-free, and it takes up water via roots by the most scrupulous filtration process, even though dissolved gasses are not excluded.

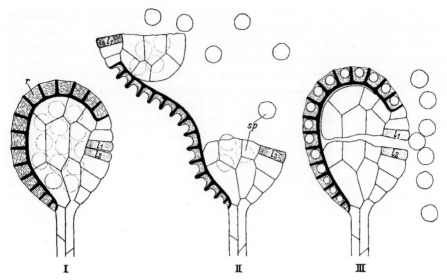

Fig. 3.5. Cohesion mechanism in the sporangium of a fern. *I* Sporangium closed. *II* Sporangium opens because transpirational loss of water from annulus cells produces tensions. *III* Increasing tension overcomes the cohesive strength of water, water vapor bubbles appear in annulus cells, the sporangium snaps closed, spores are ejected. (Stocker 1952)

Nature itself provides us with a demonstration of the tensile strength of water by the mechanism of spore ejection from fern sporangia. Certain fern sporangia possess an annulus of special cells, which reaches about two-thirds around the sporangium (Fig. 3.5). The walls of the annulus cells are thinner on the outside than on the inside. By the time the spores are mature, evaporation of water from the annulus cells decreases their volume, thus distorting their thin outer wall. The tension within the water of the annulus cells increases, thus forcing the sporangium open against the springy force of the annulus, until the tensile strength of the water is overcome. A vapor bubble appears in each of the cells, the sporangium snaps closed and the spores are thrown out (Fig. 3.5). Renner (1915) and Ursprung (1915) observed sporangia that they had enclosed in a small glass chamber with a microscope. The enclosure enabled them to control the relative humidity of the surrounding air with solutions of given concentrations. By increasing the concentrations gradually the two investigators found (simultaneously and independently) that most sporangia did not snap closed until the osmotic potential of the solution had reached ca. −35 MPa. In later experiments, Ziegenspeck (1928) and Haider (1954) confirmed that the water actually broke at the time of snapping; all cells contained bubbles, the first break seems to produce a jarring motion that triggers a break in all of them.

There are many aspects of tensile water that cannot be discussed here. Readers interested to learn more about this fascinating subject are referred to other publications (Oertli 1971; Apfel 1972; Scholander 1972; Hammel and Scholander 1976; Pickard 1981). What interests us here most is, of course, how tensile strength relates to wood anatomy. It is not at all clear whether tensile strength is related to compartment size, although the largest values were obtained with fern sporangia whose annulus cells were of the order of 10–20 µm in diameter, i.e., are the smallest test compartments used by experimenters.

Tensile strength should be a property of water, it should be high (above 100 MPa), more or less constant, and independent of the test container. In Chapter 4.3 we will see that different plants have different vulnerability curves. Why should there be a difference in tensile strength when measured with one (fern annulus) or the other (vascular bundles) plant material? One can easily come to the conclusion that it is impossible to measure the tensile strength of water experimentally. What we do measure is always the property of the enclosure. If we use an artificial material, such as glass capillaries, metal plates, etc., we may measure the dimensions of the gas nuclei extracted from the wall rather than the tensile strength of water. The somewhat erratic results obtained with artificial containers seem to support this idea. Briggs' (1950) results looked rather consistent, but one wonders whether his temperature dependence has also something to do with the size of gas nuclei. If, however, we use plant material for the measurement of the tensile strength of water, we may in fact measure the size of the pores in the walls of the water-containing plant cells. But before we get onto this topic we should first review how we measure negative pressure in plants, i.e., indirect measures and direct measures.

3.5 Indirect Measures of Negative Pressure

The C-T theory makes specific predictions about negative pressure in xylem (P_x, see Eqs. 3.10 and 3.11), but for the first 90 years after the C-T theory was proposed there was no direct way to measure negative P_x. A direct method would involve inserting a miniature pressure gauge directly into a xylem vessel. Today there are ways to do this using a cell pressure probe, which will be discussed in the next section. However, since most of our knowledge about the state of water in plants involves indirect measures of P_x, it is appropriate to review these methods.

The thermocouple psychrometer (Spanner 1951) was the first instrument capable of measuring Ψ and hence an indirect measure of P_x. This instrument used the principle that water in the liquid and vapor phase should reach thermodynamic equilibrium when enclosed in a sealed chamber with air. The equilibrium reached is given by:

$$\Psi_{liquid} = \Psi_{vapor} = \frac{RT}{V_w} \ln(H_r) \tag{3.12}$$

where H_r is the relative humidity (range from 0 to 1) and $\ln(H_r)$ represents the natural logarithm of the relative humidity. The value of $RT/V_w = 137.3$ MPa at 25 °C. The relative humidity can be defined as the vapor pressure (=partial pressure) of water vapor, p_w, divided by the saturation vapor pressure, p_o, which is the pressure that would result when air is in equilibrium with pure water, so $H_r = p_w/p_o$. Apparatus for the accurate measurement of p_w above salt solutions has been used for many years to measure the osmotic potential ($\pi=\Psi$) of different concentrations of salts. This apparatus has to measure p_w and p_o very accurately because p_w differs from p_o by less than 5% even in saturated salt solutions and this corresponds to $\Psi=137.3 \ln(0.95)=-7$ MPa. Plants in mesic climates and salt solutions rarely get to water potentials of less than –7 Mpa; however, it is worth remembering that plants in arid zones routinely fall below –7 MPa and reach –12 MPa in some cases.

An indirect way of measuring H_r in air is with a psychrometer, which consists of two thermometers one of which is kept dry and the other wet. Evaporation of water from the wet thermometer into air will make it cooler then the dry thermometer by an amount dT. The thermocouple psychrometer uses thermocouples to measure temperature. The thermocouple psychrometer is calibrated with salt solutions where the measured dT is found to be linearly proportional to Ψ_{vapor} adjacent to the thermocouple. Thermocouple psychrometers can measure Ψ only between –0.1 and –6 MPa, for technical reasons we need not discuss here. This method has been used successfully to measure quite small samples of plant tissue, e.g., leaf disks weighing just a few milligrams. However, great care must be taken to have a well-sealed chamber (so the tissue does not dehydrate while it equilibrates) and temperature has to be strictly controlled. As stated in Section 3.1, a difference in temperature of 1 °C between the sample and the thermocouple will mean Ψ_{vapor} is out of equilibrium with the sample tissue by –7.7 MPa, hence temperature has to be regulated to within ±0.001 °C to obtain measurements accurate to ±8 kPa.

The thermocouple psychrometer does not measure P_x directly in the xylem, but the argument is that tissue cells adjacent to xylem conduits are probably in approximate thermodynamic equilibrium with the water potential of the xylem, hence $\Psi_{cell}=\Psi_x=P_x+\pi_x$, and in most plants the xylem sap has a very dilute osmolal concentration making π_x fall between 0 and –0.05 MPa.

Various researchers have measured Ψ of plant tissue down to, say, –6 MPa, but does that mean that P_x can be –6 MPa in a plant? Unfortunately, we cannot draw that conclusion! Thermodynamically, P_x would have to be nearly equal to Ψ *only* if the xylem conduits are still filled with water when Ψ is measured and this condition of water-filled conduits must be proved. One way to prove that conduits are still filled, or that at least some are still filled, is to measure stem hydraulic conductivity as a function of Ψ, which is a topic we will discuss in Chapter 4.2. For now we will just say that some plants can maintain water-filled conduits down to –6 MPa and many others cannot.

About 12 years after the invention of the thermocouple psychrometer, Ted Hammel reinvented the pressure bomb (Scholander et al. 1965; see Sect. 3.2). Scholander received most of the credit for the Scholander-Hammel pressure bomb, because he went around the world lecturing about the device more than Ted did. In theory, the pressure bomb measures P_x more or less directly. The argument behind the measurement is as follows: Let us assume that a shoot is harvested from a transpiring plant. The C-T theory tells us that there will be a gradient of P_x and Ψ within the shoot at the time of harvest. Next, the shoot is placed inside a dark, humid, metal chamber (the pressure bomb) with the end of the branch protruding to the outside air through a rubber seal. Transpiration will stop and the potential gradients will dissipate because water from the 'wetter' cells will migrate to the 'drier' cells until P_x and Ψ are the same everywhere in the sample (we will ignore the residual evaporation that might occur from the portion of stem outside the pressure bomb). Water in vessels cut open at the time of harvest will also have drained up to the vessel ends, but the meniscus will hang up at the pits between vessels for reasons that will be discussed in more detail later. The cut vessels drain because when they are cut in air a new meniscus forms at the cut surface, which raises the pressure of the water in the cut vessel to something near atmospheric pressure, and the water is sucked into surrounding tissue which is at a lower water potential. For the sake of clarity, we will ignore the small increase in P_x that occurs when the meniscus withdraws into the cut vessels and hangs up on the pits. When the sample has equilibrated in the pressure bomb we have an equilibrium everywhere such that:

$$\Psi^o_x = P^o_x + \pi^o_x = \Psi^o_{cell} \tag{3.13A}$$

where the superscript 'o' represents the initial equilibrium state. Now let us suppose that we gradually increase the gas pressure, P_g, inside the metal chamber. The living cells are surrounded by a thin elastic cell wall, which is easily compressed by the gas, so the gas pressure will transmit easily to the cells increasing Ψ_{cell} such that:

$$\Psi_{cell} = \Psi^o_{cell} + P_g = P^o_x + \pi^o_x + P_g \tag{3.13B}$$

The gas pressure can be increased to a point where water is 'squeezed' out of the shoot. We are interested only in the squeezing that restores the water to the surface of the stem, i.e., the water needed to refill the vessels cut open when the branch was harvested, and the pressure needed to do this is called the 'balance pressure', P_B. At this point we have xylem water at an osmotic potential of π^o_x which must be in equilibrium with Ψ_{cell}. The xylem water is at atmospheric pressure because it is resting as a 'drop' on the surface of the stem. So at the balance pressure we have $\Psi_{cell} = \pi^o_x = P_B + \pi^o_x + P^o_x$. From this it follows that at the balance pressure:

$$P_B = -P^o_x \tag{3.13C}$$

We have gone through this rigorous derivation to dispel two common misconceptions in the literature. Firstly, many people incorrectly presume that the balance pressure measures $-\Psi$; it does not because it measures $-P_x$. Secondly, many people incorrectly presume that the balance pressure measures the 'dynamic' value of $-P_x$ that existed at the instant the branch was cut; it does not because it measures an 'average' equilibrium value *after* all the potential gradients have dissipated. We could add to this a third misconception, that balance pressure displaces the xylem fluid by squeezing the conduits to a smaller diameter; it does not because the pressure difference between the outside of the water-filled vessels and the inside remains unchanged during the return to balance pressure. For every unit of kPa P_g is increased, P_x increases by an equal amount.

What evidence is there that the equation leading to Eqs. (3.13B) and (3.13C) is correct? Putting Eq. (3.13C) back into (3.13B) tells us that the water potential measured at the cut surface of the stem, $\Psi_{stem\text{-}surface}$, should be:

$$\Psi_{stem\text{-}surface} = \pi_x + P_g - P_B \tag{3.14}$$

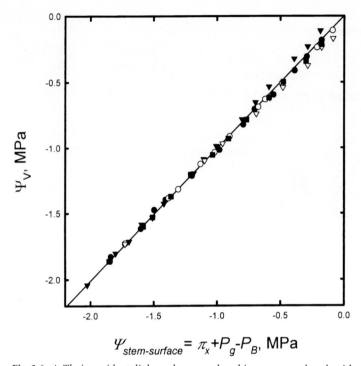

$$\Psi_{stem\text{-}surface} = \pi_x + P_g - P_B, \text{ MPa}$$

Fig. 3.6. A *Thuja occidentalis* branch was enclosed in a pressure bomb with an 8-mm diameter base protruding to the outside air. A thermocouple psychrometer was used to measure the water potential in the vapor phase (Ψ_V) and compared to the value from pressure bomb theory (Eq. 3.11). The shoot had been previously dehydrated to P_B=2.0 MPa. The agreement between methods validates the pressure-bomb theory that the pressure bomb measures xylem pressure. (Adapted from Dixon and Tyree 1984)

Dixon and Tyree (1984) attached a thermocouple psychrometer to the cut surface of a stem on a branch in a pressure bomb and confirmed the expected relationship (Fig. 3.6). More recently, Holbrook et al. (1995) have used a centrifuge to induce tension in the xylem of a branch. A branch of length 2L was mounted at the center of a centrifuge and spun at an angular velocity of ω. A branch was selected with a leaf attached at the axis of rotation. The tension developed by centrifugal force at the center of the axis of rotation is:

$$P_x = -0.5\,\rho\omega^2 L^2 \tag{3.15}$$

where ρ is the density of water. After sufficient time the leaf at the axis of rotations should reach equilibrium with P_x. The agreement was reasonably good (Fig. 3.7).

The Scholander-Hammel pressure-bomb (Scholander et al. 1965) is one of the most frequently used tools for estimating P_x. The C-T theory does not depend on the accuracy of the pressure bomb, but much of what we know about the range of P_x tolerated by different species of plants depends on the pressure bomb. Typically, P_x can range down to –2 MPa (in crop plants) or to –4 MPa (in arid zone

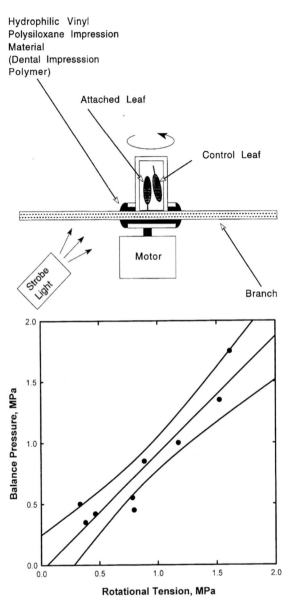

Fig. 3.7. The Holbrook experiment. The experimental method is illustrated above. A stem segment was mounted on a centrifuge motor. Centrifugation of the segment induced a P_x as given in
Eq. (3.15). The strobe light was used to measure angular velocity. *Bottom* Results showing that
the rotational tension was transmitted to the attached leaf; the x-axis is the calculated rotation
tension=P_x in Eq. (3.15) and the y-axis is the balance pressure of the attached leaf at the end of
the experiment

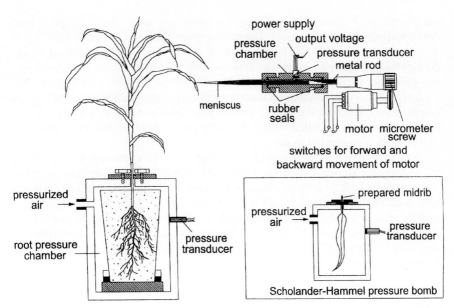

Fig. 3.8. The cell pressure probe (*upper right*) and its use in an experiment on maize plants. The roots are enclosed in a pressure bomb with a lid that can be sealed around the stem (details not shown). Pressurization of the soil mass with a gas pressure, P_g, permits manipulation of the xylem pressure, P_x, measured in a minor leaf vein. See text for more details

species) and, in some cases, –10 MPa (in Californian chaparral species). The pressure bomb is also easier to use than the thermocouple psychrometer. The pressure bomb, however, has the potential of refilling vessels emptied by cavitation events. Although it can tell us the P_x somewhere in the apoplast, it must be used in combination with other techniques to work out the limits of stability of metastable water under negative pressure in xylem.

3.6 Direct Measures of Negative Pressure: the Cell Pressure Probe

A cell pressure probe (CPP) is basically a pressure gauge that can be inserted into a single cell (Fig. 3.8) and has been available for measurement of positive turgor pressure in living cells since the early 1970s (Zimmermann et al. 1969; Zimmermann and Steudle 1974; Zimmermann and Hüsken 1979). Only since 1990 have they been applied to the measurement of negative pressure in xylem vessels.

The CPP shown in the upper right of Fig. 3.8 has an electronic pressure transducer built in a clear plastic body. A glass micropipette (1 mm OD and 0.5 mm ID) is inserted into one end of the plastic body and a movable metal rod (acting like the plunger on a syringe) enters from the other side. Both rod and pipette pass through an air-tight, rubber seal. The micropipette has been drawn to a fine tip at the other end (5 μm OD) and beveled to a 45° angle with respect to the long axis of the pipette. The sharp tip can be inserted, with the aid of a micromanipulator, into a single cell or vessel. The pipette and plastic body can be filled with water or low viscosity silicone oil. Silicone oil usually performs better than water, which tends to adhere less firmly to the plastic, rubber and metal surfaces of the CPP. The tip micropipette is always filled with 3–4 μl of water by drawing the metal rod back while the tip is immersed in water.

The CPP is much more prone to cavitations than the apparatus used to measure the tensile strength of water (Sect. 3.3). The cavitation threshold is easily measured in a CPP. The tip of the pipette is sealed off and then placed in a water bath. Heating and cooling of the water bath will raise or lower the recorded pressure because of the thermal expansion of the oil. The water-filled probes tend to cavitate at pressures of –0.7 MPa and the oil-filled probes can be taken down to –1.4 MPa (Wei et al. 1999). The principal problem with making measurements of P_x in xylem vessels is the difficulty of inserting the micropipette without seeding an embolism. The chance of air seeding can be minimized by slowly expelling water by slowly advancing the metal rod while inserting the tip into plant tissues. Surface tension probably determines when an embolism is sucked into a vessel punctured by a micropipette. To sustain an air-tight seal between the glass and the cell wall of the vessel requires that the gap between the wall and glass be less than 0.3 μm when P_x down to –1 MPa is measured. This probably explains why the lowest pressure measured so far in plants with the CPP is about –1 MPa below atmospheric (–0.9 MPa absolute).

The earliest experiments with the CPP in xylem found poor agreement with the pressure bomb in measurement of P_x but later experiments found otherwise. The experimental design has to be thought out carefully to design a fair test. The CPP measures pressure at a point with a fairly fast response time, i.e., 5–20 s. Hence the pressure probe is capable of measuring P_x at a single point in a transpiring plant where there are gradients in P_x; the pressure bomb can estimate P_x only after gradients are dissipated. The micropipette should be inserted in a leaf vein of the same leaf harvested for measurement of P_x in the pressure bomb. As will be seen later, gradients of P_x are much steeper in leaves than in stems, hence it is best to insert the micropipette into a covered, nontranspiring leaf. The rest of the plant has to be exposed to high light intensity and transpiring in order to generate negative P_x. The covered leaf will act as a pressure manometer tracking the changes in P_x in the adjacent stem. An ideal plant for the comparison is a monocot with large leaves, since the micropipette can be inserted about 15 cm from the tip while the tip distal of the micropipette is covered to prevent transpiration. This permits transpiration and pressure gradients in the basal portion of the leaf while eliminating pressure gradients where the sample is taken for the pressure bomb. The leaf tip is harvested after the CPP reading and placed in the pressure bomb.

Wei et al. (1999) have performed this kind of experiment on maize plants. A potted maize plant was placed in a pressure vessel as shown in Fig. 3.8. The lid of the vessel consists of four half-circle metal plates, which can be assembled around the stem with rubber seals. The upper two sets are offset 90° to improve the seal.

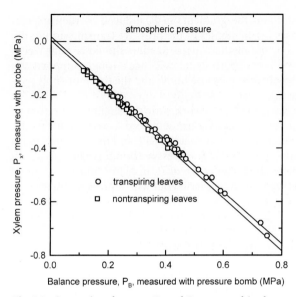

Fig. 3.9. Comparison between P_B and P_x measured in the same leaf as shown in Fig. 3.8. Each *symbol* represents measurements of P_x and P_B for a different leaf (total: 65 leaves measured on 65 plants). *Square symbols* represent leaf tips covered with aluminum foil to prevent transpiration (slope=-0.984). *Circles* represent uncovered leaf that was allowed to transpire (slope=-0.967). The results show that the pressure probe (a direct measure of P_x) and the pressure bomb (an indirect measure of P_x) measure similar values

The plates are bolted down to the body of the pressure chamber (details not shown in Fig. 3.8). The plant is exposed to high light intensity to induce transpiration and low P_x. The value of P_x is under experimental control, i.e., P_x can be decreased to more negative values by increasing light or P_x can be easily increased or decreased by adjusting the gas pressure, P_g, exerted on the soil water in the pot. Quite a good agreement was obtained between the pressure bomb and the CPP (Fig. 3.9). When the leaf tip was covered (nontranspiring), the agreement was best. When the tip was uncovered, the P_x measured by the pressure bomb was offset to slightly more negative values reflecting the fact that there were gradients in the leaf tip. A year earlier, Melcher et al. (1998) carried out similar experiments on maize and sugarcane leaves, but poorer agreement was found because the leaves harvested for the pressure bomb were not the same leaves as used for the pressure probe. Differences in light absorption on adjacent leaves can cause large differences in transpiration and gradients in P_x of adjacent leaves. Hence we would not expect as good an agreement.

Balling and Zimmermann (1990) were the first to compare P_x to P_B; experiments were mostly conducted on tobacco seedlings. They concluded that the gas pressure applied, P_g, is not transmitted on a 1:1 basis as hypothesized in the pressure-bomb theory (Eq. 3.13). The cause of most of this 'disagreement' has been explained and is beyond the scope of this book. Briefly, in some cases, I believe the authors forgot to take into account the gradients in P_x induced by their experimental methods. In other cases the dynamics of the pressure transmission in the pressure bomb were changed by the methods used. Readers should consult Wei et al. (2000) for more details. I believe experimental evidence strongly supports the notion that the cell pressure probe, the pressure bomb and the thermocouple psychrometer are all in agreement.

The cell pressure probe is very difficult to work with; the slightest vibration or movement of the leaf by just a fraction of a µm will seed a cavitation making measurement of P_x impossible. It is fortunate that the CPP and the pressure bomb are in agreement over measurements of P_x. The pressure bomb is much easier to use and most of what we know about the range of P_x in plants and most of our tests of the cohesion-tension theory can be carried out with a pressure bomb.

3.7 Testing Aspects of the Cohesion-Tension Theory

A full test of the cohesion-tension theory requires measurement of the hydraulic architecture of the plant under study. The theory predicts negative pressure, i.e., P_x below atmospheric pressure, and it predicts that there should be gradients of P_x (dP_x/dx) such that P_x declines in the direction of flow. The magnitude of $-dP_x/dx$ should increase and decrease with increases and decreases in transpiration. Furthermore, in the most exacting quantitative test, there should be an exact agreement between P_x at any point in a plant and the P_x predicted at that point. This final, quantitative test requires a detailed quantitative description of the hydraulic architecture of the plant under study. That is, we need knowledge of the hydraulic conductances of roots, stems, petioles and leaves and we need to know the hydraulic conductances with sufficient spatial resolution so that we can compare predicted P_x values to those actually measured at any given point. The quantitative basis of the comparisons requires the repeated application of Eq. (3.10) or (3.11) to each plant part, i.e., within the soil-to-leaf-surface continuum with the necessary spatial resolution. A full quantitative test has never been conducted in the entire plant continuum, but there have been reassuring confirmations of some of the predictions of the C-T theory.

However, the introduction of the Scholander-Hammel pressure bomb (Scholander et al. 1965) made measurements of gradients of negative pressures along a tree trunk possible. Like everything, the pressure chamber has its limitations. Measurements cannot be taken directly on the stem xylem; one needs a small twig or leaf. One must also realize that as one applies outside pressure to a twig, one may fill the capillary space system (discussed in Chap. 4.8), because the pressure difference between the intercellular air and the xylem water decreases, thus changing the balance pressure slightly. Boyer (1967), Kaufman (1968), and West and Gaff (1976) have given this problem special attention.

Figure 3.10 shows pressure gradients that were obtained by Scholander et al. (1965) on Douglas fir. The twigs were obtained by shooting them down with a rifle. The gradient shows the expected drift into a more negative region during the daylight hours at the time one would expect transpiration to intensify. What is

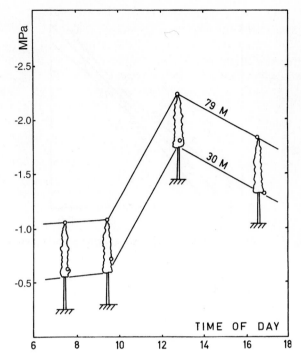

Fig. 3.10. Gradients of xylem pressure in Douglas fir. (Redrawn from Scholander et al. 1965)

surprising, however, is the fact that the gradient is always close to 10 kPa m^{-1}, indeed often less. This is physically impossible if all our assumptions are correct. During the midday hours the gradient should be steeper on account of flow resistance. Tobiessen et al. (1971) investigated this problem with a great deal of care. They mounted an elevator on a tall redwood tree and took shaded twig samples close to the trunk on a cool (11–14 °C) and relatively humid (70%) day. They sampled at five levels from ca. 20 to 80 m height and found a gradient of only 8 kPa m^{-1}, i.e., less than hydrostatic. When water flow, F, in a stem is zero, Eq. (3.10) predicts that $-(dP_x/dx)$ should equal $\rho g(dh/dx) = 10$ kPa m^{-1}. Gradients measured with the pressure chamber do sometimes appear steep enough to account for both gravity and resistance to flow (F/K_h in Eq. 3.10). Steeper gradients were measured in *Eucalyptus regnans* (Connor et al. 1977).

The solution to this problem is relatively simple. There are pressure gradients up the trunk and out into the branches. When one samples twigs, one goes quite literally out on a limb. In other words, the recorded pressure may be much lower than the pressure at the same height in the stem. Hellqvist et al. (1974) and Richter (1974) drew attention to this problem. As we shall see in Chapter 5, the xylem path may be quite constricted at branch junctions, and the conductivity of branches is less than that of the stem. Pressure chamber measurements cannot be taken as pressure values of the stem xylem without special precautions, simply because they are taken elsewhere!

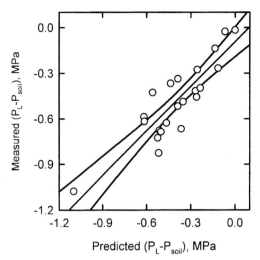

Fig. 3.11. Computed versus measured $\Delta P_x = P_L - P_{soil}$ for *Acer saccarinum* plants growing in pots. Each *point* represents a different plant. Evaporative flux density (E) was estimated from weight of water lost from the potted plants and leaf area measured at the end of the experiment. E was varied by controlling light intensity. Once a steady-state value of E had been reached, the balance pressure of a transpiring leaf was estimated (P_L). Soil water potential (P_{soil}) was estimated by the balance pressure of nontranspiring plants. So the y-axis is the measured pressure drop from soil to leaf. At the end of the experiment the whole plant was harvested and the hydraulic conductance of the root (k_R), the stems (k_S) and the leaves (k_{Leaves}) were estimated using a high-pressure flow-meter. The predicted ΔP_x on the x-axis was computed by the application of Eq. (3.10) to each conductance element (root, stem, and leaf). (Makoto Tsuda and Melvin T. Tyree, unpubl. data)

There is a relatively easy way to remedy the problem. The twigs to be measured can be identified and their transpiration can be suppressed by enclosing them in plastic bags. At the same time one has to cut off all leafy parts of the branch so that transpiration in that lateral branch is entirely suppressed. It would be interesting to repeat the measurements of Fig. 3.10 in this way, but this has not yet been done so far as I know. Hellqvist et al. (1974), Tyree et al. (1983), and Tyree (1988) found much steeper gradients in trunks and primary branches of smaller trees using the bagged shoot method, e.g., of the order of 20 kPa m^{-1} in trunks and 100 kPa m^{-1} in primary branches.

Other studies have confirmed the correctness of the Ohm's law analogue. Ewers et al. (1989) studied the hydraulic architecture of a large woody vine (lianas >20 m long) growing along the ground to avoid the effects of gravity on P_x. They measured K_h and $F = A\,E$ to apply to Eq. (3.10). They used the Ohm's law analogue to compare predicted values of $-dP_x/dx$ with values measured on bagged leaves using a pressure bomb. The predicted gradients (0.083±0.033 MPa m^{-1}) agreed with the gradients measured at midday with a pressure chamber (0.076±0.016 MPa m^{-1}). Measured and predicted gradients also agreed when $E=0$.

Tsuda and Tyree (unpubl. results) have compared the predicted drop in P_x from soil to leaf (based on gravimetric measures of E and whole shoot and root

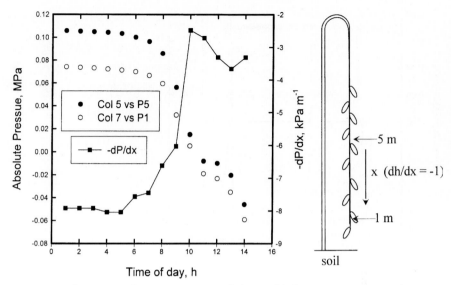

Fig. 3.12. *Right* Diagram showing orientation of vine used in long-term pressure probe measurements of P_x measured at two heights in a woody vine, *Tetrastigma voinierianum*. Note that the vine has grown about 10 m to the top of a greenhouse then begun growing down again. Measurements were made on the inverted portion of the vine. *Left* Pressure probe values read from Fig. 3.6 (Thürmer et al. 1999) at 1-h intervals and pressure gradients derived from the pressure probe readings of P_x. See text for more details

hydraulic conductances) to values of P_x measured with a pressure bomb and have found good agreement in young *Acer saccharinum* plants (Fig. 3.11). In this case, a high-pressure flowmeter (HPFM) was used to estimate root, stem and leaf hydraulic conductances to use in Eq. (3.10). This method will be discussed in more detail in Chapter 6.1.

Long-term measurements of P_x in lianas using the pressure probe have been cited as evidence against the C-T theory. However, the conclusion seems to be the result of confusion over sign conventions. Benkert et al. (1995) and Thürmer (1999) report pressure readings versus time like those reproduced in Fig. 3.12 at two different heights in *Tetrastigma voinierianum*. The pressure differences at night are about right ($\sim$10 kPa m^{-1}) but appear to decline during the day while the absolute pressure becomes more negative. However, they measured lianas in a greenhouse, which had grown up to the maximum height of the greenhouse and then grown back down towards the ground. Their measurements were done on the inverted part of the vine. Equation (3.10) is restated below:

$$-\frac{dP_x}{dx} = \frac{F}{K_h} + \rho g \frac{dh}{dx} \tag{3.10}$$

For any shoot, the positive sense of direction (x) and flow is from base to apex and this applies whether the shoot is upright or inverted. For a horizontal liana,

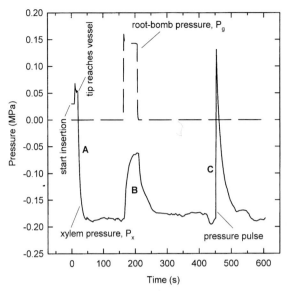

Fig. 3.13. A time sequence involving xylem pressure measurements with a cell pressure probe in maize. *A* Insertion of the micropipette is shown followed by an adjustment of the steady-state value of P_x; *B* a pulse of pneumatic pressure P_g was applied to the root which was rapidly reflected in a change of xylem pressure (half-time about 8 s); *C* typical pressure relaxation measurement, where the metal rod is rapidly moved forward into the cell pressure probe to raise the pressure. The rapid return to the original pressure indicates that the volume displaced by the rod has rapidly flowed out of the pressure probe. This proves that the pressure probe was not blocked by cellular debris. (Reproduced from Wei et al. 1999)

$dh/dx=0$ so $-(dP_x/dx)$ should start out 0 when $F=0$ and grow more positive as F increases. For a vertical liana, $dh/dx=1$ so $-(dP_x/dx)$ should start out at 10 kPa m^{-1} when $F=0$ and should grow more positive as F increases. However, for an inverted liana shoot, dh/dx has a negative value! So we would expect $-(dP_x/dx)$ to start out at -10 kPa m^{-1} at $F=0$ and grow more positive (less negative) as F increases during the day. This is exactly what happens in the data when P_x values are replotted as gradients in Fig. 3.12. So this pressure probe experiment provides strong evidence in favor of the C-T theory. An interesting aspect of this inverted-liana experiment is that if F had increased some more we would have had non-zero flow with $dP_x/dx=0$, at which point flow will be driven totally by gravitational force, i.e., $F=-K_h \rho g \, dh/dx$. In Fig. 3.12, $-(dP/dx)$ is -8 kPa m^{-1} overnight, whereas it should be -10 kPa m^{-1} when $F=0$. Other replications of this experiment gave values nearer -10 kPa m^{-1}; the calibrations on the pressure probes may have been off by 2 kPa, or there may have been some flow due to root pressure.

Quite a lot is known today about pressure gradients in large trees, some of it has been measured using the pressure bomb or thermocouple psychrometer. Much more is known from the analysis of the hydraulic architecture of trees, which will be discussed in Chapter 5.

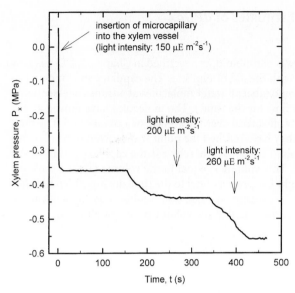

Fig. 3.14. Effect of light intensity on leaf xylem pressure in maize. Light intensity was 150 μmol m^{-2} s^{-1}, when a xylem vessel of the leaf was probed (*arrow*). Light was then increased 200 and 260 μmol m^{-2} s^{-1} (*arrows*). It can be seen that there was a rapid response in P_x that corrsponded to the measured rate of transpiration as measured (data not shown) by weight changes in the plant. (Reproduced from Wei et al. 1999)

Now let us consider the question of pressure transmission from the soil to the leaves. The C-T theory predicts that the fluid pressure at the root surface should equal the Ψ of the soil immediately adjacent to the root surface. From the root surface, the change in P_x in the plant should be determined by the following relationship: $P_x = \Psi_{soil} - F/k_p$, where F is the flow rate from the soil to where P_x is measured and k_p is the hydraulic conductance of the plant from the soil to where P_x is measured. So any change in Ψ_{soil} should be more-or-less immediately registered in P_x in the leaves. This prediction has been confirmed by the experimental setup in Fig. 3.8. The soil water potential was changed rapidly by increasing the gas pressure, P_g, on the soil mass (see Fig. 3.13).

Light intensity is predicted to affect transpiration in two different ways. First, at a given stomatal opening, an increase in light increases leaf temperature and the water vapor pressure at the evaporative surface. This, in turn, increases the force driving the diffusion of water vapor across the stomatal pores. The other effect of light intensity is to increase the stomatal width which will also increase transpiration. Hence, the C-T theory would predict a decline of P_x with increasing light intensity and this has been confirmed in maize (Fig. 3.14) and in lianas (Fig. 3.12).

For the remainder of this book we will interpret results on the presumption that the C-T theory is correct. We will, however, refer to other challenges to the C-T theory when certain new topics are introduced.

3.8 Sap Velocities and Estimates of dP_x/dx

Flow through ideal capillaries is paraboloid, as described in Chapter 1.3. Let us now visualize this in more detail with the aid of Fig. 3.15. The capillary shown here has a radius r. Imagine that we could label all water molecules of a transverse-sectional plane at A. We then let them flow for the time t. The molecules have moved at different velocities; they are now all spread over the surface of a paraboloid. The ones in the center moved fastest, they have reached the point B and covered the distance h. The volume of the paraboloid is $r^2\pi h/2$, whereby r is the capillary radius and h the height of the paraboloid. This volume is equal to the volume flowing during a given time t whereby the height h is proportional to the time during which flow has taken place. The flow rate, according to Hagen-Poiseuille, has been given in Eqs. (1.1) and (1.2). We can set this equal to the volume of the paraboloid:

$$\frac{Q}{t} = \frac{\pi r^4}{8\eta}\frac{dP}{dl} = \frac{r^2\pi h}{2t} \tag{3.16}$$

whereby Q/t is the volume flow rate, η the viscosity, and dP/dl the pressure gradient. The peak velocity is therefore:

$$\frac{h}{t} = \frac{r^2}{4\eta}\frac{dP}{dl} \tag{3.17}$$

Velocity is directly proportional to the pressure gradient and to the square of the capillary radius (Fig. 3.15). The fact that flow follows a paraboloid is important conceptually because it means that velocities range from zero at the vessel walls to a peak in the center of the vessel. When measuring velocities and when talking about them, we must be aware of this. Peak velocity is a useful concept, which we shall discuss further. Another concept is average velocity. This is based on the notion that liquid moves like a solid cylinder = Q/t divided by the cross-sectional area of the cylinder (πr^2). This average velocity turns out to be exactly 1/2 the peak velocity, i.e., Eq. (3.16) divided by πr^2 compared to Eq. (3.17). This never happens, of course, but it is nevertheless a useful concept when dealing, for example, with specific mass transfer in sieve tubes (Zimmermann and Brown 1971). If we inject a dye into a conduit, what velocity is measured? Given that velocities range from zero at the sides to a peak in the center, the situation was not at all clear to many plant physiologists.

Sir Geffory Taylor was the first to address this problem. He reasoned that diffusion must be averaging out the different velocities. Laminar flow will make the concentric sheet of water move at different velocities. Lets say that all the sheets start out with the same concentration of dye behind a plane and zero concentration in front of the plane. If the peak sap flow is 2 m h^{-1} in the center of a 60-μm diameter pipe then, in 1 s, the center will advance about 0.28 mm or 280 μm. However, while the dye is advancing down the axis it is diffusing radially. How long would it take for diffusion to carry the average molecule from the center to the edge of the cylinder (30 μm)? It turns out that the time required is about 0.5 s (Tyree and Tammes 1975). Taylor (1953) solved the mathematical problem of what happens in this situation. Theory predicted that the dye would move at the aver-

Fig. 3.15. Flow paraboloid in a capillary of radius r. If, at zero time, we could label all water molecules on a transverse-sectional plane at A, they would be lined up on the surface of a paraboloid at time t. The fastest ones, in the center of the capillary, would have covered a distance h and reached the point B. A parabola is usually shown as $y^2=2px$; the x- and y-axes are indicated, but the drawing would be rotated 90° counterclockwise. In reality, diffusion of molecules in the radial (or y-) direction would make this mental experiment impossible (Taylor 1953); the labeled molecules would move at the average velocity of the water if diffusion occurred in narrow tubes

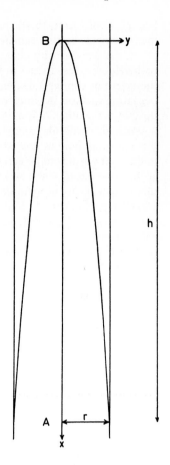

age velocity, which is equal to one half the peak velocity. Experiments have been performed to show that the theory is correct.

Velocities can be measured with a tracer (dye or isotope) only on cut shoots after the cut end has been immersed in water and xylem pressures are relaxed. If the tracer is injected into an intact plant, a point of +100 kPa is introduced into the xylem stream and the existing pressure gradient is thereby totally changed. This can be seen from the fact that injected dyes always move down as well as up in the stem. Acid substances move fastest, alkaline substances get absorbed on the wall. The latter are useful to mark a track, but cannot be used to obtain even a rough velocity measure.

The most elegant method to measure flow velocities is the thermoelectric or heat-pulse method. It was first used by Rein (1928) to measure flow velocities of blood in animals, and adapted for the measurement of xylem sap velocities by Huber (1932). The principle is to heat up the liquid briefly at a point along the stem and time the arrival of the heat pulse downstream. A little wire loop underneath a bark flap is heated up by an electric current of a very brief duration. The arrival of the heat wave is recorded by a thermocouple 4 cm downstream from the heating wire. Using this arrangement, Huber and Schmidt (1936) made many measurements on cut branches and on standing trees and reported diurnal as well as annual flow velocity changes. However, this arrangement does not work well with velocities less than 1 m h^{-1} (0.3 mm s^{-1}). For such slow movements (e.g., in the xylem of conifers), Huber and Schmidt (1937) developed the so-called compensation method in which one of the thermocouple junctions is located 16 mm upstream, the other 20 mm downstream from the heat source.

Methods used today sometimes differ slightly in configuration (in some procedures a small hole in drilled into the wood, thus interrupting the xylem at that point) and one has to be careful when trying to obtain absolute values (Marshall 1958; Cohen et al. 1981; Swanson and Whitfield 1981). In fact, absolute velocity measurements are quite useless (see Chap. 4.3). Velocities range from zero at vessel walls to a peak in the center of the widest vessel. Vessel diameters usually vary widely in any plant part. The only well-defined velocity is therefore the peak velocity in the center of the widest vessels, but it is doubtful if this is ever recorded (Rouschal 1940, p. 230). In addition, ecologists are often not really interested in velocities but in volume flow that can be calculated from heat-pulse measurements with proper calibration.

The heat-pulse method has provided us a great deal of *comparative* information about xylem sap velocities in plants. Velocities decrease to almost zero at night and usually reach a peak shortly after noon. The curves shown in Fig. 3.2 show three major points. First, velocities in wide-vessel trees like ash are considerably greater than those of narrow-vessel trees like birch. Second, in some trees (e.g., ash), velocities diminish from the bottom to the top of the trunk, and in others (e.g., birch) they increase. Third, the diurnal plots of velocities at the bottom vs top of a tree show a hysteresis loop, which indicates that water is pulled up by transpiration (Sect. 3.1). Huber and Schmidt (1936, 1937) also reported peak velocities of many different species. Trees with wide vessels (diameters of 200–400 µm) showed midday peak velocities of 16–45 m h^{-1} (4–13 mm s^{-1}). Trees with narrow vessels (50–150 µm) had slower midday peak velocities, in the range 1–6 m h^{-1} (0.3–1.7 mm s^{-1}). To this latter group belong two species of the dry Mediterranean region, namely *Fraxinus ornus* and *Quercus ilex*, whose northern relatives belong to the first (wide-vessel) group (Rouschal 1937). Slowest peak velocities were recorded on conifers with the compensation method; they corresponded to the slowest velocities of narrow-vessel trees.

It is useful to discover what pressure gradients are required to achieve such midday peak velocities. Even if we consider that vessels are only about 50% efficient (Chap. 1.3), it can be seen that pressure gradients in tree trunks, due to flow resistance, should be of the order of 10 kPa m^{-1}. However, we must take this value of dP/dx as a minimum because the velocity of heat flow is likely to be less than the average velocity of the sap. This follows because the heat will diffuse into the surrounding wood tissue as the water column advances. This leakage of heat will likely be more than the leakage of a chemical tracer since plasma membranes of living cells will retard the leakage into adjacent cells. When molecules spend part of their time in slowly moving water layers near the conduit walls, Taylor's theory (1953) predicts that the average velocity is half the peak. The surrounding tissue is equivalent to a larger non-moving region that will reduce velocity even more. This seems to be generally true for trees.

However, consider now a very short herbaceous plant. If conditions in leaves and roots are similar to those of trees, then the plant can afford very much steeper gradients and very much greater velocities. Begg and Turner (1970) found pressure gradients of 0.8 bar m^{-1} in tobacco. Rouschal (1940) found a maximum velocity of 108 m h^{-1} in *Impatiens parviflora*. More recently, Passioura (1972) has found in wheat plants that flow rate measurements indicated velocities up to 900 m h^{-1} (25 cm s^{-1}) through a single vessel of the taproot. By forcing plants to obtain water from a single root, velocities could be increased experimentally to 2880 m h^{-1} (0.8 m s^{-1}). Comparable findings (i.e., steep pressure gradients in small plants) were reported by Woodhouse and Nobel (1982).

3.9 Another Mechanism of Sap Flow (Sugar Maple)

The objective of this section is to review the present state of knowledge concerning the mechanism to transport sugar-rich sap in early spring when ambient temperatures hover around 0 °C. This sap exudation is the basis for an industry for production of maple syrup and maple sugar in northeastern North America. Sap flow occurs in commercial quantity in March and April in the northeast, the start date depending on climate. Holes about 6 mm in diameter and 10 cm deep are drilled into the sapwood and a tap is inserted to guide sap flow to a bucket or a tubing system. Sap flow is spontaneous and occurs under positive pressure and occurs before bud expansion begins. The sap is typically 2% sugar by weight and 40 l of sap generally comes from one tap hole in a season. This 40 l is concentrated by boiling down to 1 l of syrup containing 66% sugar. Although the direction of sap flow in a tapped tree is physiologically artificial, the mechanism inducing the flow must somehow be active in the intact tree.

Despite an extensive literature of several hundred papers, the mechanism of maple sap exudation is still not fully understood. Various hypotheses involving both vitalistic (Wiegand KM 1906; Johnson 1945) and physical (Jones et al. 1903; Stevens and Eggert 1945) mechanisms have been proposed but none as yet satisfies all observations. Root pressure is not considered a significant driving force since exudation may occur in excised stem segments whereas little exudation is found from root-stocks after removal of shoots. It has been established that sap flow is temperature-dependent with maximum yields occurring during the period when the wood temperature fluctuates above and below 0 °C (Marvin and Greene 1951; Marvin and Ericson 1956; Wiegand 1906). Uptake of soil water occurs during the cooling period and has been termed a "conditioning" phase by Marvin (1958). This conditioning is required for large sustained volume flows during the warm period.

The literature can be divided into ideas and research prior to 1980 followed by a 'watershed' in concepts beginning with the PhD dissertation written by O'Malley (1979). Reports begin from the early part of the seventeenth century. Space does not permit a full and critical review of even the most important papers. Although there are a few inconsistencies in the literature regarding experimental facts, most of the disagreement pertains to mechanistic interpretations, and the early discussions of mechanism are often closely tied to the disputes common in the nineteenth and early twentieth century literature regarding the now-understood mechanism of sap ascent in trees. From the past literature, the following basic facts emerge regarding sap flow rates from tap holes or from excised branches (see Marvin 1958 and literature cited therein).

- Sap flow rate decreases over the long term if several days pass without subzero temperatures.
- Over the short term, sap flow rate increases as the wood temperature increases, and flow rate decreases or becomes negative (=sap uptake) as the wood temperature decreases. The flow rate changes are much larger in magnitude when the wood temperature fluctuates above and below 0 °C, but flow rates are a function of temperature even without the occurrence of freezing.
- Sap flow rates in standing trees are most closely correlated with temperature changes in the small branches in the canopy (Marvin and Ericson 1956).
- When excised trees or stems are provided water during a freezing cycle, very rapid water uptake rates are noted at about the time subzero temperatures arise in the wood (Stevens and Eggert 1945; Marvin and Greene 1951). Roots probably supply water to the leafless canopy of intact trees during the freezing cycle.
- Sap exudation from the stumps of felled trees is negligible in the spring, but copious exudation persists from the excised shoots provided they are supplied with water during freezing conditions (Stevens and Eggert 1945).
- The distribution of gas and sap in the stem is an important feature of maple stems that make them unique. Wiegand (1906) observed the distribution of gas and sap in the fibers and vessels of a number of species during the winter. Vessels in maple stems were generally filled with sap and very little gas, whereas fibers contained large quantities of gas. Wiegand reported that *A. camprestre* had a considerable amount of gas in vessels and exuded only slightly when brought into the laboratory in winter. All other species of maple exuded in the laboratory. *Juglans cinerea* (butternut) had little gas in vessels, but large quantity in fibers. It also exuded when warmed. Species of all other genera (*Salix, Populus, Ulmus, Vitis, Fraxinus,* and *Quercus*) contained large quantities of gas in vessels, little gas in fibers and none exuded when warmed.

It has been suggested that the mechanism of maple sap exudation is at least partly related to physical events occurring during the cooling process: Marvin (1958) writes: "During the cooling part of the temperature cycle there is an absorption of water... This absorption is independent of the composition of the vessel solution and may be in part a physical phenomenon. It apparently does not depend entirely upon the activities of living cells." In spite of the above knowledge, a precise correlation in time between the rate of sap flow from the stumps of excised branches and the duration of ice crystal growth or melting as measured by freezing exotherms and thawing endotherms was not reported prior to 1980. Marvin and Greene (1951) came closest to reporting these relationships in excised stem segments, but they emphasized only semiquantitative relationships and the time resolution of their flow measurements and temperature measurements did not permit them to take full advantage of their apparatus.

Marvin et al. (1967) showed a requirement for osmotically active substances in the vessel solution for the exudation process to continue. When stem segments were perfused with hexoses no exudation occurred during the thaw, but when perfused with sucrose exudation occurred and the volume of exudation seemed to be positively correlated with the sucrose concentration and the amount of ray cell tissue (Marvin et al. 1967; Morselli et al. 1978). Past explanations for exudation from maple have been based on activities of living cells (Wiegand 1906; Johnson 1945), osmotic effects resulting from freezing of sap (Stevens and Eggert 1945), or the role of CO_2 generation in maple stems forcing sap out of vessels (Sauter et al. 1973; Sauter 1974)

The PhD dissertation of P.E.R. O'Malley (1979) provided a major refocus on the mechanism of maple sap exudation. O'Malley's results were published in two papers (O'Malley and Milburn 1983; Milburn and O'Malley 1984); their work centered primarily on uptake and exudation of sap during freezing and thawing of sycamore maple stem segments (Acer pseudoplatanus). In their first paper, excised stem segments were connected via water-filled tubing to a mercury manometer; a mechanical float on the outlet side of the manometer actuated movement of a pen on a clock-driven, rotating-drum recorder. Because of the substantial volume displacement required to cause a movement of the mercury column in the manometer, their measurements were a blend of volume displacement and pressure measurement. The manometer required a volume displacement of 3.6 ml/100 kPa of pressure compared to a typical exudation volume of 1.5 ml from their typical stem segment during exudation. The maximum pressure decline they observed during a freeze was –16 kPa and the maximum pressure increase upon the thaw was 15 kPa. No positive pressures above atmospheric pressure were observed because of limitations of their experimental design. Although the pressure changes observed were less than occur in trees (–80 kPa during the freeze vs. +200 kPa during the thaw), their measurement did give relative estimates of volume displacement. Their work demonstrated that: (1) some sap uptake occurred during stem cooling from +15 to 0 °C; (2) about 2.5 times more volume uptake occurred during the freezing exotherm when the temperature fell from 0 to –5 °C,; and (3) volume exudation started upon rewarming and most of the exudation occurred during the endotherm.

In their second paper (Milburn and O'Malley 1984), they demonstrated that: (1) the volume of water uptake and exudation was a linear function of % water content (g water per g dry weight) for water contents from 42 to 65%. At a water content of 60% no uptake occurred upon freezing, for larger water contents exudation occurred upon freezing and for lower water contents uptake occurred during freezing; (2) killing stems by heat or by perfusion with KCN did not prevent water uptake upon freezing; and (3) stem segments collected during the summer contained <0.2% sucrose and yet could be made to take up sap during the freeze and exude sap after a thaw provided they were incubated for >36 h at 2 °C prior to the freeze-thaw cycle that took an additional 24 h and provided the stem water content was <60%. Based on all the above findings, Milburn and O'Malley concluded that sucrose was not needed for maple sap uptake and exudation and thus they proposed the purely physical model described below.

Based on the information above, Milburn and O'Malley (1984) proposed a new hypothesis to describe sap flow in dormant maple trees. The model provided an explanation of what happened during the cooling sequence when water is absorbed and during the warming sequence when water is exuded. During the

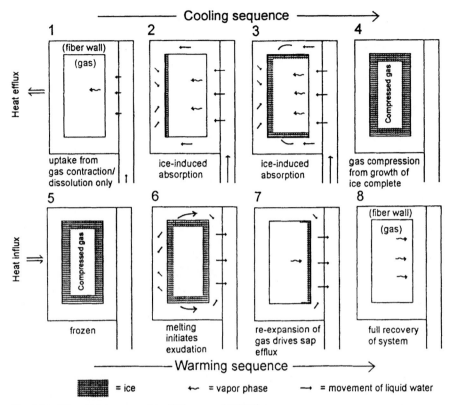

Fig. 3.16. The model proposed by Milburn and O'Malley

cooling sequence, the absorption of sap is described as a biphasic process and is entirely apoplasmic moving from the vessels through wall material into gas-filled fibers (Fig. 3.16). Initial absorption of sap (as the temperature falls toward 0 °C) can be explained entirely as a consequence of gas dissolution and contraction within the stem, presumably in embolized fibers surrounding vessels. Absorption associated with freezing occurs as a direct result of ice crystal formation (on the inner wall of fibers) and proceeds as long as ice crystal growth occurs. Growth of ice crystals does not progress into cell wall capillaries because the freezing point is depressed due to surface absorption effects. Continuous flow of liquid water occurs up to the site of crystal formation. Gas entrapped in the fibers becomes compressed as ice crystal growth continues and contributes to the positive pressure driving sap out of the stem during the thaw. This system is similar to the process used to describe water flow in frozen soils (Everitt 1961; Vignes and Dukema 1974; Biermans et al. 1976; Loch and Kay 1978).

At the end of the cooling sequence, air is left compressed by ice in fiber lumina. During the warming sequence, the compressed air remains after the ice has melted. Exudation out of the stems is driven by the force of the compressed air (in horizontal stems) and exudation is enhanced by gravitational potential forces.

Sap, which was sucked up into the crown of a sugar maple during the freeze, is now free to fall down towards the base of the tree.

The O'Malley-Milburn model is a physical one which requires neither living cells nor sucrose. Experimental evidence in support of these negative requirements was obtained in both field and laboratory freezing experiments on *Acer pseudoplatanus*. In these experiments, Milburn and O'Malley (1984) observed absorption of sap on freezing and exudation during thawing in the absence of living cells. Stems collected in summer, which presumably contained no sucrose in the vessel sap, performed similarly. Most importantly, there was a clear demonstration that the onset of absorption was simultaneous with the freezing exotherm (O'Malley and Milburn 1983).

Using improved methods, Tyree (1983) studied uptake and exudation on whole excised branches of sugar maple about 2 m long and 20 mm in diameter at the base. Electronic pressure transducers were used that permitted pressure measurement with minimal volume flow and an electronic flowmeter that permitted volume flow rate measurements with negligible pressure. Tyree (1983) demonstrated the following facts not previously reported by O'Malley (1979): (1) a phase of rapid uptake was coincident within a few seconds of the first detectable freezing exotherm (improved time resolution over O'Malley's work). (2) Branches could be supercooled to –2 °C and held at that temperature without freezing. During cooling, there was a small flow rate into the branch that ceased once temperature became constant. Thus, uptake during the freezing exotherm was independent of cooling uptake. (3) The volume of uptake during the freeze was highest during a slow freeze and lowest when rapidly frozen. (4) If a branch was allowed to take up sap during the freeze but exudation was prevented during the thaw, then a positive pressure of 50–100 kPa was generated. (In unpublished results, Tyree found this pressure would remain more or less constant for >24 h if exudation was prevented.) (5) The pattern of uptake and exudation of excised branches in the forest was comparable to that measured from the taphole of an entire tree. Thus, all the above results appeared to be consistent with the O'Malley-Milburn model.

Johnson et al. (1987) returned to studies on excised stem segments to address a conflict in the literature, e.g., that Milburn and O'Malley (1984) said the presence of sucrose was not necessary for the sap-flow mechanism versus the work of Marvin (1958) that indicated that sucrose was necessary. In this work segments were perfused with different concentrations of sugar and different kinds of sugar all in 10 mM NaCl or perfused with a control solution of 10 mM NaCl alone. This work confirmed and extended the work of Marvin (1958) and demonstrated that: (1) stem segments perfused with NaCl or NaCl plus glucose, or fructose, or mannitol would take up sap upon freezing and take up more sap during the thaw, i.e., no exudation could be induced; (2) stem segments perfused with sucrose, maltose, lactose, or raffinose would take up sap upon freezing and exude sap upon the thaw; (3) the volume of exudation increased with increasing sucrose concentration in the perfusate; and (4) sugar maple and sycamore maple had essentially identical behaviors, thus O'Malley's results could not be attributed to species differences. They concluded that the exudation mechanism probably did not involve simple osmotic effects with living cells because exudation did not occur with low-molecular-weight solutes. However, no specific role of sucrose in the exudation process was suggested.

While the O'Malley-Milburn model satisfactorily explained sap uptake during the cooling phase and during the freezing exotherm, it was in conflict with the above findings. O'Malley's negative evidence for sucrose requirement came only from work done on stems collected in summer. Such stems contained <0.2% sucrose when harvested and O'Malley presumed it remained that way throughout his experiments. The conflict regarding the requirement for sucrose can be resolved by noting that incubation of stem segments at 2 °C caused an enzymatic shift from starch storage to conversion of starch to sucrose in sugar maple wood (Sauter et al. 1973), and therefore O'Malley's stem segments may have contained sucrose following exposure to low temperatures for the duration of their experiments.

Johnson and Tyree (1992) extended the above work by investigating the role of wood water content on the uptake and exudation process in sugar maple and butternut (*Juglans cinerea*). They demonstrated that the volume of exudation depended not only on low wood water content, but also on *how* the low wood water content was achieved. Stem segments were collected in winter and sap uptake and exudation volumes measured during a freeze-thaw cycle. The segments were then put into storage and dehydrated further. Stems stored and dehydrated an extra 10% at +4 °C exuded only about half as much initially, but stems stored and dehydrated an extra 10% at −12 °C exuded more (about twice as much) as initially. They explained this in terms of *where* air bubbles formed depending on how dehydration proceeded. Dehydration while water is in the liquid state (at +4 °C) will cause embolisms in vessels that will block uptake and exudation during the freeze-thaw cycle (Tyree and Sperry 1989a). Dehydration while water is frozen (at −12 °C) might cause ice crystals in wood fiber lumina to migrate to ice crystals in vessels. The physical basis of this has to do with the thermodynamics of ice that favors large crystals over small crystals because the surface energy of large crystals is less than of small crystals. Thus a thermodynamic driving force exists allowing vapor distillation from small to large ice crystals. This freeze-dehydration step seems to be a necessary preconditioning step for air bubbles to form in wood fibers before the events described in the O'Malley-Milburn model can take place.

So what is wrong with the O'Malley-Milburn model and what is the role of high-molecular-weight sugars? An important clue comes from recent studies of the physics of the stability of air bubbles in wood (Tyree and Yang 1992; Yang and Tyree 1992). An air bubble in a confined space (such as a fiber lumen) will be at a higher pressure than the surrounding fluid because of surface tension. This means that the air bubble is unstable and will eventually dissolve, as discussed in greater detail in Chapters 4.1 and 4.6. While the O'Malley-Milburn model predicts that an air bubble may be compressed to a very high pressure (say 200 kPa) during a freezing event, theoretical calculations reveal that bubble dissolution and diffusion away will reduce the bubble pressure to initial values within 1 h or less and, over a period of about 12 h, the bubbles will totally disappear if the xylem fluid remains pressurized. Some unpublished experiments easily confirm the lack of stability of air bubbles. Bubble pressurization can be decoupled from freezing by perfusing stem segments with pure water to remove sucrose. Then the stem is pressurized at say 100 kPa for a few minutes to infuse water under pressure and compress the air bubbles. If we stop perfusion and prevent water exudation while

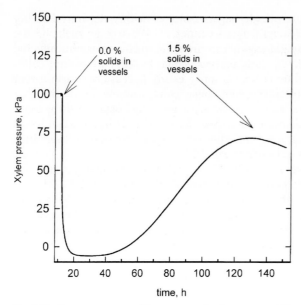

Fig. 3.17. Pressure measured in sugar maple stem segment 10 mm diameter and 0.5 m long kept constantly at 2 °C. At time zero the segment was perfused from base to apex with pure water for 1 h until sap emerging from the tip had 0.0% solids. Then the stem was infused from both ends at a pressure of 100 kPa while measuring flow in from both ends. Air bubbles were being compressed by the infusion. When the flow nearly stopped after a further 2 h, both ends were sealed to a pressure transducer to measure pressure with no volume change (at *left arrow*). At the end of 150 h, analysis of the sap revealed that 1.5% solids (sucrose) had returned to the xylem sap. See text for more interpretation. (M. T. Tyree, unpubl. data)

measuring fluid pressure, we find that the fluid pressure drops to zero in a few hours because the air bubbles slowly dissolve (Fig. 3.17). On the other hand, if the stem is first perfused with a sucrose solution, with π=−100 to −200 kPa, and then pressurized to compress the bubbles, the air pressure remains stable. So there is something about the sucrose (perhaps osmosis) that keeps the bubbles from dissolving and hence the observed pressure may indeed be of osmotic origin.

But where is the osmotic membrane and how does it interact with air and water in wood fiber cells? One hypothesis that has still to be tested is that the lignified cell walls of wood fibers may be semipermeable to sucrose solutions, i.e., water can pass through but not sucrose molecules. Hence we could have osmosis without living membranes. A revised model for the exudation process that does explain the osmotic role of solutes can be put forward with predictions. The hypothesis can be divided into a number of related statements: (1) lignified cell walls in maple xylem are semipermeable to large solutes but permeable to small solutes. So the osmotic effect of a solute will be given by $\sigma_i RTC_i$, where σ_i is the reflection coefficient of solute i and C_i is the solute concentration. Tentatively, I propose that high-molecular-weight sugars have a value of σ nearly equal to 1 and lower-molecular-weight sugars have a much lower value of σ. (2) During the freezing phase, low-molecular-weight sugars can pass with water into the fiber lumina, but high-molecular weight sugars cannot. (3) When water melts during the thaw, high-molecular-weight sugars can draw the water out osmotically and can generate a maximum hydrostatic pressure equal to $\sigma_i RTC_i$. Low-molecular-weight sugars do not generate an exudation pressure because σ is nearly zero and/or because the low-molecular-weight sugar concentrations inside the fiber and vessel lumina are almost equal. The pressure generated osmotically by high-molecular-weight sugars will persist for many hours without bubble dissolution and explains how stable pressures persist for >20 h (Tyree, unpubl. results). Working with this hypothesis, let us return to Fig. 3.17.

In Fig. 3.17 the pressure returned after 50–100 h. What is the interpretation? Let us assume the wood fiber cells contained x ml of air after the perfusion with pure water. Then, after infusion with water at 100 kPa above air pressure (200 kPa absolute), the bubbles will have been compressed to half their original volume (based on the ideal gas law). Hence the fibers now contain $0.5x$ ml of compressed air and $0.5x$ ml of extra water (=the water needed to displace the air bubbles to half the volume). The stem is now sealed off so there can be no more net volume change but the ratio of air volume to water volume can change slightly if needed. So, over the next 3 h, the air bubble pressure drops (with no volume change) because air molecules go into solution (based on Henry's law of solubility of gas in water). When the bubble pressure reaches atmospheric pressure, the air pressure in the bubble is in solubility equilibrium with the outside air. However, to remain stable at atmospheric pressure, the water pressure has to be slightly below atmospheric to counter the capillary pressure at the air/water interface (see Eq. 4.2, Chap. 4). So why does the pressure return? Over a much longer time scale, starch is converted to sugar and dumped into the xylem. If we assume the sucrose does not permeate the lignified cell walls between the vessels and wood fibers, then we have to suppose it will osmotically draw some water out of the wood fibers to raise the pressure of the sap above atmospheric. This explanation also works if we assume the water is drawn osmotically out of living cells (such as ray cells in the wood) but it does not explain why low-molecular-weight sugars do not raise the pressure unless we suppose the low-molecular-weight sugars pass easily into the ray cells. Unfortunately, there has been insufficient time and money to complete these studies and I hope that someone will be interested in the future to resolve these questions.

4 Xylem Dysfunction: When Cohesion Breaks Down

Xylem is an unusual tissue. The conduits (vessels or tracheids) are alive and filled with water and cellular organelles from the time they divide through growth and differentiation. When they are mature they die to become functional. Conduits can remain functional for just a few days or for more than 100 years, but the first step towards a state of permanent dysfunction probably involves a state of embolism. Freezing can induce embolism because air comes out of solution as water turns to ice. Drought stress can induce embolism because xylem pressure, P_x, becomes negative enough during drought to induce a cavitation. Simple mechanical damage can also induce this first stage of xylem dysfunction. However, embolism dysfunction is not always permanent. Sometimes embolisms can be repaired daily and sometimes by the next growth season. In addition, embolized conduits are not totally dysfunctional, since even embolized conduits retain some water and the amount of water retention increases and decreases with P_x. So embolized conduits still function as water-storage organs. In this chapter we will review the mechanisms of xylem dysfunction, vulnerability curves, and mechanisms of embolism repair (removal).

4.1 Xylem Functionality and Cavitation

A cavitation event in xylem conduits ultimately results in dysfunction. A cavitation occurs when a void of sufficient radius forms in water under tension. The void is gas filled (water vapor and some air) and is inherently unstable, i.e., surface tension forces will make it spontaneously collapse unless the water is under sufficient tension (negative pressure) to make it expand. We must take a necessary diversion to explain why this is true.

The chemical force driving the collapse is the energy stored in hydrogen bonds, the intermolecular force between adjacent water molecules. In ice, water is bound to adjacent water molecules by four hydrogen bonds. In the liquid state, each water molecule is bound by an average of 3.8 hydrogen bonds at room temperature. In the liquid state hydrogen bonds are forming and breaking all the time permitting more motion of molecules than in ice (Slatyer 1968). However, when an interface between water and air is formed, some of those hydrogen bonds are broken and the water molecules at the surface are at a higher energy state because of the broken bonds. The force (N=Newtons) exerted at the interface as hydrogen bonds break and reform can be expressed in pressure units (Pa) because pressure is dimensionally equal to energy (J=Joules) per unit volume of molecules, i.e.,

J m^{-3}=N×m m^{-3}=N m^{-2}=Pa. Stable voids in water tend to form spheres because spheres have the least surface area per unit volume; thus, a spherical void has the minimum number of broken hydrogen bonds per unit volume of void. The underlying principle is that matter tends to assume a geometry that minimizes energy. The pressure tending to make a void collapse is given by $2\tau/r$, where r is the radius of the spherical void and τ is the surface tension of water (=0.072 Pa m at 25 °C).

For a void to be stable, its collapse pressure (2t/r) must be balanced by a pressure difference across its surface or meniscus that is equal to $\bar{P}_v-\bar{P}_w$, where $\bar{P}_w$ is the absolute pressure of the water and $\bar{P}_v$ is the absolute pressure of the void.

$$\bar{P}_v - \bar{P}_w = \frac{2\tau}{r} \tag{4.1}$$

$\bar{P}_v$ is usually above absolute zero pressure (=perfect vacuum) since the void is usually filled with water vapor and some air. Relatively stable voids are commonplace in daily life, e.g., the air bubbles that form in a cold glass of water freshly drawn from a tap. An entrapped air bubble is temporarily stable in a glass of water because $\bar{P}_w$ is a relatively constant 0.1 MPa and $\bar{P}_v$ is determined by the ideal gas law, $\bar{P}_v=nRT/V$, where n is the number of moles of air in the bubble, R the gas constant, T absolute temperature, and V the volume of the bubble. So the tendency of the void to collapse ($2\tau/r$) makes V decrease, which causes $\bar{P}_v$ to increase according to the ideal gas law because $\bar{P}_v$ is inversely proportional to V. The rise in $\bar{P}_v$ provides the restoring force across the meniscus needed for stability. However, an air bubble in a glass of water is only temporarily stable because, according to Henry's law, the solubility of a gas in water increases with the pressure of the gas. So the increased pressure exerted by $2\tau/r$ makes the gas in the bubble more soluble in water and it slowly collapses as the air dissolves, i.e., as n decreases.

Air bubbles are rarely stable in xylem conduits because transpiration can draw $\bar{P}_w$ to values <0. As $\bar{P}_w$ falls towards zero, the bubble expands according to the ideal gas law, but V can never grow larger than the volume of the conduit, so $\bar{P}_v$ can never fall to or below zero to balance $2\tau/r$. Once the bubble has expanded to fill the lumen, the conduit is dysfunctional and no longer capable of transporting water. Fortunately for the plant, a dynamic balance at the meniscus is ultimately achieved. This stability will be discussed first in the context of a vessel and its pit membranes.

Fig. 4.1. Diagram of the wall structure between adjacent xylem vessels showing intervessel pit structure. The porous pit membrane develops from the primary cell wall of the two vessels and middle lamella. It is overarched by thick secondary walls to form a pit chamber that opens to the vessel lumen via a pit aperture. When a vessel is embolized, air is prevented from spreading to adjacent functional vessels by the capillary force or surface tension of the air-water meniscus spanning the pit membrane pores. As the pressure difference increases across the pit membrane pores, the meniscus is gradually pulled through. At 0 MPa the menisci are flat, at –1 MPa the menisci have a radius of curvature of 0.14 μm but still not small enough to pass through the largest pore. At –2 MPa the radius of curvature is 0.07 μm and is capable of passing through pores <0.14 μm diameter. As pressure difference increases the pit membranes are also going to be stretched (deflected) toward the water-filled side

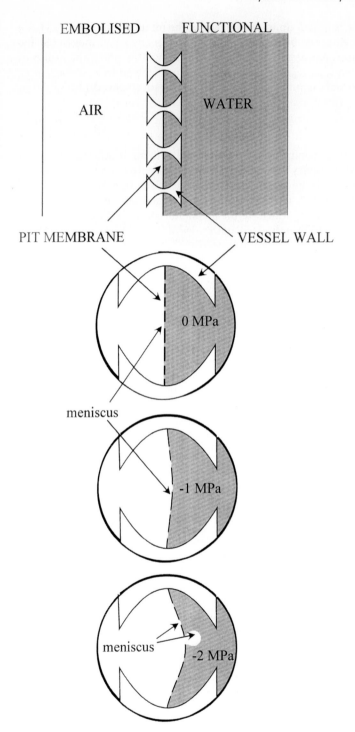

As an air bubble is drawn up to the surface of the pit membrane (Fig. 4.1), the pores in the pit membrane break the meniscus into many small menisci at the opening of each pore. As the meniscus is drawn through the pores, the radius of curvature of the meniscus, r_m, falls towards the radius of the pores, r_p. As long as r_m exceeds r_p, the necessary conditions for stability are again achieved, i.e.,

$$\bar{P}_v - \bar{P}_w = \frac{2\tau}{r_m} \tag{4.2}$$

Usually, a dysfunctional conduit will eventually fill with air at atmospheric pressure (as demanded by Henry's law) so $\bar{P}_v$ eventually approaches 0.1 MPa as gas diffuses through water to the lumen and comes out of solution. When $\bar{P}_v$ equals 0.1 MPa, the conduit is fully embolized. As $\bar{P}_w$ rises and falls as dictated by the demands of transpiration, r_m adjusts at the pit-membrane pores to achieve stability. When the conduit is fully embolized, both sides of Eq. (4.2) can be expressed in terms of xylem pressure,

$$P_x = \bar{P}_w - \bar{P}_v = -\frac{2\tau}{r_m} \tag{4.3}$$

The minimum value of P_x that can be balanced by the meniscus is given when r_m equals the radius of the biggest pit-membrane pore bordering the embolized conduit. If the biggest pore is 0.1 or 0.05 μm, then the minimum stable P_x is −1.44 or −2.88 MPa, respectively. So the porosity of the pit membrane is critical to preventing dysfunction of vessels adjacent to embolized vessels (Sperry and Tyree 1988). When P_x falls below the critical value then the air bubble is sucked into an adjacent vessel seeding a new cavitation. The same kind of surface tension forces can arise at the air/water interface of spongy mesophyll cells where water evaporates from leaf surfaces. The effective pore radius of mesophyll cell walls can be of the order of 0.01 μm meaning that theoretical differences in pressure of −14 MPa can be generated. Hence, the cohesion-tension theory can postulate quite negative pressures. The conduits are the weak link because of their vulnerability to cavitation.

Although the explanation given above is more or less what is commonly found in the literature, it is an oversimplification of the real situation regarding how air/water interfaces interact with solids of various composition and fine structure, and therefore we will take an aside to consider surface tension in a little more detail. One effect of surface tension in cylindrical tubes is a phenomenon called capillary rise, as illustrated in Fig. 4.2. The pressure difference across the meniscus is still determined by r_m, but when a meniscus is contained in capillary tube of radius r, the value of r_m can be greater than or equal to r depending on the contact angle, α, the meniscus makes with the wall of the capillary. In general, $r = r_m \cos \alpha$. Some surfaces can be more hydrophilic than others because they enter into stronger hydrogen bonding with the water. Hydrophilic surfaces have contact angles between 0 and 90°, whereas hydrophobic surfaces have contact angles between 90 and 180°, i.e., the meniscus bends the other way, i.e., convexly.

The surface tension equation (Eq. 4.3) becomes inaccurate at r_m values around 20 to 60 nm. There is still substantial surface tension and large pressure differences can be sustained, but the magnitude of the surface tension cannot be calcu-

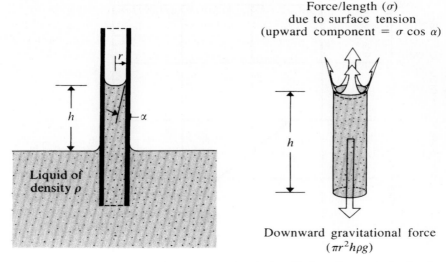

Force/length (σ)
due to surface tension
(upward component = $\sigma \cos \alpha$)

Downward gravitational force
($\pi r^2 h \rho g$)

Fig. 4.2. Capillary rise of a liquid: **a** parameters involved and **b** force diagram indicating that surface tension projected in the upward direction is balanced by gravity acting downward. (Nobel 1983)

Fig. 4.3. Artist's interpretation of the spacing of cellulose fibers in a primary cell wall or pit membrane. Note the very irregular interstices, i.e., inter-fiber spaces. (Reproduced from Nobel 1983)

Primary
cell wall

Interstices

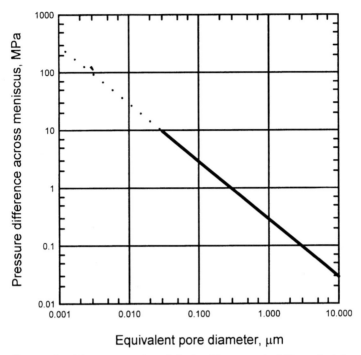

Fig. 4.4. Graphic representation of the 'capillary equation' (Eq. 4.2) giving pressure drop ΔP across a meniscus versus diameter of an equivalent pore that can sustain the meniscus assuming a contact angle of zero degrees. The *dotted line* indicates the region beyond which the equation applies, but a substantial capillary effect will still be present. (Schneider et al. 2000)

lated. In addition, there are no circular pores in pit membranes or cell walls. If the cellulose fibers of primary cell walls could be enlarged they would probably look like Fig. 4.3 and would contain very irregular, water-filled spaces. Nevertheless, it is very useful to talk in terms of equivalent pit-membrane porosity as defined by the radius in Eq. (4.3). The equivalent porosity helps us explain air-passage at a given pressure difference. The total range for the pressure drop across a meniscus versus equivalent pore diameter ($2r$) is shown on a log-log scale in Fig. 4.4.

Consequently, the genetics that determines pit morphology and pit-membrane porosity must be under strong selective pressure. A safe pit membrane will be one with very narrow pores and thick enough, and thus strong enough, to sustain substantial pressure differences without rupturing. The cost, however, is that narrow pores and long pores (in thick pit membranes) do not conduct water efficiently (see the Hagen-Poiseuille law above). So a balance must be achieved between safety and efficiency of water transport. Hydraulic efficiency could be increased, by increasing the surface area of pit membranes, but this would weaken the conduit and make it more liable to collapse under negative pressure.

4.2 Tension Limits: "Designed Leaks" and Vulnerability Curves

Plant cell walls might be "*designed*", i.e., predetermined, to allow water columns to break at a certain "suitable" tension. This statement may appear unreasonable and contradict the experimental results of Ziegenspeck (1928) and Haider (1954), but Martin believed it does not. The concept of designed leaks, otherwise called the air-seeding hypothesis for cavitation, was a very insightful idea suggested by Zimmermann (1983) years before the first experiment was conducted to test the idea. The basic idea for air seeding is the sequence of events shown in Fig. 4.1. Martin provided few details about where the pores were located that provided designed leaks but he probably would not have excluded pit membranes and may well have suggested other cell wall imperfections in vessels as other loci of designed leaks.

The mere fact that the "tensile strength" of water is consistently different in different plant cells indicates that the notion of "designed" leaks may not be so far-fetched. Let us look first at fern sporangia again. Both Ursprung (1915) and Renner (1915) found independently a tensile strength of water in annulus cells between 30 and 35 MPa. In a more recent paper, Haider (1954) confirmed this for a number of species of the fern family Polypodiaceae. However, in the fern *Aneimia rotundifolia*, he found tensions up to 44 MPa. *Selaginella and Equisetum* have similar spore ejection mechanisms which snap at lesser tensions (Ziegenspeck 1928). It is obvious that it is not the property of the water, but the structure of the cell wall that must be different. The question then is whether annulus cells of ferns of the Polypodiaceae have wall perforations of ca. 10 nm diameter and those of *Aneimia* of ca. 7 nm. Renner (1925) and his students Holle (1915) and Frenzel (1929) worked on this problem. Annulus cells are permeable to potassium nitrate and glycerol but not to sucrose. It was therefore concluded that the pores can not be much wider than ca. 1 nm, the size of the sucrose molecule. However, permeability even of dead cells depends not only on pore sizes, but also on the respective chemical nature of the wall and the permeating molecule. It is therefore somewhat questionable whether permeability studies can be interpreted so "mechanistically", even though glycerol and sucrose are chemically not so different. Furthermore, pores may well be wider when the cells are under stress (Fig. 3.5 at II).

Let us now explore the question of "designed leaks" further. Renner (1925) was very interested in the nature of cell wall pores and he considered that after a cell's death plasmodesmata deteriorate and the spaces they had occupied in the wall become accessible to air penetration. Holle (1915) and Frenzel (1929) found that air enters dead pith cells of *Sambucus* at a tension of 0.5–0.6 MPa. This corresponds rather well with the dimensions of plasmodesmata openings. Frenzel (1929) measured pore sizes in all sorts of cells with dye molecules of known size, gold suspensions, etc., and found ranges of pore sizes which become available to air penetration after the death of cells.

4.3 Vulnerability Curves and the 'Air-Seeding' Hypothesis

A vulnerability curve (VC) is a plot of percent loss hydraulic conductivity (PLC) in stem segments versus the water stress (measured as Ψ or P_x) that induced the PLC. PLC is measured by collecting a shoot and excising a segment from it under water. The initial hydraulic conductivity, K_i, is measured in a conductivity apparatus. The initial conductivity is usually less than the maximum possible because of some air embolism. The air emboli are dissolved or displaced by flushing the stem segment with degassed water at a pressure of 0.1–0.2 MPa. After each flush the conductance is measured until a maximum value ($=K_{max}$) is reached. PLC is calculated from $100(1-K_i/K_{max})$. A vulnerability curve is constructed by dehydrating replicate plants to different known P_x values and then measuring PLC at each stress level (Sperry et al. 1988a). Dehydration is usually performed on excised branches. Consequently, air emboli are prone to be sucked into conduits from the cut base via pit membranes. It is generally presumed that this procedure does not bias the VC and this has been confirmed in some instances by comparing the VC from dehydrating whole rooted plants versus excised branches of the same species (Tyree et al. 1992). Some variations on the method are needed for conifers because tori often remain aspirated after dehydration (Tyree and Dixon 1986; Sperry and Tyree 1990).

Extreme care must be taken in the sampling of stem segments for the measurement of PLC. It is undesirable for the dehydration procedure or sampling method to induce 'unwanted' loss of conductivity. When excised branches are dehydrated, air will be sucked the full length into vessels cut open at the excised surface. So the stem segments must be excised from stems held under water, otherwise more air bubbles will be sucked in when they are cut. Prior measurements of maximum vessel length also have to be made to insure stem segments are cut far enough away from wounds exposed to air during the dehydration process. Hydraulic conductivity must be measured with water free of particles that might plug up pits during the flow measurements or during the flushes. Water pressure can also displace bubbles from vessels cut open on the downstream side during the measurement of K_i. Bubble displacement can be avoided by measuring K_i with an applied pressure less than 3 or 4 kPa. In some species conductivity cannot be restored by flushes. The torus of conifers seals (aspirate) against the bordered pit after a cavitation; hence, K_{max} cannot be restored. Gums and/or latex released by cuts plug up the stems very quickly in other species. In these cases, K_i values of unstressed samples can sometimes be used in place of K_{max}.

The VCs for a number of species are reproduced in Fig. 4.5. These species represent the range of vulnerabilities observed so far, i.e., 50% loss of conductivity occurring at P_x values ranging from –0.7 to –11 MPa. Vulnerability curves have probably been reported on >100 species to date; a paper in 1994 reviewed 60 known curves (Tyree et al. 1994). The vulnerability curves have agreed very well with field ranking of 'drought tolerance' reported from sylvicultural observations. Many writers have asserted the hypothesis that vulnerability to cavitation is the single most important physiological parameter explaining drought tolerance. However, this hypothesis has never been tested! We will review this hypothesis in

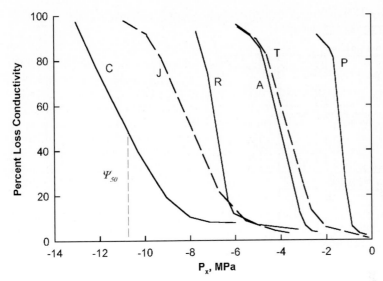

Fig. 4.5. Vulnerability curves for various species. *y*-Axis is percent loss of hydraulic conductivity induced by the xylem pressure, P_x, shown on the *x*-axis. *C Ceanothus megacarpus; J Juniperus virginiana; R Rhizophora mangle; A Acer saccharum; T Thuja occidendalis; P Populus deltoids.* Only the trend-lines are plotted. Individual points have very high standard deviations when PLC is between 20 and 80%. The points are excluded to make the trends clearer. (Adapted from Tyree et al. 1994)

Section 4.5. The issue now is what seeds a cavitation, i.e., what 'fails' when a cavitation event occurs?

Four mechanisms for the nucleation of cavitations in plants have been proposed and these are illustrated in Fig. 4.6. This figure shows for each mechanism the sequence of events that might occur as P_x declines in the lumen of a conduit.

Of these four mechanisms, homogeneous nucleation is the most studied in the physical sciences (Pickard 1981). Homogeneous nucleation can be quantitatively predicted from the statistical mechanics of molecular motion in water. A calculation is done to predict the probability of a void spontaneously occurring in the 'center' of a water container, i.e., far enough away from the container walls that the molecular bonds between the wall and water can be ignored. The factors that have to be taken into account are the kinetic energy of thermal motion of water molecules and the probability that several water molecules will be moving away from a common locus with sufficient kinetic energy to break the hydrogen bonds holding them together. If the water is under tension (negative pressure), the hydrogen bonds will already be stretched and thus weakened making the statistical probability of a void forming more likely. The size of the void that must form to be unstable and continue to expand is given by Eq. (4.3) and is a function of P_x. Readers may consult Pickard (1981) for details and other references for the statistical formulations; all the reader needs to know is that the probability, P_c, of a cavitation occurring will increase with: (1) the volume of the water in the container (conduit

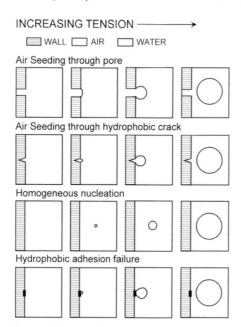

Fig. 4.6. Diagram illustrating four mechanisms of cavitation nucleation in xylem conduits. (Tyree et al. 1994)

volume), (2) tension applied to the volume, and (3) with the time (duration) the tension is applied. In short, P_c increases with the time–tension–volume domain.

Heterogeneous nucleation occurs when water pulls away from the container wall (top two and last example in Fig. 4.6). The nature of the forces binding water to wall material differs from that holding water to water. Heterogeneous nucleation is less well understood because it depends on the nature of wall defects. The best understood mechanisms are nucleation at a hydrophobic crack and nucleation by air seeding. Pickard (1981) discusses at length nucleation at hydrophobic cracks. Briefly, air bubbles can remain stable in hydrophobic cracks without dissolving because, at the hydrophobic interface, the angle of contact at the wall–water–air interface is altered forming a reversed meniscus. Both the angle of contact and the size of the crack determine the radius of curvature of the meniscus, but an equation similar to Eq. (4.3) can be used to predict the P_x needed to pull a bubble out of a crack based on its radius of curvature. A second kind of heterogeneous nucleation might occur at the interface of water and hydrophobic patches without the presence of air bubbles. No predictive equations are available for this case. The third kind of heterogeneous nucleation is the air-seeding hypothesis of Zimmermann (1983), which has already been discussed. Air seeding occurs when an air bubble is sucked into a water-filled lumen via a pore from an adjacent air space. One interesting side question is: are there pathways for air entry that communicate directly between the outside atmosphere and a xylem conduit? Or do all air-seeding events originate from adjacent air-filled conduits?

Of the four hypotheses, air seeding is unique because it predicts that xylem embolism could be induced without xylem tension. Air seeding depends only on the difference in pressure between the xylem and the outside air, $\bar{P}_y - \bar{P}_w$, the same pressure difference of, say, 1 MPa results if $\bar{P}_y = +1.1$ MPa and $\bar{P}_w = 0.1$ MPa or if $\bar{P}_y = 0.1$ MPa and $\bar{P}_w = -1.1$ MPa. In the other three mechanisms, the value of $\bar{P}_y$ is unrelated to the value of $\bar{P}_w$ that induces cavitation.

A number of earlier papers estimated the air pressure needed to displace water in cut stems (Crombie et al. 1985; Sperry and Tyree 1988, 1990). Stem segments were placed with one end in a pressure bomb and the other end outside. The pressure difference across the stem segment needed to push air through pit membranes was similar to what was expected from vulnerability curves, and these studies certainly lend support to the notion of air seeding. However, they did not test Zimmermann's idea (Zimmermann 1983) that there can be a pathway for air movement from outside a shoot to vessels. Cochard et al. (1992) were the first to test the air-seeding hypothesis in a way I consider convincing. They enclosed a flexible willow branch in a large pressure bomb; the branch, with leaves attached, was bent so that both ends protruded through a rubber seal outside the pressure bomb. In this way, the hydraulic conductivity, K_h, could be measured while keeping P_x continually above atmospheric pressure except for the downstream side of the branch that was at atmospheric pressure (Fig. 4.7). As the air pressure inside the bomb, P_g, was increased, air could be seen emerging from both cut woody surfaces of the willow thus showing that there were direct air pathways from the leaves and/or bark and the wood. As P_g was gradually increased from 0 to 1 MPa, there was little or no change in K_h but, as the pressure was increased from 1 to 2 MPa, there was a gradual loss of hydraulic conductivity of up to 70–90%. As the air pressure was decreased, there was no further loss of K_h. This experiment measured a vulnerability curve without P_x ever dropping below atmospheric pressure. Vulnerability curves were measured on similar shoots by two independent means. Some branches were excised and dehydrated in the laboratory. This caused a gradual decline in P_x, which was measured in a pressure bomb on excised leaves. Then PLC was determined in the normal way. In another experiment, excised shoots were placed inside a pressure bomb and dehydrated to desired balance pressures $P_B = -P_x$. According to pressure-bomb theory (Chap. 3.5), P_x was above atmospheric during the dehydration but became negative when the bomb pressure was returned to zero prior to harvesting stem segments for the measurement of PLC. Since the resulting vulnerability curves were the same by all three methods, the air-seeding hypothesis must be correct.

Experiments to test the air-seeding hypothesis have been repeated by others with similar results, e.g., Sperry and Saliendra (1994). Today, double-ended pressure bombs (called pressure collars) are used, so that stems can be passed through two rubber seals without bending the stems. The advantage of the pressure collar is that a vulnerability curve can be constructed on a single stem segment in much less time than in the traditional method. Pressure collars have also been fabricated so that the collar can be assembled around intact branches in the field. Pressure collars have also been used to induce an incremental increase in $\bar{P}_y - \bar{P}_w$ to see what effect this has on loss of hydraulic conductivity and other physiological events (Salleo et al. 1996; Tyree et al. 1999a; Rood et al. 2000).

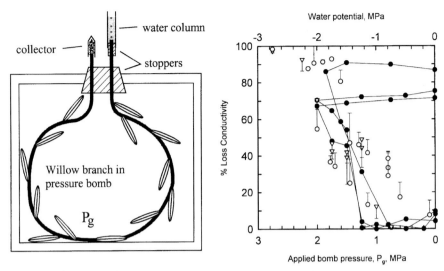

Fig. 4.7. *Left* A willow branch is coiled around inside a pressure bomb and both ends pass through a rubber stopper to the outside. One end is connected to a water column, which provides the hydrostatic pressure to drive water flow through the branch to a collector with absorbent paper positioned on the downstream end. *Right* Vulnerability curves of willow obtained by three different methods. The *open circles* are vulnerability data obtained by air dehydration of excised branches causing xylem tension (negative P_x). The *open triangles* were obtained from shoots dehydrated in a pressure bomb. The *solid circles* connected by *lines* were generated by the apparatus shown on the *left*. The bomb pressure, P_g, was increased in steps of 0.3 MPa and held for 1 h at each pressure. The applied pressure caused a loss of hydraulic conductivity without xylem tension

4.4 Sealing Concepts

Since cavitations are induced by air seeding, it seems appropriate to review the structural features of plants that keep air bubbles away from xylem conduits. The direction of flow in transport systems is normally regulated by demand. If the pressure drops at the receiving end, flow intensifies. It is possible that an accidental sink, caused by an injury, drains the system and thus represents a grave danger. Animals can bleed to death, but there is a blood-clotting mechanism preventing such a fatal result most of the time. Injured phloem or xylem could be drained by exudation, but there are mechanisms that can seal the injured system. Eschrich (1975) has described mechanisms of sealing of the phloem.

Xylem pressures are rarely above and far more commonly below atmospheric. In the case of an injury suffered while pressures are less than atmospheric, the injury is not a sink but a "source" (to use the terms of phloem physiology). Air enters the xylem, displacing the liquid, until the air/water interface has reached a wet membrane, namely the tracheary wall, where its movement is stopped automatically by the surface tension of the water. Even though water flow is stopped, the injured vessel is now open to the outside world and represents a potential entry site for harmful microorganisms. Plants have therefore evolved secondary means of sealing the air-blocked vessel with gums (carbohydrates) or living cells (tyloses; Chap. 8.3). In addition, injured xylem is often isolated by suberization of the cell walls next to the injury. This seals the extra-fascicular pathway between the damaged area and the living plant part. These secondary-sealing mechanisms will be discussed in Chapter 8.3. Let us now return to the first seal, which is provided by capillarity.

Occasionally, pressures in the xylem are above atmospheric. This happens in some species in early spring as mentioned above. It also happens in certain herbs during nights when absorption conditions are favorable and transpiration is minimal. Excess water is then bled from hydathodes at the margins of leaf blades. If the plant is injured when pressures are above ambient, it bleeds. This does not appear to be harmful, because it only happens when there is a surplus of xylem sap. The rather special case of aquatic angiosperms will be discussed later (Chap. 7.4). Most of the time, pressures are below ambient; when the xylem is injured, air is drawn in and it is thus self-sealing by its very structure. We could even speculate that low operational pressures were selected during evolution, because they not only make xylem self-sealing, but also provide an intercellular air-duct system, by decreasing capillary meniscus radii.

In certain plant parts we find that the cell walls are suberized and thus free apoplastic water movement is blocked. This barrier is usually called a Casparian strip. The endodermis of the root with its Casparian strip has been known since the last century. Haberlandt (1914) provided a review of the older literature and Clarkson and Robards (1975) have summarized more recent papers and described the histological development of the Casparian strip. Peterson and Cholewa (1998) have shown that some plants even have two Casparian strips! The aspect that concerns us here is the fact that the cell walls of the endodermal layer in the root provide an effective seal for apoplastic water movement in a radial direction. "Outside" and "inside" apoplasts are separated into different compartments by the barrier. This means that pressures in the two can be substantially different. This pressure-seal concept was experimentally tested many years ago (de Vries 1886). Positive pressures in the xylem of the root system would not be possible without such a seal. Transport between the two compartments is metabolically controlled and goes via the symplast, i.e., the plasmodesmata (Läuchli and Bieleski 1983). The endodermis is mechanically well developed in plants of dry areas and in marsh plants that dry up periodically (Haberlandt 1914). In extremely dry desert habitats, for example, the Casparian strips are much wider, presumably because the pressure difference between the soil and the stele is greater (Fahn 1964; see also Chap. 6.1).

Xylem pressures are normally highest in the roots and rhizomes where they are more often above atmospheric than anywhere else. As pressures in aerial parts are normally below atmospheric, an apoplastic seal is only necessary at the plant surface in order to prevent water loss by evaporation; this is provided by the epidermis or the periderm. However, in desert shrubs, suberization can be found in the stem phloem parenchyma (Jones and Lord 1982), the rays, between xylem layers, etc. (Fahn 1974), thus sealing the xylem more effectively against water loss.

Vascular bundles of some leaves are surrounded by a bundle sheath, containing a suberized layer comparable to that of the Casparian strip in the roots (Schwendener 1890; O'Brien and Carr 1970). This seal separates the apoplast into two compartments, one inside and the other outside the bundle sheath. The two areas are only connected by plasmodesmata that connect living cells (O'Brien and Carr 1970). It may well be that the bundle sheath and/or presence of hydathodes prevent flooding of the photosynthetic tissue in those plants that experience positive root pressures regularly. That flooding can actually happen was found by Stahl (1897), who grew plants that lacked hydathodes in warm, humid soil and covered them with glass jars to prevent transpiration. The intercellular space of the photosynthetic tissue was thereby infiltrated. These intercellular spaces are often protected by the poor wettability of their interior cuticle (Häusermann 1944; Scott 1950).

Sperry (1983) studied water relations of the Mexican fern *Blechnum lehmannii*. Measuring root pressure on petioles with bubble manometers, Sperry recorded xylem pressures up to about 20 kPa above ambient in the plants growing near the water, and progressively lower pressures in plants growing higher up on the stream bank. Only the hydathodes of the young leaves guttated (in intact plants); those of older leaves were sealed and did not guttate. This appears to be a mechanism to direct the (nutrient-containing) xylem stream toward the younger leaves when transpiration is low and positive root pressure exists. We shall encounter this mechanism again when we discuss aquatic angiosperms (Chap. 7.4).

In a previous section (Sect. 4.1) we looked at the quantitative properties of the seal provided by a wet membrane. Smaller vessel-wall pores of desert plants seal outside air against greater negative pressures (e.g., Scholander et al. 1965). As discussed in Section 4.2, air leaks in tracheary elements may be provided by the wall perforations of plasmodesmata left behind after protoplasmic degradation. In this respect, it is perhaps of interest to consider Braun's (1963, 1970) *Hydrosystem*. Braun recognizes five levels of evolutionary development of xylem. The lowest level is that of gymnosperms where most of the axial elements are dead tracheids. In the next level, some vessels are embedded in a matrix of dead tracheids, fiber tracheids, and fibers. Further evolutionary development gradually surrounded the vessels with (living) parenchyma. In the wood of the most advanced stage, vessels are entirely enveloped by paratracheal parenchyma (Fig. 4.8). This is certainly a very effective seal, which protects the vessels from accidental air leaks.

Sealing mechanisms of coniferous tracheids deserve special attention. Tracheids are connected by bordered pit pairs, which are located primarily (but not exclusively) on radial walls. Their peculiar structure is shown by one of Bailey's (1913) drawings, which is reproduced in Fig. 4.9. Two features characterize the coniferous, bordered pit: (1) the pit membrane is rather porous, and (2) it contains a central thickening, the torus. Many electron micrographs have been published during the past 50 years and these have been obtained by many preparative techniques such as thin sectioning parallel to the wall to remove the pit border, carbon replication, etc. They all show the pit membrane as a rather drastic modification of the original primary wall. It looks as if the torus is suspended by more or less radially oriented bundles of microfibrils (Fig. 1.14; see also Frey-Wyssling and Bosshard 1953; Frey-Wyssling et al. 1959). The gaps between these fibrillar networks appear to be rather large, of the order of a fraction of a micrometer.

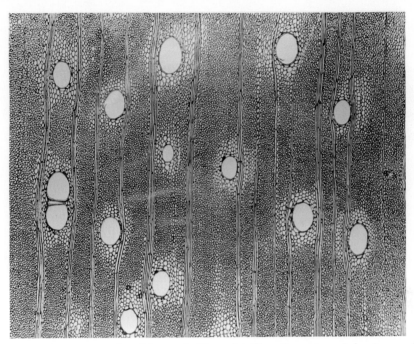

Fig. 4.8. Transverse section of the wood of *Prosopis juliflora* (Leguminosae) showing vessels surrounded by living parenchyma cells (paratracheal parenchyma). The parenchyma cells can be recognized by their lighter color, the darker-colored (i.e., thick-walled) cells are fibers. (Photomicrograph by I.W. Bailey)

Functionally, this means that the membrane alone could not contain an embolus against high tensions. The suspicion could arise that this structure is an artifact of preparation, but permeation experiments by Liese and Bauch (1964) showed that the gaps are real. When wood of various species was perfused with metal suspensions, particles up to 0.2 μm diameter often filtered through. This implies that an embolus could spread in coniferous wood at tensions of only 1.5 MPa. However, the torus is easily displaced if the pressure difference between the two tracheids becomes significant. When a tracheid is injured, its lumen is filled with air at ambient (+101 kPa) pressure. The pressure in the intact, water-containing neighboring tracheids may still be negative; a considerable pressure drop therefore exists across the pit membranes. Their tori are therefore pulled against the pit borders of the neighboring, water-containing tracheids (Fig. 4.9, right). The air-blocked tracheid is thus sealed off from the water-conducting tracheids not only by capillarity, but also by the valve action of the torus. Dixon (1914) showed this situation rather nicely (see his Figs. 17 and 18). Tracheids are so much longer than wide that a naturalistic drawing would not be very clear. Figure 4.10 is an attempt to explain the situation at two levels of simplification.

How well does the valve action of coniferous bordered pits work? The question is particularly interesting in the cases where the surface of the tertiary wall of the

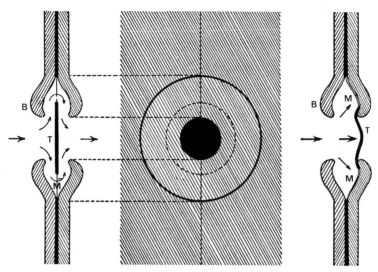

Fig. 4.9. *Center* Surface view of the radial wall of a coniferous tracheid, showing a bordered pit. *Left* The same pit in section, *arrows* indicating the path of water from one tracheid into the next. *Right* Section showing the valve-like action of the torus. *T* Torus; *M* pit membrane; *B* pit border. (Bailey 1913)

tracheids contain warts. Krahmer and Côté (1963) published such electron micrographs (e.g., their Fig. 11), and the seal looks quite effective. Nevertheless, pressures in coniferous xylem rarely drop below –3 MPa, in other words, we must assume that the seal fails at this level. This indicates residual leaks of around 0.1 μm diameter.

An interesting indication of the function of coniferous bordered pits was reported by Hudson and Shelton (1969). When they pushed liquid through freshly cut stems of southern pines, they increased the flow rate by a factor of 20–30 by first cutting off a 3–5-cm-thick disk from the application end. When, after initial perfusion, they cut off a second disk, the flow rate increased 400-fold. This can be explained as follows: After felling, the xylem of the stems was still under tension, the bordered pits at the cut ends therefore closed. This made perfusion difficult. Successive removal of these sealed xylem regions improved xylem conductance, particularly when the stem xylem was first pressurized, i.e., the tension released.

Finally, most conifers possess a mechanism to seal off injured xylem parts by resin impregnation. Even those species that do not normally have resin ducts in the wood develop traumatic ducts upon injury. Similar mechanisms exist in some dicotyledons. This seal is very effective in that it closes the entire apoplast including the cell walls, a phenomenon which will be taken up in Chapter 6.3.

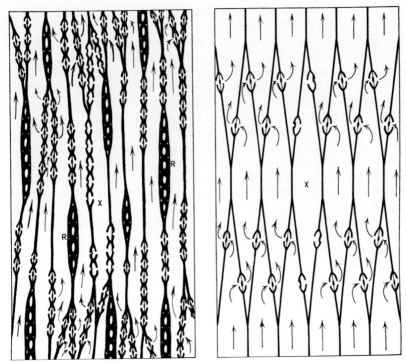

Fig. 4.10. *Left* Diagrammatic view of a (tangential) section of coniferous wood. The tracheid in the center (marked *X*) is vapor-blocked and does not function; water flows around it. The negative pressure in the conducting tracheids has pulled the pit membranes away from the vapor-blocked cell to seal it off. Note that in reality tracheids are much longer in relation to their diameter. *Right* The same situation is shown in a still more diagrammatic manner. Note that coniferous tracheids are never exactly stored as this diagram may imply. (Certain dicotyledonous wood have their cells stored, i.e., arranged all at the same height like books on a bookshelf)

4.5 Vulnerability Curves and Adaptation to Drought

Wide conduits are much more efficient water conductors than narrow conduits, so evolution has driven plants towards wider conduits. However, it is strange that the upper limit on useful vessel diameter is approximately 0.5 mm (Putz and Mooney 1991). This limit has been reached in parallel in many diverse taxa in trees, vines and monocots. The range of conduit diameters is from 0.01 to 0.5 mm (a factor of 50 from minimum to maximum) whereas the range of functional unit lengths is much wider, i.e., a factor of 50,000 from 1 mm for tracheids to 50 m for vessels (Chap. 1.2). A vessel, as a functional unit, is defined here as an catina of vessel elements closed at all surfaces by pit membranes so that vessel-to-vessel transport of water requires passage through pit membranes of simple or bordered pits in the same way that tracheid-to-tracheid transport is through pit membranes of bordered pits. However, the pit membranes of most conifers are specialized comprising a central torus and a fibril or margo (Chap. 1.5).

The reason for the upper limit of conduit diameter cannot be traced to a loss of mechanical strength. The strength of large plants (which determines their maximum possible mechanical height) is determined by density-specific stiffness and strength. The former is the ratio of Young's modulus to density (E/ρ) and the latter is the ratio of tissue-breaking stress to density (φ/ρ). The ideal mechanical tissue with which to construct a vertical stem is one for which both of these ratios are maximized because this maximizes the extent to which a stem can grow vertically before it bends or breaks under its own weight (Niklas 1993a,b). It is always possible to enhance hydraulic conductance and maintain strength by surrounding a few very large conduits (which decrease local strength but increase conductance) with a large number of small, lignified cells such as fibers (to preserve a suitable φ and E).

It is presumed by many that there is some kind of tradeoff between large conduits with high hydraulic efficiency and the vulnerability of xylem to dysfunction. This presumption is borne out of many anatomical surveys using both systematic and floristic approaches (see, e.g., Carlquist 1975, 1977a,b, 1988, 1989; Baas 1976, 1982b; Carlquist and Hoekman 1986). These studies have shown a general trend in conduit diameter, as shown in Fig. 4.11. Wet-warm environments tend to favor species with wide conduits whereas cold or dry environments tend to favor species with narrow conduits. Our purpose here is to discuss the possible benefits and costs of large conduits from a biophysical perspective.

The difficulty with looking for a relationship between vulnerability to cavitation and vessel diameter is to devise some kind of measure of 'average' vulnerability and 'average' vessel diameter. It is easy to show with dyes that large conduits are more vulnerable than small conduits within a given stem. A stem at maximum conductivity can be perfused with dye and nearly all conduits are stained by the dye. But if a stem if first dehydrated before a dye perfusion, then it is clear that the larger conduits are embolized first (Fig. 4.12). The water potential at 50% loss of hydraulic conductivity (Ψ_{50}, Fig. 4.5) is a reasonable measure of vulnerability. However, some extra thought is needed for an estimate of 'average' vessel diameter, because the Hagen-Poiseuille law tells us that hydraulic conductivity increases with the fourth power of the radius or diameter. Hence wide vessels contribute disproportionately to the stem hydraulic conductivity. This is illustrated for the conduits of *Gnetum microcarpum* in Fig. 4.13. The vessel diameters were measured and grouped into diameter size classes of 25 μm (25 to 50, 50 to 75, etc.) and the percentage of conduits in each size class plotted as a bar histogram. Also plotted is the percentage contribution to the whole stem conductivity based on the Hagen-Poiseuille equation. The two largest size classes, which account for only 6% of the conduits, account for 48% of the hydraulic conductivity!

The problem is to find an appropriate measure of diameter for all taxa. A few large conduits contribute much more to conductivity values than do many small ones, as discussed above. One way around the problem is to measure 200–400 conduit diameters (D) at random and tabulate a column of D^4 ranked from largest to smallest value. The D^4 powers are summed for all conduits and then summed again until the sum equals 95% of the total stem conductance. This approximates the number of conduits likely to be responsible for about 95% of the total stem

Fig. 4.11. Diagram showing the ecological gradients in temperature and moisture and the direction of evolution of vessel diameter in response to these gradients. The *double-headed* arrows indicate that plants evolve in both directions along the indicated gradients

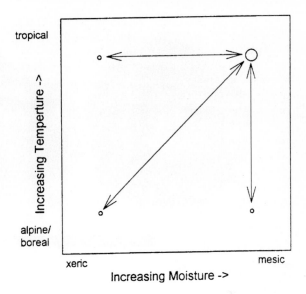

conductance. The mean diameter, D_{95}, is computed for these conduits. When D_{95} values were compared to means of all, D_{100}, the D_{95} values were generally 5 or 10% more than D_{100} for diffuse porous trees. This is because D values varied over a fairly narrow range. In ring-porous trees, where there are many small vessels and only a few large ones, D_{95} was 50–100% more than D_{100}.

Another possible estimator is to take the mean hydraulic diameter, D_H, which is defined as $[(\Sigma D^4)/N]^{1/4}$, where N is the number of conduits. D_H is the mean diameter that vessels in a stem would have to be if it is to have the same total conductivity for the same number of conduits as the sampled stem. This estimator generally produced a D_H between D_{100} and D_{95}. Another diameter that could be used was recently defined by Sperry et al. (1994) as $D_S=\Sigma D^5/\Sigma D^4$. D_S values were close to D_{95} values for ring-porous species, but were deemed no less arbitrary then D_{95} values. D_S has been used frequently but is probably the least justified measure for 'effective' vessel diameter; it is simply a statistic that weights large vessels but not in a physically meaningful way. For most purposes, D_H is a more meaningful measure in terms of the physics of water transport.

A plot of Ψ_{50} versus D_{95} for 60 species was compiled by Tyree et al. (1994) and is reproduced in Fig. 4.14. A log-log transform was used to look for both linear and non-linear correlations, e.g., if Ψ_{50} should increase with D^x, then the log-log transform would linearize it. The regression line is shown in bold in Fig. 4.14. The dotted lines are 95% confidence intervals on the regression. The weak, but statistically significant, regression accounts for only 22% of the variation ($R^2=0.22$, $p=0.007$). The weak correlation may be of use to evolutionary biologists but is not of sufficient accuracy to be of predictive value to a comparative physiologist, i.e., a physiologist cannot predict the vulnerability of a species by measuring the mean conduit diameter. It is still true that large conduits cavitate at a lower water stress

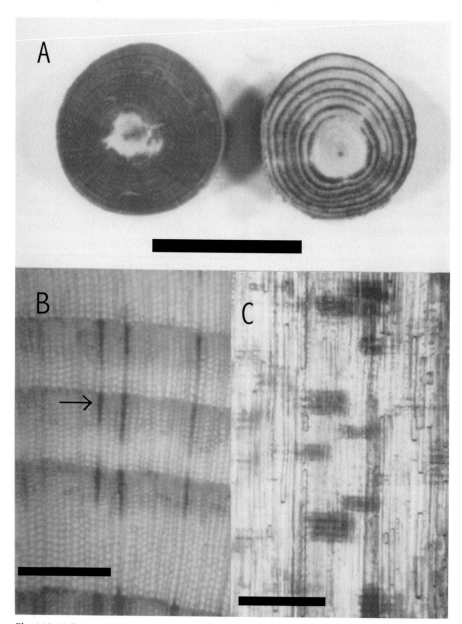

Fig. 4.12. A Dye-staining patterns of branches before (*left*) and after (*right*) dehydration. More vulnerable non-stained xylem is arranged in concentric bands. **B** Embolized bands correspond to larger diameter earlywood tracheids. *White and black parts of vertical bar* spanning the central growth ring correspond to non-stained earlywood and dye-stained latewood, respectively. Dye staining is also evident by the concentration of dye in ray parenchyma (*arrow*). *Scale bar* 0.5 mm. **C** Longitudinal section of freshly stained branches show non-stained, air-filled, earlywood tracheids (corresponding to *white parts of bar across top of photograph*) and dye-stained and water-filled latewood tracheids (*black bar*). *Scale bar* 0.25 mm. (Sperry and Tyree 1990)

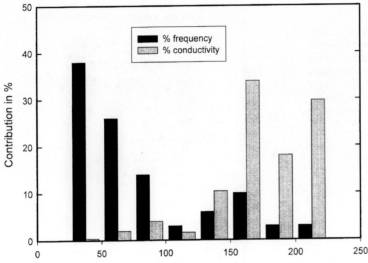

Fig. 4.13. *Black bars* are % of total vessel count in vessel diameter size classes of 25-µm increments. *Gray bars* are % contribution of vessels of each size class to total stem con-ductance based on the Hagen-Poiseuille law (computed from conduit size distribution of *Gnetum microcarpum* in Fisher and Ewers 1995)

than small conduits within a given stem (Tyree and Sperry 1989a; Hargrave et al. 1994).

The reader might ask why we do not correlate Ψ_{50} with D_{50}, i.e., the average diameter accounting for 50% of the hydraulic conductance. This correlation is not any better perhaps because so few of the largest conduits account for 50% of the conductance of stems, hence it is hard to obtain an accurate mean (Fig. 4.13). Figure 4.14 does not clearly answer the question of whether Ψ_{50} is proportional to volume ($=\pi D^2 L/4$) or surface area ($=\pi D L$), since the conduit lengths, L, were not generally reported. Conduit length varied from 1 mm for tracheids to 1.4 m for *Quercus agrifolia*. (Large conduit lengths were not represented since vulnerability curves were generally measured on young stems where conduit lengths are less than in tree trunks.) We must rely on smaller subsets of data to examine dependency of Ψ_{50} on volume or surface area. All the conifers in this study had tracheid lengths in the range 1.1–1.3 mm. No significant correlation was found between Ψ_{50} and volume or surface area for the conifer subset. Sperry and Sullivan (1992) sought a correlation between Ψ_{50} and conduit volume in five diverse species (2 conifers, 2 diffuse porous and 1 ring porous species). Among the species, diameters ranged from 10 to 100 µm and L from 1 to 400 mm. No correlation was found between Ψ_{50} and volume.

So why is there little or no correlation between conduit size and vulnerability to cavitation? One possibility is that the mechanism for cavitation is related to pit porosity rather than conduit size. Alternatively, the 'cost' of making 'safe' conduits may be so slight that there is little selective tradeoff between size and safety.

Fig. 4.14. A log-log plot of xylem tension ($=-\Psi_{50}$) causing 50% loss hydraulic conductance in stems versus the mean vessel diameter of the vessels that account for 95% of the hydraulic conductance (D_{95}). Each *point* represents a different species

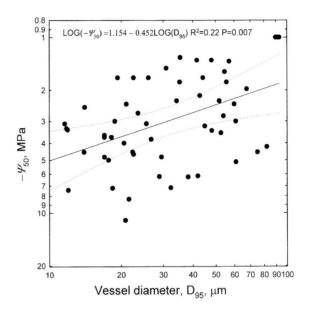

Reduced vulnerability to cavitation will still be strongly selected against in xeric or cold environments, but not in warm, mesic environments. In mesic environments we would expect genetic mistakes for high vulnerability to be tolerated and would thus eventually come to represent a large fraction of the gene pool.

It is unlikely that Ψ_{50} would ever be found to correlate with volume since this depends on a homogeneous nucleation mechanism that can be discounted on purely physical grounds. Pickard (1981) has shown that the probability of homogeneous nucleation is minuscule at biological pressures.

We are left with only heterogeneous mechanisms, and the question comes down to whether the wall defects causing nucleation are random defects or under genetic control. Pit-membrane porosity might be under genetic control because this involves the orderly synthesis of cellulose microfibrils and the regulation of secondary wall growth and lignification around the pit followed by partial enzymatic hydrolysis of the pit membrane (Butterfield and Meylan 1982). It is harder to imagine how the creation of hydrophobic cracks or patches might be controlled genetically and easier to imagine that these are random mistakes. If the air-seeding hypothesis is correct, there is no reason to believe there ought to be a strong correlation between pit-membrane porosity and conduit size.

Many papers have been published on drought-induced xylem dysfunction (loss of conductivity) since the early 1980s, and one of the most frequently stated paradigms is that vulnerability to cavitation is the most important factor determining the drought resistance of plants. As stated previously, there certainly is a correlation between Ψ_{50} and sylvicultural impressions of drought tolerance of woody plants. However, no one has actually tried to test this hypothesis. We could restate the question by asking: Is xylem dysfunction by drought-induced embolism the first step towards the death of a plant subjected to drought?

Rood et al. (2000) have recently studied the dieback in the crown of *Populus* trees. Years of repeated drought episodes seem to induce branch dieback throughout the crown in a patchwork fashion. The leaves on certain branches turn yellow 2 or 3 weeks early in the fall of any given year, and those same branches are seen to be dead in the next growth season. Rood et al. (2000) used a pressure collar to induce extra embolisms in a short length of intact stem in June or July. The leaves did not immediately die distal of the treatment zone, but the leaves did turn yellow early that fall and branches did not flush new leaves the next spring. A patchwork fashion of dieback can be predicted on theoretical grounds and will be discussed more fully in Chapter 6.2.

However, in general, clear proof that xylem embolism (PLC, percent loss conductivity) results in death of plants is hard to establish. This will be a fascinating area of study for future investigations. I have been involved in several studies of drought stress in small woody shrubs and young trees in Panama for several years. Central Panama is a seasonally dry rainforest, where the dry season starts when the direction of the trade winds changes. The dry season lasts an average of 135 days receiving an average of 2.2 mm of rain a day (293 mm for the dry season) and the wet season is an average of 230 days long, with an average of 10.2 mm of rain per day (2340 mm for the wet season). However, some dry seasons can be quite extreme; for example, some *el niño* years can be very much drier, e.g., in 1962 there was a 68-day period with 0 mm of rainfall, and in 1992 there was a 93-day period with only 15 mm of rain and an overlapping 108-day period with only 19 mm of rain. Little of the small rain showers in the 1992 drought penetrated the over-story canopy. Probably, most of the rain was intercepted by leaves of the over-story and evaporated away without ever hitting the ground. Moist-forest tropical species are often poorly adapted to such long periods of low rainfall and high death rates occur among the least drought-tolerant species. In studies we have conducted on *Psychotria* spp. we have found no correlation between native state embolism (PLC observed in field plants) and survivorship in 1997, when there were 74 days with 5.5 mm rain, and 2000, when there were 70 days with 22 mm rain (Wright, Pearcy and Tyree, unpubl. observ.).

In another study in Panama, B. Engelbrecht (pers. comm.) subjected 3-week-old seedlings of *Ochroma pyramidale* to a progressive drought. At various times subsets of plants were irrigated to score survivorship and this was related to leaf water potential measured with a psychrometer just prior to irrigation. A statistical (probit) analysis of the survivorship data revealed a Ψ for 50% death of −6 MPa. However, in adult trees, the vulnerability curves revealed a 50% loss of conductivity and >95% loss at $\Psi=-1$ and −1.5 MPa, respectively (see Fig. 4.15). Does this mean that seedlings are less vulnerable to embolism than adult trees? Or does it mean that seedlings can refill embolisms after irrigation as long as some meristems survive? In Section 4.7 we will review the mechanism of embolism reversal.

In general, the relationship between survivorship and water stress is very difficult to document because of the pattern of dieback we have observed in tropical plants. Leaf blades often die back from the distal to the proximal regions. So it is very common to see leaves with brown-dry-crisp distal areas and areas near the base or midrib that are still green and pliable. When these plants again see water, the brown areas are shed but the green areas survive. Sometimes all leaves can die and drop off but the stem meristems and, in some cases, only the roots remain

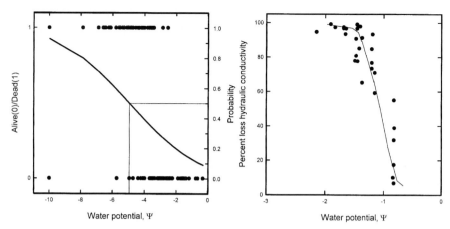

Fig. 4.15. *Left* Survival curve for 3-week-old seedlings of *Ochroma pyramidale*; the original data are survival (live/dead) versus leaf water potential measured with a psychrometer. The probability curve is from statistical analysis (Probit analysis, unpubl. results provided by B. Engelbrecht). *Right* Vulnerability curve measured on stems from a mature tree of the same species. (Adapted from Machado and Tyree 1994)

alive. During a drought there are very steep gradients in Ψ in leaves and stems, hence it is difficult to find a representative sample of living tissue to measure in a thermocouple psychrometer and often the Ψ is beyond the range that can be measured by these instruments. The pressure bomb also fails at extreme drought because it measures *only* xylem pressure, P_x. As leaf cells die the solutes are released into the xylem fluid making π_{xylem} very negative. There is still a local approximate equilibrium between Ψ of the xylem and the remaining living cells, but now P_x will be reduced ($=\Psi_{living\ cells}-\pi_{xylem}$).

Why is there such poor correlation between survivorship and PLC? In Chapter 5 we will see that there is a strong correlation between onset of xylem embolism and stomatal closure. Once stomates are closed, further water loss and progressive embolism depend on minimum leaf conductance to water vapor and water storage. High water storage and low vapor conductance will reduce the rate of embolism accumulation. Perhaps the main impact of xylem dysfunction is on long-term carbon gain. Plants seem to survive drought by shutting down gas exchange and delaying desiccation. So a plant with low vulnerability may be able to maintain a higher rate of carbon gain longer into the dry period. Hypotheses such as these need to be checked in future research.

We cannot leave this section on vulnerability curves without a mention of phenotypic plasticity in vulnerability curves. There is a growing body of evidence that the growth conditions of plants cause a shift in vulnerability curves. In beech trees, Cochard et al. (1999) found a 1-MPa shift in the water potential for 50% loss of hydraulic conductivity for shade branches, which were more vulnerable than sun branches. Several *Populus* clones grown in high-N soils showed an increased vulnerability to cavitation whereas K-enhancement of soils showed no effect on vulnerability curves (Harvey and van den Driessche 1999). The interesting suggestion is that xylem vulnerability is not static but can adjust if growth conditions are changing.

4.6 Freezing-Induced Embolism

The above discussion of drought-induced xylem dysfunction still does not answer the question of why conifers with small conduits tend to dominate in boreal and alpine habitats. Do small conduits confer some advantage in cold environments? There is a growing body of evidence that small conduits are less vulnerable than large conduits to freezing-induced embolism.

The first question that must be answered is whether xylem water of trees does freeze in the winter. Any forester who has cut trees in the winter knows that the answer is yes. But the recording of the release of latent heat of fusion is certainly the most reliable detection of freezing. One occasionally encounters the myth that xylem water supercools substantially. This idea probably arose from the fact that it is difficult to freeze fast-moving water if the stem is chilled locally, because the chilled water continuously "escapes". Scholander et al. (1961), for example, were unable to freeze water in *Calamus* (a vine of the palm family) with dry ice unless they stopped the movement artificially. Zimmermann (1964) spent a summer studying low-temperature effects in a number of North American tree species and found that supercooling amounted to no more than a few °C in all tested species. The freezing point of xylem water is only a few hundredths of a degree below 0 °C, because the solute content of xylem water is rather low. While water in tracheary elements freezes rather easily, water outside the tracheary elements may supercool substantially (Burke et al. 1976; Hong and Sucoff 1982). Such "bound" water, some 30–40% of the dry weight, may remain unfrozen (Lybeck 1959). (Note that the bound water shown in Fig. 8.1 amounts to ca. 30% if based on dry weight.)

The second question is whether xylem water contains dissolved air. The answer to this question is also yes. Scholander et al. (1955) found that the xylem sap of the grapevine is fully saturated with atmospheric nitrogen and partly with oxygen; there is no reason to believe that this is not the case in other plants.

Lybeck (1959) spun water in a Z-shaped capillary like Briggs (1950) (Chap. 3.4), at a speed which produced a tension of 0.18 MPa in the center. When a piece of dry ice was brought into contact with the central part of the spinning capillary, the water column broke immediately. Scholander et al. (1961) studied the effect of freezing of xylem water in *Calamus,* the rattan vine, a genus of the palm family containing many species. The lower end of the vine had been cut and was taking up water from a burette. When a stem section was briefly frozen and thawed, water in the previously frozen section had evidently cavitated, because when the section was raised to a point 10 m above the burette, the water drained back to the burette instead of being transpired by the leaves. When the frozen-and-thawed section was lowered to ground level and the supply burette raised, full transpiration resumed, as indicated by the rate of water uptake from the burette. It is quite obvi-

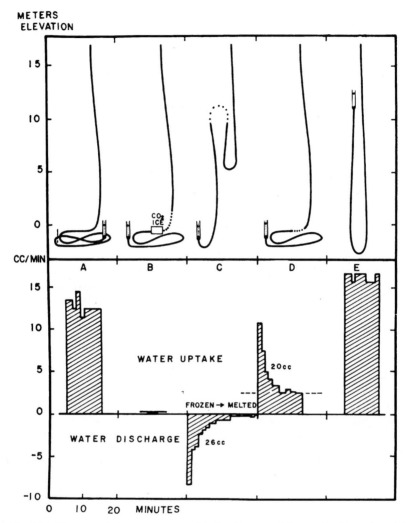

Fig. 4.16. Water uptake by a rattan vine: *A* before freezing; *B* frozen; *C* with nucleated loop elevated 11 m; *D* with cavitated section on ground; *E* with burette elevated 11 m. *Dotted line* cavitated section (Scholander et al. 1961)

ous that this arrangement had refilled the embolized vessels with water by positive pressure (Fig. 4.16).

How do trees from cold climates cope with the problem of winter freezing? There seem to be at least three entirely different methods. First, there is the "throw-away" method. Ring-porous tree species have such large and therefore efficient vessels that a single growth increment of the trunk is sufficient to provide the crown with water. This has been discussed repeatedly in previous chapters. In this case, vessels are made before leaves emerge in the spring, a phenomenon that had been known to Coster (1927), and was discussed at some length by Priestley et al. (1933, 1935). The second method is the refilling method, which is most clearly seen in the grapevine. Although the grapevine has rather large vessels (see Fig. 1.8), which embolize during the winter, they are refilled in the spring by positive root pressure (Sperry et al. 1987). The third method is perhaps the most difficult one to understand. It seems that wood that contains only tracheids or only relatively small vessels is not easily embolized by freezing.

Hammel (1967) froze 2–3-cm-long stem sections of twigs experimentally on standing trees with dry ice on summer days. He monitored the diurnal change in xylem pressure on nearby control twigs with a pressure chamber. Some time after freezing, he also measured pressures in the distal part of the frozen stem. In the four coniferous species he tested, freezing did not affect subsequent diurnal pressure changes. In other words, flow resistance had not been increased substantially by freezing, i.e., not much cavitation had taken place. In dicotyledons, temporary freezing of the tensile water did increase flow resistance to greater or lesser extent, i.e., part of the frozen xylem path was inactivated by cavitation. When air-saturated bulk water freezes fast, air is trapped in the form of many small bubbles; when freezing is slower, fewer and larger bubbles appear (Carte 1961). Sucoff (1969) carried out experiments similar to those of Hammel. He estimated that ca. 9% of the water irreversibly migrated into the unfrozen part of the plant. Thus, 9% of the water-filled tracheids would be lost to water conduction every time freezing occurs. This estimate is rather high (as Fig. 8.1 indicates), especially if we consider that xylem may freeze and thaw repeatedly during the winter. The safety feature of smaller tracheary compartments is here again evident, the smaller the compartment, the more confined bubbles remain. To what extent, if at all, is the valve action of the torus involved in conifers? As Hammel (1967) suggests, we do not know. Both freezing and the displacement of dissolved air (bubble formation) cause a local volume increase. The process of freezing therefore brings about a local pressure increase. When thawing begins, this higher pressure may help dissolve bubbles. The smaller they are, the faster they will be dissolved. It is probable that the last compartments in which ice thaws are the most likely ones to suffer cavitational damage. Thus, a certain percentage of the xylem may be lost each winter, the percentage being smaller the smaller the compartments. The chance of recovery is better the greater the effectiveness with which bubbles are prevented from coalescing within a vessel (Chap. 7.3).

In contrast, the first frost event is known to induce >90 PLC in oaks (Cochard and Tyree 1990; Sperry and Sullivan 1992; Cochard et al 1997) and some wide-vessel chaparral shrubs (Langan et al. 1997).

Freezing should induce embolisms, because air is 1000 times less soluble in ice than in water (Scholander et al. 1953). So when water freezes, air comes out of solution. If water is saturated with air at 0 °C, when it freezes then approximately 2.8 ml of air will come out of solution for every 100 ml of water frozen. What happens to this air when the ice melts? If the ice melts slowly and no tension develops in the tissue, then the air will dissolve. However, if tensions develop beyond the critical values in Eq. (4.3), then the bubbles will expand to make the conduit fully embolized and dysfunctional. Apparently, this happens in oak with vessels 100 μm in diameter, less in maple with vessels 30 μm in diameter, and least in conifers with tracheids 10 μm in diameter. Sperry and Sullivan (1992) have demonstrated a

Fig. 4.17. Relationship between xylem tension and loss of conductivity. *Open circles* are tension (negative P_x) needed to induce 50 PLC ($=\Psi_{50}$), *closed circles* are tension before a freeze required to produce 50 PLC ($=\Psi_{50F}$). The *upper plot* shows the relationship to conduit volume and the *lower plot* shows the relationship to specific conductivity. The *lower curve* suggests a tradeoff between vulnerability to freezing-induced loss of conductance and specific conductivity of wood. (Reproduced from Sperry and Sullivan 1992)

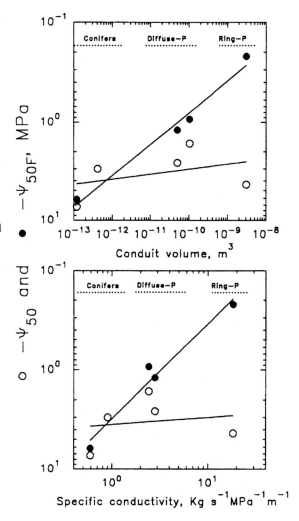

strong correlation with vulnerability to freezing-induced embolism and conduit volume. Figure 4.17 compares the Ψ_{50} values (open circles) of five species versus conduit volume or specific conductivity; it can be seen that there is no tradeoff between conduit size and drought-induced dysfunction (as stated previously); nor is there a tradeoff between increased stem hydraulic efficiency and drought-induced dysfunction.

The solid circle (Ψ_{50F}) values were obtained in the following way: Shoots were dehydrated to various initial water potentials, Ψ_i, and then wrapped in a bag, frozen to –20 °C overnight, and then thawed to >20 °C. This procedure induced a certain PLC. What is plotted is the value of Ψ_i that induced a PLC of 50 after the freeze–thaw cycle. While there is a strong correlation, it is not clear why Ψ_i measured before a freeze should influence PLC after a freeze–thaw cycle.

The Ψ_i value does not influence the amount of air coming out of solution upon freezing; the same amount of air will come out regardless of the initial conditions because the solubility of air in ice is very low compared to water at 0 °C. Since the samples frozen were not frost-tolerant, many living cells must have died and this would alter Ψ_i by an unknown amount. However, the value of P_x will certainly have risen from a value initially close to Ψ_i to something much less negative. It is the magnitude of P_x after the freeze and the capacity of the previously frozen tissue to take up water from conduits after the freeze that will determine how much the embolisms expand and how much PLC results. If no air dissolved immediately after the thaw, only 2.8% of the tissue volume would be occupied by air (=the volume fraction of air in solution at the time of the freeze). Only if these bubbles expand to fill the entire conduits would we expect PLC values of 50 or more.

It is difficult to know if PLC measured by tension-freeze experiments will induce the same level of PLC as in the field because the additional tendency of transpiration to cause bubble expansion is not duplicated. Trees certainly freeze under tension and some tension remains after the thaw. These objections have been well addressed by more recent studies (Langan et al. 1997; Davis et al. 1999) in which xylem tension was maintained in excised segments by spinning them in a centrifuge while freezing and thawing. Embolism after a freeze-thaw cycle was clearly a function of vessel diameter and tension; vessels >30 µm were very prone to freezing-induced cavitation even at modest P_x values <-0.5 MPa.

Why should large conduits be more prone to freeze-induced dysfunction? It probably has something to do with how long it takes air bubbles to dissolve rather than the tension when the ice first forms. This is because bubbles have to dissolve before the onset of the critical tension that causes them to expand. The physics of air-bubble dissolution is now well understood (Pickard 1989; Tyree and Yang 1992; Yang and Tyree 1992). Analysis of the kinetics of bubble dissolution reveals that the time it would take for a bubble to dissolve increases approximately with the square of its initial diameter. If many small bubbles were formed when ice melted and if the bubbles were the same size regardless of size of the conduit, then conduit size may not influence freezing-induced dysfunction. However, Ewers (1985) studied bubble formation while freezing and thawing water in small glass capillary tubes and observed that bigger bubbles formed in large diameter tubes than in small tubes and that they took longer to redissolve in large versus small tubes. It seems likely that the same will happen in xylem conduits.

4.7 Embolism Repair

What happens to a vessel once it is embolized? Does it remain dysfunctional forever or can it refill? If it does refill, how long does it take and what mechanisms are involved? These are the topics to be discussed in this section. It has long been known that the fractions of wood occupied by air, water and solids can change with time. The solid volume is relatively constant but the ratio of air to water can change on a daily and on an annual basis. This is a topic that will be discussed in

Fig. 4.18. A Percentage loss in hydraulic conductivity beginning in May 1986 for segments ≥5 mm (*solid circles*) and <5 mm diameter (*open circles*). Means for six trees are shown with 95% confidence intervals. **B** Minimum daily temperature during the study period. **C** Monthly precipitation as percentage departure from normal during the study period. Weather data from National Weather Service Office within 3.5 km of study site. (Reproduced from Sperry et al. 1988b)

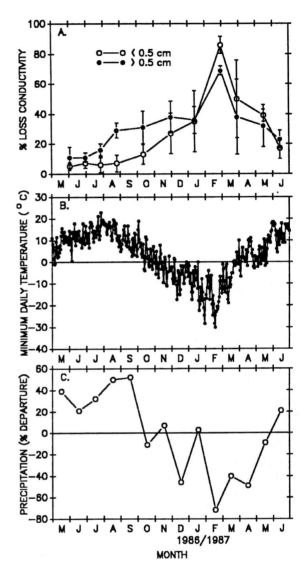

Fig. 4.19. A February main axis segments from five trees perfused with dye. All show southern location of non-stained, embolized xylem (*notch* marks north, *scale bar* 5 mm. **B**, *upper row* February twig segments perfused with dye showing extreme proportion of non-stained, embolized xylem; *lower row* adjacent segments from the same tree as in *upper row*, but perfused with dye after having been flushed to maximum conductivity. All the xylem is stained, confirming that flushing eliminates embolism. *Scale bar* 5 mm. **C** Longitudinal section from partially embolized segment such as those in **A**; safranin-stained vessels identified by dark-staining contact cells in black and white (*arrows*) are water filled, whereas the non-stained vessels contain air bubbles. *Scale bar* approx. 30 μm. (Reproduced from Sperry et al. 1988b)

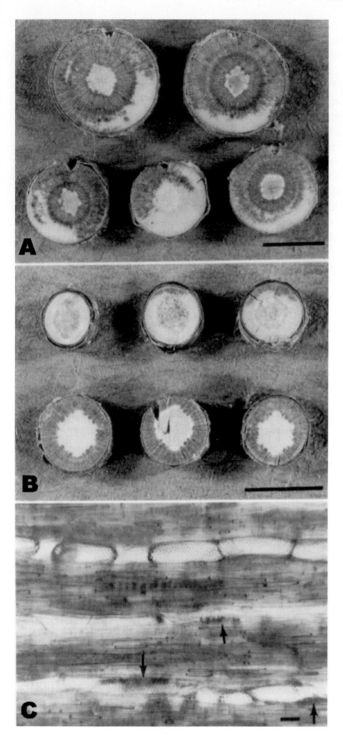

Section 4.8. Since air/water ratios can change in wood fiber cells that are not very conductive and in more conductive vessels or tracheids, the best way to investigate xylem dysfunction (embolism formation) and embolism repair (bubble removal) is to monitor short- and long-term changes in hydraulic conductance as well as percent loss of conductivity (PLC).

The first examination of seasonality of embolism formation and repair was conducted in sugar maple, *Acer saccharum* (Sperry et al. 1988b). There was a gradual increase in PLC in stems >5 mm diameter throughout the summer with little change in smaller stems (Fig. 4.18). The biggest changes occurred in winter. When temperatures remained below freezing for weeks, the stems seemed to freeze-dry with preferential loss of water from the southern exposures (Fig. 4.19) in the bigger stems and complete embolism in smaller stems. The recovery of hydraulic conductance (reduced PLC) occurred over a 6–7 day period when temperatures were persistently above freezing (Fig. 4.20).

As discussed in Chapter 3.9, sugar maples have a stem pressurization phenomenon in March to April in the northeasteern USA. It seems very likely that this stem pressurization is responsible for the refilling of the vessels in Figs. 4.18 and 4.19. Three possible mechanisms for xylem conductivity recovery from embolism have been suggested: condensation of water vapor, dissolution of gas, and expulsion of gas (Sperry et al. 1987). Embolism removal by the condensation of water vapor is rare and must happen immediately after a cavitation when the embolism is comprised mostly of water vapor.

The expulsion of gas occurs only in conduits with ends exposed to the air through open vessels or dry pit membranes. This mechanism accounts for about half the embolism removal in wild grape in spring. Over winter, grape stems become freeze-dried much like the small maple stems in Fig. 4.19. Root pressure in wild grape is sufficient to push the water column up several meters and expel air bubbles through the dry leaf scars (Sperry et al. 1987). When the advancing front of water reaches the leaf scars, some pits become wet before all the air has been expelled from adjacent vessels. When pit membranes are wet, it is unlikely that plants can generate sufficient root pressure or stem pressure to push air through the tension of water at the air/solution interface in pores on pit membranes bordering embolized and non-embolized conduits. Since conduits with ends open to the outside air are likely to embolize under only slightly negative P_x, the refilling of these conduits by gas expulsion can be of no functional value.

Bubble dissolution is the primary mechanism of hydraulic conductivity recovery from embolism in plants except in the branches of plants that initially are extremely dry (Sperry et al. 1987). The mechanism of bubble dissolution for embolized conduits encompasses three processes: (1) the dissolution of gas in air bubbles into the liquid at the surface of the bubbles; (2) the movement of dissolved gas to the outside surface of the stem, where it comes out of solution and enters the air; and (3) the replacement of the volume of the bubbles with xylem fluid (mostly water). There are two ways in which the air can be transported in solution from the inside to the outside of stems: (1) air diffuses through liquid from the surface of bubbles inside the wood to the surface of stem, driven by the air-concentration gradient; this air is then released from liquid at the stem surface into the atmosphere; and (2) the air is transported by a mass flow of water and

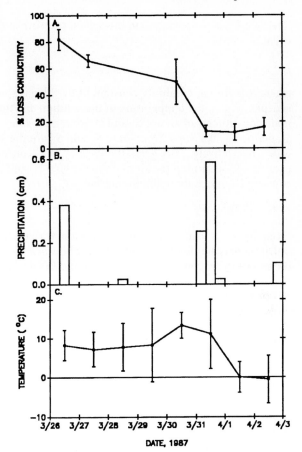

Fig. 4.20. **A** Mean percentage loss in conductivity (with 95% confidence intervals) for each of six trees in March vs. date. **B** Precipitation vs. date. **C** Mean temperature with range vs. date. (Reproduced from Sperry et al. 1988b)

carried to the evaporative surfaces (stem or leaf surfaces). These two processes occur simultaneously, but we believe that diffusion is the primary mechanism for movement of dissolved gas during conductivity recovery in most cases.

Gas will diffuse away from bubbles inside lumina of xylem conduits only if the concentration of dissolved gas in the water at the air/water interface of the bubble is more than the concentration at another place in the stem, e.g., in the water at the outside surface of the stem. Henry's law states that the equilibrium solubility of gas at an air/water interface is proportional to the gas pressure. Gas bubbles trapped in lumina are frequently at higher-than-atmospheric pressure because of surface tension of bubbles trapped in capillaries, as expressed by the 'capillary equation' (Eq. 4.2). This insures that there can be concentration gradients from gas bubbles to the outside surface of stems. The rate of diffusion away from bubbles is determined by the concentration gradient through Fick's law. These laws and equations are explained below.

Henry's law predicts the equilibrium solubility of a gas between gas/liquid interfaces. The gas concentration of the ith species in the liquid at the gas/liquid interface, C_{li}, is proportional to the partial pressure of the gas (absolute pressure), $\bar{P}_i$, or

$$C_{li} = S_i \bar{P}_i \qquad (4.4)$$

where S_i is the gas solubility constant of the ith gas species in the solution, which is dependent on the temperature of the solution and the gas species. Since the two main constituents of air are N_2 and O_2 we can consider an average solubility of air, S_a, defined as $0.8\ S_N + 0.2\ S_O$, where S_N and S_O are the solubilities of N_2 and O_2 in water, respectively. This gives the correct "air" concentration, C_{la}, from the air pressure, P_a, assuming the air bubbles are 80% N_2 and 20% O_2. In this case, the air concentration in bubble/liquid interfaces, C_{la}, is:

$$C_{la} = S_a \bar{P}_a \qquad (4.5)$$

From Eq. (4.5) it can be seen that in a system with two air regions separated by liquid, if there is a difference in air pressure between these air regions, there will be a difference in air concentration in the liquid between the borders of those two air regions.

The capillary equation is a special case of the surface tension equation that can predict the air pressure of a bubble trapped in a cell lumen. The air pressure of bubbles in a stem can vary from a value near to 0 Pa (vacuum) to any higher pressure depending on the stem xylem water pressure. This air pressure is calculated by:

$$\bar{P}_a = \frac{2\tau}{r_m} + \bar{P}_x \qquad (4.6)$$

where $\bar{P}_x$ is the absolute xylem water pressure ($= P_x +$ the current atmospheric pressure), $\bar{P}_a$ is the absolute air pressure of the bubbles (which will contain N_2 and O_2 as principal gas species), τ is the surface tension of the solution, and r_m is the radius of the curvature of the meniscus.

The air concentration in the liquid at the air/liquid interface of bubbles, C_{la}, can be obtained by substituting $\bar{P}_a$, the right side of Eq. (4.6) into Eq. (4.5),

$$C_{la} = S_a \left(\frac{2\tau}{r_m} + \bar{P}_x\right) \qquad (4.7)$$

Similarly, if the air pressure at the outside surface of a stem is given as $\bar{P}_{ao}$, the air concentration of the liquid on the surface of the stem, C_{lao}, will be:

$$C_{lao} = S_a \bar{P}_{ao} \qquad (4.8)$$

According to Eqs. (4.7) and (4.8), as long as $C_{la} > C_{lao}$ or $\bar{P}_a > \bar{P}_{ao}$, air will diffuse in the liquid from the surface of bubbles to the surface of the stem.

Fick's law predicts the rate of diffusion from a bubble surface from the concentration gradient,

$$F_a = -D_a \frac{dC_a}{dx} \qquad (4.9)$$

where D_a is the diffusion coefficient of air in the stem (e.g., wood), F_a is the diffusional flux across a unit area normal to the x-direction, and dC_a/dx is the air concentration gradient in the x-direction at a fixed time. The diffusion coefficient for air in wood, D_a, should be less than the diffusion coefficient of air in pure water because a certain fraction of wood is occupied by cellulose solids and air would have to diffuse around the cellulose fibers (Fig. 4.3). Yang and Tyree (1992) discuss how D_a should be estimated.

The air concentration in the liquid at the surface of bubbles must be higher than that at the surface of the stem for air to diffuse out of the stem. The only way $\bar{P}_a$ can exceed atmospheric pressure is when $P_x > -2\tau/r_c$, where r_c is the radius of the conduit containing the bubble. We have discussed already how r varies with P_x (Sect. 4.1). $\bar{P}_a$ changes because of the movement of the air/water interface in the lumen of the conduits. When xylem pressures are more negative than $-2\tau/r_c$ (which is most of the time during the day), the gases in embolized conduits will equilibrate, given enough time, with the air concentration in the xylem fluid, which should be in approximate equilibrium with the atmosphere. If embolism recovery is by a purely physical means, then the value of $-2\tau/r_c$ is the critical P_x for conductivity recovery in plants with conduits of the diameter equal to $2r_c$. For a typical conduit 50 µm in diameter, this critical pressure is about –5.8 kPa.

Equations (4.4) through (4.9) can be used to write a system of equations used in an iterative, computation-intensive calculation to predict the rate of bubble dissolution and the concentration gradients and fluxes anywhere in a predefined geometry of space within a stem divided into gas and liquid phases. The derivation of the model is given in Yang and Tyree (1992). Most readers can skip the derivation of the iterative equations and need only understand that they can be used to predict certain easily measurable parameters and others not so easily measured. The easily measurable parameters are: (1) the time required for all bubbles to dissolve from an initially defined state; (2) the tempo of recovery of hydraulic conductivity; and (3) the tempo of net water gain as water replaces bubbles. The less easily measurable parameters are: (1) gas flux rate versus time and place; and (2) concentration versus time and place in the liquid phase.

Tyree and Yang (1992) have used computer solutions of the system of equations that describe the rate of bubble dissolution according to the capillary equation and Henry's and Fick's laws and compared the predictions to the rate of hydraulic conductivity recovery in maple stems. Calculations based on this model predict that the time for the hydraulic conductivity to reach its maximum (i.e., when no embolisms remain) is 2.47 and 9.03 times faster when P_x increases from 0 kPa to 13 and 150 kPa, respectively. Based on the typical diameter of maple vessels, the model predicts that embolisms in stems will not dissolve and hydraulic conductivity will not increase when P_x is <-6 kPa; these predictions were tested by Tyree and Yang (1992).

Fully embolized (non-conductive) maple stem segments were connected to water-filled tubing. The tubing at both ends was inserted into beakers of water set on top of digital balances. The balance connected to the base of the stem segment was higher than the balance connected to the apex. The pressure difference ΔP driving flow across the segment stem segment equaled 10 kPa per m difference in height between the water level in the balances. The basal-balance measured the flow into the stem segment and the apical-balance measured the flow out of the

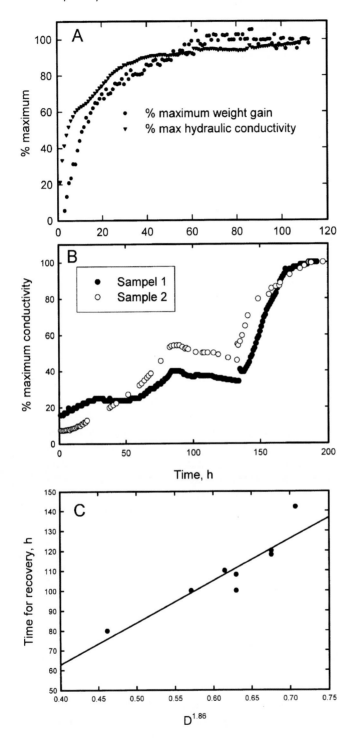

stem segment. The differences in flow gave the net increase in water content of the segment, which should equal the net decrease in air volume. By moving the stem segment either above or below the two beakers of water, the value of xylem pressure could be adjusted between negative and positive values (relative to atmospheric pressure), respectively.

Figure 4.21 shows the experimental results from several experiments conducted by Tyree and Yang (1992). The time course of hydraulic conductivity recovery is similar to the time course of net weight gain of the stem segment. Generally, about 100 h is needed for complete recovery from compete embolism in stems 5–7 mm in diameter. The rate of recovery is very sensitive to changes in P_x so, when P_x is reduced from +11 to –6 kPa, recovery of hydraulic conductivity stops. Finally, the time for recovery is a function of stem diameter to the power of 1.86 as predicted by theory. The theory also predicted that the time for recovery should be approximately proportional to the inverse of $(P_x+2\tau/r_c)$ for values >0 or otherwise no recovery should occur; this was confirmed by experimental results.

Hence, repair of embolisms appears to be a purely physical process requiring that the emboli be compressed by some force so that the pressure of the emboli is above atmospheric. This notion has been borne out by other experiments. For example, Lewis et al. (1994) observed 0.1-mm-thick slices of *Thuja occidentalis* wood under a microscope and were able to observe the gradual formation of emboli as wood samples dried and the dissolution of emboli after the samples were flooded with water. The time for all emboli to dissolve was predicted very well from Eqs. (4.4)–(4.9) for the geometry appropriate for these experiments.

Several theories of refilling have been investigated by Borghetti et al. (1991), who concluded that there is no evidence for active secretions by ray parenchyma (the so-called 'vital theory' of refilling), and that temperature changes could not induce sufficient condensation to account for the quantities of water involved in their limited experimental conditions. Borghetti et al. suggested that tracheids may be chemically active and that this, in combination with surface tension forces exerted by the water menisci at the ends of the embolized tracheids, may be sufficient to induce re-dissolution of gases when water potential rises above the value at which cavitation is caused. The nature of the 'chemical activity' of the tracheid walls was never defined in physical terms, making these suggestions unconvincing.

Fig. 4.21. A Percent hydraulic conductivity recovery and weight gain versus time of a maple stem segment 7 mm in diameter. The stem was kept at 1 °C to reduce microbial growth and perfused with a pressure of 14 kPa at the base and 12 kPa at the apex for an average pressure of 13 kPa. **B** Percent hydraulic conductivity recovery versus time of a maple stem segment versus time. The pressure difference was always held at 2 kPa but the average pressure was changed from +11 kPa to –6 kPa by moving the branch either 1.1 m above or –0.6 m below the source and sink reservoirs, respectively. **C** Time for complete recovery from complete embolism to maximum conductivity versus stem diameter. The theory in Yang and Tyree (1992) predicted the recovery time should be proportional to stem diameter to the power of 1.86 so the x-axis shows diameter to that power. Each point is a different stem and the *solid line* is the linear regression (adapted from Figs. 2, 3 and 4 in Tyree and Yang 1992)

In an important subsequent study, Edwards et al. (1994) designed experiments to investigate the relationship between water content and stem hydraulic conductance and looked specifically for unexplained cases of bubble collapse under negative pressure. Balances were used to measure water flow into and out of *Pinus sylvestris* stem segments under a pressure difference of 20 kPa while maintaining negative pressures down to –40 kPa. Experiments were conducted on stem segments 100–200 mm long, 5–25 mm in diameter. Evaporation from the stem surfaces was prevented by placing the segments in tight-fitting, polyethylene tube-sleeves. Water flow through the segments gave a measure of hydraulic conductance and accumulative differences in flow in versus flow out gave the net water uptake, which was equated with the volume of bubbles displaced. These experiments were very similar to those of Tyree and Yang (1992) except that greater care was taken to prevent evaporation from the stem segments. Bubble collapse was observed with and without net flow through the segments and with and without de-aeration of perfusate solution.

Tyree and Yang (1992) argued that bubble collapse should continue as long as $P_x > -2\tau/r$ and demonstrated that bubble collapse ceased in maple when P_x was more negative than $-2\tau/r$. In *P. sylvestris*, the tracheid diameters were 20–30 μm so collapse should cease when $P_x < -10$ to –15 kPa. Although Edwards et al. (1994) observed bubble collapse at $P_x = -40$ kPa, there is no reason to alter our understanding of the physics of bubble collapse because these observations were made when perfusing stems with de-aerated water. The authors correctly observe that Henry's law would predict that the air content of emboli should fall to zero when stems are perfused with air-free water (and radial diffusion of air is prevented which would re-aerate the water). Under these conditions, a pressure difference was established between the air-free emboli and the source reservoir of water on the balance that could push water to a height of 10 m (=a tension of about 100 kPa). When capillarity and pressure difference are taken into account, bubble collapse should not stop until P_x falls below –110 to –115 kPa.

The results of Edwards et al. (1994) provide an interesting confirmation of our existing theories, but under artificial conditions not likely to be encountered in nature because water moving from the soil to stems of trees is likely to be nearly air-saturated. Edwards et al. (1994) demonstrated that collapse was greatly reduced when the stems were perfused with aerated water at –40 kPa. There are other reports of partial bubble collapse under *apparent* tension (e.g., Salleo and LoGullo 1989). These interesting experiments have been repeated under better-defined, biophysical conditions during the partial reversal of embolisms in *Laurus nobilis* stems (Tyree et al. 1999a) and therein lies a substantial paradox!

If we conclude that positive pressures are a means for dissolving embolism, then we should ask how commonly positive pressures might arise in xylem. One mechanism for generation of positive pressure is root pressure or root exudation. In some species, e.g., grape (Sperry et al. 1987), bamboo (Cochard et al. 1994) and maize (Miller 1985), very substantial root pressures can arise. The origin of the root pressure is thought to be the buildup of solute in the fine roots by active uptake of nutrients from the soil water. Recently, Enns et al. (2000) have challenged the osmotic origin of root pressure but not the concept that roots can push water forward under positive pressure generated by some means. Root pressure events usually occur early in the morning or during rainstorms when transpirational pull, which would nullify root pressure, is minimal. How commonly does root pressure occur in plant taxa? Ewers et al. (1997) surveyed the occurrence of root pressure in 26 species of tropical liana in Panama and found instances of it in only three species. Fisher et al. (1997) surveyed 109 tropical vines and woody species and measured root pressure above 2 kPa in only 50 species (maximum 148 kPa, median 38 kPa). In terms of growth habit, most of the species studied were herbaceous vines and liana and about half the species exhibited exudation from cut shoots. Interestingly, of the 6 species of trees and 11 species of hemiepiphytes studied, all had exudation. Kursar (pers. comm.) carried out a survey of Panamanian species. In canopy and subcanopy trees (13 spp. in 12 families), no root pressure was found. In understory shrubs, root pressure was more common, being found in Rubiaceae (in 8 species of *Psychotria* and *Palicorea*) and in 7 species of *Piper* (Piperaceae). In seven other families of understory shrubs, root pressure was found only in *Acalypha diversifolia* (*Euphorbiaceae*). In monocots, root pressure was much frequently found in all species examined but was absent in all three species of palms examined. Monocots may rely more on root pressure to dissolve embolisms than dicots because they lack cambium. In woody dicots, growth of new functional vessels is a very effective means of replacing embolized vessels. Given that root pressures are commonly only 10–30 kPa, i.e., enough to push water up 1–3 m above ground, it seems unlikely that root pressure could refill vessels in tall plants with a few notable exceptions (Sperry et al. 1987). The paucity of root pressure in woody lianas with wide vessels and limited secondary growth is particularly surprising.

Do embolisms dissolve in plants without root pressure? Recent studies (Salleo and LoGullo 1989; Canny 1995a,b, 1997) have reported embolism repair under conditions when the threshold xylem pressure was much less than $-2\tau/r$. Cryo-SEM methods have been used recently to look at the state of vessels in roots, petioles and stems frozen in liquid nitrogen (LN$_2$) at various times of the day. Most frozen vessels appear to be filled with ice in early morning but gradually a large number of vessels become air-filled by noon. Then the frequency of air-filled vessels decreases gradually until late afternoon, when the number of ice-filled vessels is about the same as in the morning (see, e.g., McCully et al. 1998; Pate and Canny 1999). This work was interpreted to mean that vessels start refilling with water in the early afternoon, when xylem water pressure is very negative. However, this interpretation has been discounted definitively by ingenious experiments by Cochard et al. (2000). Briefly, they have shown that freezing walnut petioles in LN$_2$ induces embolisms at tensions that do not induce embolisms without freezing. Canny's recent reply (Canny 2001a) to Cochard et al. (2000) is very flimsy because of the lack of controls and in my opinion is totally without merit because it missed the main point. Cochard measured hydraulic conductance of walnut petioles while spinning them in a centrifuge. When spun to induce tensions of –0.7 MPa, no loss of conductivity occurred but, when frozen in LN$_2$ at the same tension, air bubbles were observed (Canny et al 2001b; Cochard et al 2001; Richter 2001).

Holbrook et al (2001) recently looked at embolism formation and reversal in grapevines using magnetic resonance imaging (MRI) techniques. Grape vines were grown in 8-l pots and periodically pruned to yield unbranched shoots 4 m in length. The shoots were passed through the magnet of an MRI machine so that most of the leaves passed through the machine and could be illuminated to maintain transpiration. Water was withheld until xylem pressure potential fell to −1.2 MPa at which point MRI began. A number of air-filled vessels were observed and this number increased by 10 over the next 24 h while lights were kept on continuously and the xylem pressure potential fell to −2 MPa. At this point the pot was watered. Upon watering the xylem pressure potential increased to −0.2 MPa and no refilling was observed over the next 9 h. However, when the lights were turned off, approximately 30 gas-filled vessels refilled over the next 12 h without positive pressure in the xylem. During the dark period, a short side branch at the base was cut off and observed for evidence of root pressure and water exudation. At the end of the experiment the whole shoot was excised from the roots and still no evidence of root pressure or water exudation was observed. Hence, the 30 vessels appeared to refill without water pressure reaching high enough pressure to account for refilling by gas dissolution according to Henry's law. Because of the expense of MRI research and the many demands on time for use of the MRI facility, the experiment on grape vine described above was not replicated, i.e., only one experiment was done.

Salleo et al. (1996) reported the most replicated example of embolism recovery under water stress. Studies were conducted on laurel (*Laurus nobilis*) in dry soil (predawn, Ψ_{leaf} equal to −1 MPa). Therefore, the threshold xylem pressure should have been less than −1 MPa before embolisms were induced. One-year-old twigs were cavitated by air injection in a pressure collar and recovered from embolism 20 min after pressure release. Xylem refilling was greatly increased in twigs treated with 50 mM KCl or with KCl plus 1 mM indoleacetic acid (IAA) solutions placed into contact with the exposed phloem of subsequently cavitated twigs. In the same study, it was shown that xylem refilling was prevented or strongly reduced by girdling the stem before inducing cavitation. The shorter the time interval between girdling and measurement of PLC, the larger the PLC recovery.

The results reported above could fit with the existing paradigm if the threshold xylem pressure were locally greater than −2τ/r in refilling vessels while full vessels were at −1 MPa. These conditions could be obtained if refilling involved the excretion of osmotica (salts and/or organic molecules) from living cells surrounding the cavitated vessel. If the π_x of the sap in the refilling vessel were less than −1 MPa, the threshold xylem pressure of the sap would be at a pressure greater than 0. Grace (1993) made similar suggestions. These experiments were repeated by Tyree et al. (1999a): potted *Laurus nobilis* plants were dehydrated to a predawn water potential of −1 MPa, which is not enough to induce much embolism. Then a pressure collar was built around a section of stem and pressurized to +1.26 MPa, which was enough to cause a substantial amount of embolism because of the pressure differential of −2.26 MPa between the air and the xylem. Some branches were harvested and PLC measured within 2 min and others after 20 min. Even though P_x was always −1 MPa, some embolism recovery occurred and the recovery was even more pronounced when stems were pretreated with a mixture of KCl+IAA (Fig. 4.22).

Some stems were frozen in LN_2 10 min after the pressure collar treatment and the samples were analyzed in a cryoscanning electron microscope. Frozen xylem fluid was analyzed by elemental microanalysis. No detectable quantities of C, K, Ca, Na, P or Cl were found. Direct analysis of xylem perfusate revealed an osmolality of 43 mOsm kg^{-1} and, as the microanalysis could have overlooked a maximum

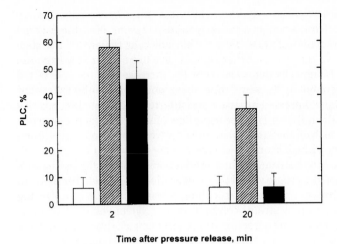

Fig. 4.22. Percent loss conductivity ±SD ($n=10$) measured in 1-year-old twigs of *Laurus nobilis*. *White columns* control twigs; *hatched columns* untreated twigs but induced to cavitate with a pressure collar; *black columns* KCl/IAA-treated twigs but induced to cavitate with a pressure collar. Values are 2 and 20 min after release of the pressurization that induced the cavitations. All plants were at a predawn water potential of –1 MPa

osmolality of 100 mOsm kg^{-1}, π_x was >-0.34 MPa and recovery could not be explained by localized osmotic effects, so that recovery of embolisms can occur at $P_x < 2\tau/r$, i.e., at values of at least –0.6 MPa.

Clearly, a new paradigm is needed. Canny (1995b, 1997) introduced the concept of "tissue pressure" to explain how changes in the π of living cells in stems might exert a prolonged amelioration of the xylem pressure in adjacent xylem vessels. We feel that this theory is thermodynamically and mechanically impossible. Surrounding tissues cannot effect a lasting change in xylem pressure in the presence of continued transpiration. Canny proposed that starch-to-sugar conversion in stem tissues would cause a swelling of the living cells (true) and that the swelling of these tissues would apply a tissue pressure against the cell walls of vessels (also true). The stress of the additional tissue pressure would cause a transitory release of pressure in the vessel lumen because of the contraction of the vessel diameter under the stress of the increased tissue pressure. However, eventually (in seconds or less), the stress of the tissue pressure would be totally balanced by the strain (reduction in vessel wall dimensions). The water pressure in the lumen would soon return to the same or lower pressure value as additional water was withdrawn from the vessel by transpiration in the attached leaves. Hence tissue pressure cannot cause a permanent increase in xylem pressure (Tyree 1999a).

Could transitory increases in xylem pressure be sufficient to cause refilling of vessels? Transitory changes in tissue pressure, if large enough, could return the xylem pressure of cavitated vessels to values above atmospheric for a brief period. In Canny's paradigm (1995b, 1997), some living cells undergo a starch-to-sugar conversion; this conversion lowers their π and draws in water, which causes them to swell. These swelling (living) cells then compress other living cells, causing them to release water. This paradigm is unlikely for two reasons. First, there would have to be a countercurrent of water flow within adjacent regions of the stem cross-section. For living cells to swell, there would have to be water flow from full xylem vessels to the living cells that swell after the increase in π because of the starch-to-sugar conversion. At the same time, there would have to be water flow from other living cells, compressed by tissue pressure, to the cavitated vessels that might be adjacent to the full vessels. Since mass flow of water cannot occur in two directions simultaneously in the same space, the flow would have to occur in adjacent but separate places, which is unlikely. Second, there would be a large mechanical disadvantage in Canny's tissue pressure mechanism. Some of the mechanical swelling would be wasted, as some would be in a direction away from the cavitated vessels and might increase stem diameter without releasing water. Some swelling pressure would be taken up by strain (compression) of xylem vessels and would not release water. Therefore, only a fraction of the mechanical advantage would be exerted where it is needed, i.e., on living cells squeezed by tissue pressure.

As a counterproposal, a more efficient paradigm could be considered. Canny (1995b, 1997) proposes a decrease in π in some living cells to release water from other living cells. It makes much more sense to propose an increase in π (e.g., by sugar-to-starch conversions) in all living cells surrounding a cavitated vessel. This would cause a rise in cell Ψ and a corresponding release of water and decrease in turgor pressure of the affected living cells. There would be no countercurrent in water flow (toward swelling cells and away from shrinking cells) and therefore no mechanical disadvantages.

Unfortunately, there are four serious problems with this more efficient paradigm and that of Canny (1995b, 1997). The first problem is that in woody tissue the modulus of elasticity of cell walls is very high, so volume changes are quite small; as little as 0.1% volume change MPa^{-1} (Irvine and Grace 1997). In woody stems, vessel lumina can account for up to 20% of the tissue volume (Tyree and Yang 1992). Vessel lumina occupied 10.3% (SD=2.1; n=5) of the stem volume in our laurel twigs. Therefore, if 80% of the stem is filled with living cells and they all release 0.3% of their volume for a 3-MPa change in turgor pressure, that would be enough to refill only 2.4% of the vessel volume (=0.3%×[80%/10%]). The potential volume of water that could be released by shrinking would therefore be too small.

The second problem is that the water released would not flow preferentially from shrinking (living) cells to cavitated vessels, it would flow to all vessels simultaneously. Actually, more water would flow from shrinking (living) cells to full vessels than to embolized vessels because the pressure drop from the shrinking (living) cells to the full vessels would be more than from the shrinking (living) cells to the embolized living cells. Based on these first two problems, it appears that tissue-volume changes would not be enough to account for the volume flows required to refill embolized vessels.

The third problem is that water would flow from living cells to cavitated vessels only if the Ψ of the living cell became more than the Ψ of the water in the filling vessel. For example, in the experiments in Fig. 4.22, the predawn value of Ψ_{leaf} was −1 MPa, so it is likely that all living stem cells were at an initial Ψ of −1 MPa, and for every living cell, $\Psi = P_t - \pi$, where P_t is turgor pressure. Let us assume that the π

of the living cells is the same as that in leaves ($\pi=-2.4$ MPa); the turgor pressure would then have to be 1.4 MPa when the living cells are in equilibrium with a threshold xylem pressure of -1 MPa in functioning vessels. For water to flow from the living cells to the cavitated vessels, where xylem pressure is approximately 0 during filling, the cell π would have to rise rapidly by more than 1 MPa to raise the value of Ψ above 0 (e.g., from $\pi=-2.4$ to -1.3 for Ψ to reach $+0.1$ MPa). A similar argument applies to Canny's tissue-pressure hypothesis (1995b, 1997). Tissue pressure caused by swelling (living) cells would have to compress other living cells to raise their P_t by more than 1 MPa for Ψ to be above 0. Consequently, the swelling cells would have to lower their π by more than 1 MPa (e.g., from -2.4 to less than -3.4 MPa).

The final problem with the two paradigms is that filling vessels would sometimes be adjacent to full vessels. In the pit membrane between the full and empty vessel there would be a water meniscus that sustains the large pressure difference between the vessels. The problem is that during filling of the embolized vessel the water might reach the pit membrane in some of the bordered pits before it reaches all of the pits. If the meniscus in a full vessel is rejoined with the water in a partly filled vessel, why isn't the water sucked out as fast or faster than it enters? One possibility is that the cell walls of the over-arching bordered pits might be slightly hydrophobic (lignin in secondary walls is known to be hydrophobic). Water in the filling vessels might not pass through the bordered pits until the vessel lumen is completely full. At that point the water in the lumen might rise a few kPa above atmospheric pressure and thus push through all pit borders simultaneously; a similar idea was independently proposed by Holbrook and Zwieniecki (1999).

The problems discussed above suggest that our proposed paradigm and the existing paradigm by Canny are both very unlikely. The biggest problem might be the speed with which changes in π would have to occur. While π is increasing, water would be simultaneously flowing out of the living cell; so for Ψ to reach a value greater than 0, the halftime, $t_{1/2}$, for the π increase would have to be less than the $t_{1/2}$ for water equilibration between living cells. Pressure-probe studies on leaf cells reveal the $t_{1/2}$ to be 1 to 10 s. Since $t_{1/2}$ is inversely proportional to bulk modulus of elasticity, we would expect the $t_{1/2}$ of woody cells to be even less. It seems improbable to me that π could change more than 1 MPa in less than 1 s.

In conclusion, I feel that data confirm the refilling of embolized vessels in plants when some adjacent vessels are at a threshold xylem pressure much less than $-2\tau/r$. However, to explain this refilling, a new paradigm may be needed since all existing paradigms seem improbable. I are reminded of a quote from a famous fictional detective, "... when you have eliminated the impossible, whatever remains, *however improbable,* must be the truth" (Doyle 1986). It seems that we have eliminated some of the possible and some of the improbable paradigms, so for the moment we should be open to new suggestions, however improbable they may seem. Clearly, more work on the mechanism of xylem refilling will be required to answer the issues raised above.

4.8 Storage of Water

This section is not meant as a general review of water storage in plants, but concerns primarily storage mechanisms that are more-or-less directly related to water movement.

The following mechanisms come to mind.

1. *Elasticity of Tissues.* As xylem pressures fluctuate, the plant body shrinks and swells. When pressures drop, water becomes available to transpiration by a volume decrease of the plant body, and when they rise, water is stored. This mechanism is based on osmotic water movement, on imbibition and on cell-wall elasticity.

2. *Capillarity.* Any water that is held by capillarity must be in equilibrium with the apoplast pressure. This can provide considerable storage potential in rigid stems. Strangely enough, it is hardly ever considered.

3. *Cavitation.* Some plants have special water storage cells, giving up water by "air seeding" at a given tension.

The first mechanism is practically unavoidable and must function at all times, because of the elasticity of the plant body. Osmotic water movement can provide considerable water-storage capacity in living tissues. Some plants have special water-storage cells near the photosynthetic cells. Their anticlinal walls crumple when they give up water, but they are still alive and refill when apoplastic pressure rises (Fig. 4.23 A,B). The discovery of diurnal swelling and shrinking is ascribed to Kraus (1881) and his student Kaiser (1879). Leaves, fruits, nuts, and other plant parts shrink and swell diurnally as a result of water-content changes (MacDougal 1924; Tyree and Cameron 1977, etc.). Friedrich (1897) designed a device for the accurate measurement of tree growth and recorded diurnal swelling and shrinking of tree trunks due to water stress. However, according to MacDougal et al. (1929, p. 8), his efforts were still much hampered by the high-temperature coefficient of the metal out of which his device was constructed. Instruments of this nature, dendrometers or dendrographs, have subsequently been used by numerous investigators, perhaps most extensively by Nakashima (1924) and MacDougal (1924). A more recent description has been published by Fritts and Fritts (1955). A survey of the devices that have been designed over the years has been given by Breitsprecher and Hughes (1975). Part of the expansion and contraction of a stem resides in the bark (Kraus 1877; Klepper et al. 1971), but it has been rather well documented that xylem also undergoes diurnal diameter changes (MacDougal 1924). Even individual vessels contract under stress without rupture of the water columns in them (Bode 1923).

Holbrook and Sinclair (1992a,b) and Holbrook et al. (1992) report a very elegant study of water storage in palm trees. They used both time-domain reflectometry and electrical capacity measurements to estimate non-destructively stem and soil water content in *Sabal palmetto* as trees of different size progressed into a drought period. Elastic (tissue) water storage per unit leaf area increases with palm tree size. Consequently, large palm trees can survive drought better than small palms (Fig. 4.24). Hence it is common sylivacultural practice to transplant large palm trees rather than small ones. The trees are transplanted with almost no

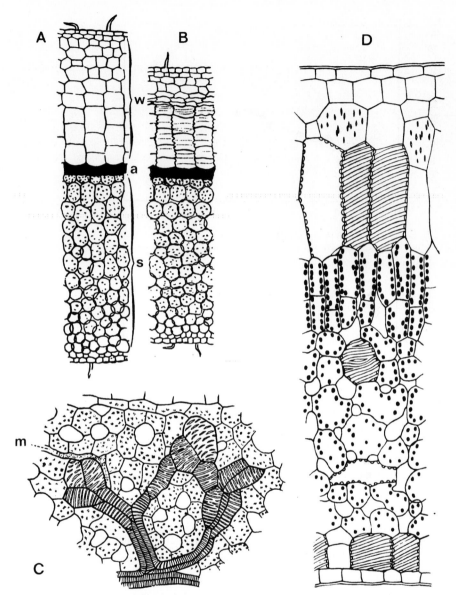

Fig. 4.23. Example of water storage by osmotic water movement, i.e., elastic volume change (**A**, **B**), and by cavitation (**C**, **D**). **A** and **B** are water-storage cells in a leaf of *Peperomia tricho-carpa*. **A** The leaf in fresh condition; **B** severed leaf that has been transpiring for 4 days; *w* water storage tissue; *a* photosynthetic tissue; *s* spongy mesophyll. **C** Bundle ends with storage tra-cheids in the leaf of *Euphorbia splendens*; *m* portion of a latex tube. **D** Transverse section through a leaf of *Phyosiphon Landsbergii* showing storage tracheids. (Haberlandt 1914)

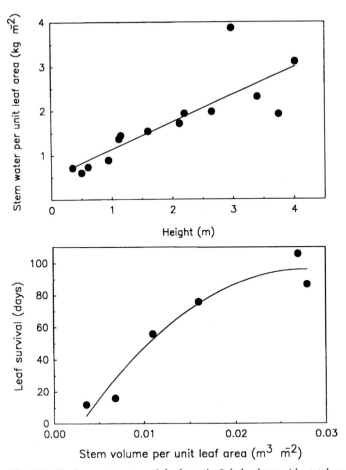

Fig. 4.24. *Top* Stem water per unit leaf area in *Sabal palmetto* (the total amount of water within the aboveground portion of the stem divided by the leaf area of the entire crown) as a function of plant height. Each *point* is an individual tree and the *solid line* is the linear regression $y=0.51+0.62x$, $r^2=0.83$ (from Holbrook and Sinclair 1992a). *Bottom* Survival time of leaves attached to felled trunks as a function of the stem volume divided by the leaf area attached to that stem. The *solid line* is a second-order polynomial regression. (Holbrook and Sinclair 1992b)

roots attached, a practice that rarely works for large dicot trees. There is enough water storage in large palms to meet transpirational needs until new roots grow. During an 8-day dehydration a large palm growing in a pot transpired 58 kg of water of which approximately 30 kg came from stem-water storage and the rest from the soil (Holbrook and Sinclair 1992b).

Our discussion will now concentrate on the second and third of the mechanisms listed above. The second, capillarity, has been almost entirely ignored in the past, whereas the third, cavitation, has been much discussed. Let us begin with the rather puzzling question of how tree stems can be relatively "dry' in the summer and "refill" in the fall. By far the most detailed information about this phenomenon has been provided by Gibbs (1958).

In many papers R.D. Gibbs reported the seasonal changes in water content of tree stems. This work was originally started to supply the practical information of when to float logs to the mill and avoid losses by sinking. However, it soon acquired theoretical interest in relation to the mechanisms of water storage and sap ascent. Gibbs (1958) summarized these investigations in the First Cabot Symposium, which was held at the Harvard Forest in April 1957. Many aspects of Gibbs' work are interesting, but we need to concern ourselves here only with the controversial question of water storage in xylem. The water content of wood drops throughout the summer and reaches a low value some time before leaf fall. For about a month after leaf fall, some of the tree species appear to "refill" their stems with water, others do not. It is primarily the ring-porous species that do not refill, perhaps because the wide earlywood vessels become embolized in the fall. The timing of cavitation in ring-porous species has never been recorded to my knowledge, although Hermine von Reichenbach (1845) found that tylosis formation in *Robinia* takes place in the fall. As embolism normally precedes tylosis formation (Chap. 8.3), we may assume that at least some vessels embolize in ring-porous trees before winter (Zimmermann 1979).

The interesting and puzzling finding of Gibbs (1958) was that *diffuse-porous* species showed an increase in water content after leaf fall, amounting to roughly 15% (weight per volume of wood). In *Acer saccharum* and other maple species, it was closer to 25%. This phenomenon is puzzling because it is hard to conceive that vessels that have embolized during the summer could refill in the fall in the absence of positive pressure. Where is this water storage space?

Some space is provided by the elasticity of the stem, i.e., mechanism (1) above. Some of the tissue contraction is released by cutting while air is drawn into the cut vessels. This, however, is not enough. In addition it may well be capillarity, a mechanism neglected in the past, although it has been known for very specialized structures. Schimper (cited in Haberlandt 1914) described well-developed intercellular spaces for water storage in petioles of an epiphytic *Philodendron* species. When the external water supply fails, water moves from these spaces to the leaves. The amount of gas space present in wood is not very well known, but it could be considerable. A wood section on a microslide is entirely liquid-saturated, comparable to a log that sinks in the river. But the wood of a living tree is not that waterlogged, it contains a considerable amount of gas. Martin first became aware of the extensive gas-duct system when he made paint infusions for the vessel-length distribution measurements described in Chapter 1.2. He used paint applicators made of metal (Zimmermann and Jeje 1981, Fig. 1), which were attached to the cut end of the log. The applicator was first filled with distilled water for vacuum infiltration of the cut vessels. Air that had entered the cut vessels after felling the tree and cutting the log could thus be removed. To his great surprise, water in the applicator kept bubbling no matter how long he applied the vacuum. It was quite evident that air was entering the cut end of the log beside the applicator, was drawn through the intercellular spaces and into the applicator. We can assume that intercellular spaces are necessary to provide ducts for air movement in wood, but it was a surprise that there was so many of them.

Let us now explore capillarity of these spaces with a very simplified model in order to see how it could work and how effective it might be. The purpose of this exercise is to show how the changing radii of the air/water interface can provide storage space in any crack between solid surfaces. Let us assume that wood consists of bundles of perfect cylinders (representing the fiber matrix of wood). These cylinders can be packed in a tight, hexagonal pattern. Intercellular spaces are probably under atmospheric pressure (+101 kPa). Let us assume that the xylem pressure during a summer day is –1.4 MPa and in the fall –50 kPa. In the fall, therefore, extracellular water sits in cracks of a width less than 2 μm (see Eq. 4.3), as illustrated in black in Fig. 4.25 (left). The two large circles represent the cylinders (the fibers) with a radius R of 5 μm. The circle with the radius $r_1 = 1$ μm, drawn to scale, represents the water meniscus at –50 kPa pressure, i.e., the fall condition. In the summer this meniscus is smaller; at –1.4 MPa it is 0.1 μm ($=r_2$). To the right of Fig. 4.25 these menisci are shown at higher magnification. During the summer at greater tension, water withdraws to form the smaller meniscus r_2; in the fall, as the pressure in the xylem rises, water (ascending from the roots) must leave the conducting elements to form the larger meniscus r_1. The hatched area shows the capillary storage area that is refilled after leaf fall. If the cylinders are tightly packed and endless, there are six such water-storage cracks per cylinder. We can now calculate the dimensions of this space and find that for the above conditions it represents ca. 6% of the total volume. This value may be too optimistic, as capillarity may depend on hydraulic, rather than measured radii. The purpose of the model is to show the principle of capillary storage. Wood does not consist of perfectly cylindrical cells; potential storage locations have yet to be discussed.

Intercellular spaces are rarely seen between axial elements, although air penetration tells us that they must exist (see, e.g., p. 152 in Baas and Zweypfenning 1979). It appears that there are more spaces along rays than along axial tissue. Bolton et al. (1975) reviewed our knowledge about interstitial spaces in wood and described their own observations in the Araucariaceae. MacDougal et al. (1929) made extensive measurements of the permeability to air of intercellular spaces in wood and reported that they are exceptionally tight in *Sequoia*. *Sequoia* may depend less on oxygen in the undissolved gas form; it is perhaps this property that makes it more resistant to flooding than other trees. Hook et al. (1972) examined the entire question of aeration in trees. The important message of the extremely simplified model shown in Fig. 4.25 is that the changing meniscus radii provide very effective water storage in wood; however, it does not show the location of the space.

We now have to consider two very important points. First, intercellular air spaces are very small in diameter. Vacuum could not pull air through any wet pore with a diameter smaller than 3 μm. If, however, we visualize that the surfaces of these intercellular spaces are partly or entirely lined with a water-repellent coating, it could be done. Second, it is very likely that the largest gas volume in the wood is present inside dead fibers, fiber tracheids and tracheids, and perhaps vessels where the water columns have cavitated (see Fig. 6.1). Vapor pressures in intracellular spaces are anywhere between 2.3 and 100 kPa. The slender tips of these cells can store and release considerable amounts of water as the meniscus radii of the capillary water vary. This situation is shown in Fig. 4.26, which illustrates another capillary-storage model, namely, a single wood fiber. Its outside diameter is 10 μm, its transverse section circular. For mathematical simplicity it is

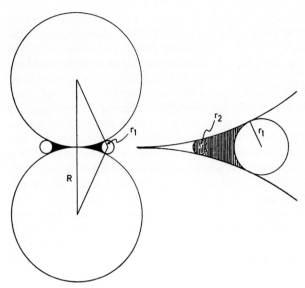

Fig. 4.25. Capillary storage of water (*black*) between cylinders. The radius of the cylinders, representing wood fibers, is shown as R=5 µm, r_1 is the radius of the meniscus at –0.05 MPa pressure, and r_2 the radius of the meniscus at –1.4 MPa pressure. Dimensions are drawn to scale. On the *right* details are shown at higher magnification. The *hatched area* is the waterstorage capacity when the pressure increases from –1.4 to –0.05 MPa

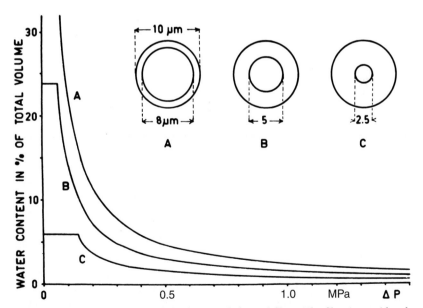

Fig. 4.26. Capillary storage inside dead, air-seeded wood fibers. The fiber is considered to be a 0.4-mm-long double cone; storage capacities are shown for three different wall thicknesses (*A—C*). Δp is the pressure difference between the fiber lumen and the apoplast water. For further explanation, see text

assumed to be a double cone of 0.4 mm overall length. Three wall thicknesses are considered, corresponding to a central lumen diameter of 8, 5, and 2.5 µm, respectively (A, B, and C in Fig. 4.26). The fiber is considered dead and air-seeded, i.e., it may contain water vapor. We can now calculate the amount of water it must hold by capillarity at given differences of pressure between the lumen (water vapor) and the apoplast. When the apoplast (xylem) pressure is very low (Δp large), capillary water sits only in the very tip of the pointed fiber lumen. As Δp decreases, the radii of the water menisci increase and water is taken up into the fiber lumen. In Fig. 4.26 the amount of this water is expressed in percent of the total fiber volume (i.e., it would be 100% at $\Delta p=0$ and an infinitely thin wall). The curves show a peculiar flat segment near $\Delta p=0$; this is where capillarity has filled the cell completely. Obviously, once the cell is full, it cannot take up more water. The model ignores complications such as elasticity.

In considering this model, we see that the greatest storage capacity is at small Δp, i.e., at xylem pressures that are not very low. This is exactly what happens in autumn when the leaves drop: xylem pressures increase to values a little below atmospheric and the stem is able to store a considerable amount of water. This volume of water, together with the volume taken up by elasticity and imbibition, probably provides the water storage described by Gibbs (1958).

We now have to consider a complication. Some fibers may contain water vapor only (and an infinitesimally small amount of air from the air-seeding process), others may contain additional gasses such as carbon dioxide. Gas-filled fibers are known from a number of species, such as maples, apple trees, etc. In this case the inside gas pressure changes according to the gas equation whenever the xylem pressure changes, i.e., Δp may not quite reach zero, even at positive xylem pressures, because the gas pocket in the lumen is merely compressed. In time it may become dissolved, of course.

When pieces of fresh sapwood are vacuum-infiltrated and left submerged for a day or so, they take up a considerable amount of water, increasing in weight by 50% or so. This increase is essentially the storage capacity at atmospheric pressure, but of course we know that xylem pressures are rarely that high in a standing tree. However, we do not know how much of this storage is due to elastic expansion (imbibition and osmosis), how much is extracellular storage in cracks between cells, and how much is capillary storage inside vapor-filled cells. Wood contains too many irregularly shaped cells, and does not provide us with neat spherical surfaces for easy mathematical exercises.

Tyree and Yang (1990) investigated the water-storage capacity of 2-cm-diameter stem segments of *Tsuga canadensis, Thuja occidentalis*, and *Acer saccharum*. Stem segments, initially at full hydration, were dehydrated in air while measuring water loss by weight change; simultaneously, the stem water potential was measured with a psychrometer and cavitation events were detected acoustically using ultrasound detectors. In all three species, three phases of dehydration could be discriminated (Fig. 4.27). The first phase was apparently capillary dehydration because no cavitations were detected. The second transition phase may have been water coming mostly from the bark but there were also a few cavitations detected. Water from cavitated vessels becomes new capillary water and is lost rapidly by evaporation from the stem surface. The third phase was water released by extensive cavitation events.

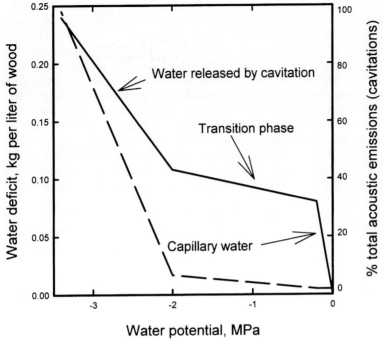

Fig. 4.27. Conceptualized data showing the three phases of dehydration of woody stems. *Solid line* is water deficit (weight loss per liter of wood) versus stem water potential. *Dashed line* is cumulative acoustic emissions (cavitations) observed during the dehydration of the same sample. (Extracted from Tyree and Yang 1990)

Capillary storage poses a serious problem with pressure measurements in a pressure bomb (Scholander et al. 1965). Pressure-bomb measurements assume only storage by elasticity (imbibition and osmosis). With respect to these parameters, negative pressure can be replaced by positive pressure. In the case of capillary water storage, this is not possible. Any embolized tracheid, for example, that is *emptied* by negative pressure, is filled by positive pressure, unless the applied positive pressure has access to the cell lumen (which it probably rarely has). However, if the emptied conduits are filled with air, the volume of water entering the embolized space will be limited to just a few percent of the air-bubble volume, because surface tension will raise the pressure of the bubble a few kPa above atmospheric pressure and then become quasi-stable in size for the short duration of the measurement of balance pressure. Such errors have been presumed to be the cause of disagreement between readings of water potential measured with a pressure bomb and a psychrometer (Boyer 1967, Kaufman 1968). These authors considered these discrepancies to be caused by the filling of non-xylem tissues and recommended that routine pressure-bomb readings be calibrated for each species by psychrometer measurements. The magnitude of the error will be most when balance pressures of leaves are measured while still attached to massive stems, but values in individual leaves are not likely to be affected much. A number of other sources of error can explain the discrepancy between the methods, so the exact cause is uncertain (Talbot et al. 1975).

West and Gaff (1976) ascribed excessive pressure-bomb readings to cavitation of xylem water. While leaves were permitted to transpire after excision and before insertion into the pressure bomb, cavitation clicks could be recorded (Chap. 3.2). Subsequent pressure-bomb readings increased. If the xylem contains large numbers of embolized tracheary elements, pressure-bomb readings can become dramatically increased. Julian Hadley (pers. comm.) found in March 1982 that in conifers of the timberline in Wyoming, needles of individual twigs gave pressure-bomb readings of –3 MPa on the sheltered side, and exceeded –7 MPa on the exposed side of the same twig, while the entire twig gave a reading of the order of –3 MPa. Here we are undoubtedly dealing with widespread embolization of tracheids on the exposed side.

The third mechanism, water storage in certain cells and retrieval by "air seeding", is probably a very special mechanism that does not play an important role in trees. *Sphagnum* spp. are supposed to store water in special cells which are air-seeded at very slight tensions (Huber 1956). Herbs whose xylem experiences positive pressures nightly are another possible case (see below).

Haberlandt (1914) described water-storing tracheids in leaves. They are round-ish in shape, and located either at the tip of the veins or even detached from the transporting xylem (Fig. 4.23C,D). In more recent publications, they have been called "tracheoid idioblasts" (Foster 1956; Pridgeon 1982). These can operate in various ways, depending on their submicroscopic structure, i.e., the dimensions of their wall pores (Sect. 4.2). If air-seeded like ordinary tracheary elements, they could refill completely during a rainfall event, and thus serve as water-storage compartments. In the absence of atmospheric pressure, and thus in the absence of complete refilling, the space between the spiral thickenings and the tips of the cells holds variable amounts of water, depending on the dimensions and the xylem pressures. The spacing in Pridgeon's illustrations is about 2 μm, a very suitable dimension. In leaf tissue one also encounters occasionally large, branched, almost star-shaped sclereids (e.g., Figs. 5 and 6 in Foster 1956). The shape and dimensions of these are ideal for capillary storage when embolized. In general, air-seeded fibers are ideal for water storage, because their slender tips can hold a considerable amount of water.

Could tracheids of coniferous leaves have "designed leaks" as the above-mentioned observation by Hadley seems to imply? Tensions exceeding some 3 MPa can cause dieback in some coniferous twigs. When stress becomes too great, they dry out and turn brown (e.g., in trees at the timber line). If "designed leaks" are located at needle tips, water could be "retrieved" into the stem. We might well have a clue here to the function of the bundle sheath in the coniferous leaf. If the bundle sheath in coniferous leaves is open at the distal end, it could favor distal air seeding of the tracheids, and water would be drawn toward the stem if tensions become excessive. This would be an extreme safety measure; the leaf would thus be sacrificed. However, we do know that coniferous needles turn brown and die under extreme stress conditions. Even in dicotyledonous leaves, it is sometimes only the margins that dry up. This is a fascinating area for future research.

When water columns break in regular tracheary elements, they give up water to other plant parts and thus constrict the flow path. Milburn (1973a,b) suggested that this might be an important mechanism to regulate water flow and distribution in herbaceous plants that produce positive root pressures and guttate nightly. This would refill the xylem again at night.

Let us briefly summarize the concept of capillary storage of water. The meniscus radius between the cell wall and the gas space is dictated by the pressure difference between the two phases. The difference in space occupied by capillary water at different xylem pressures appears to be quite large. A small part of this may be located in the intercellular space system of wood, mostly along rays, unless these surfaces are water repellent. However, most of the capillary storage space is probably located in the slender tips of embolized fibers and tracheids. It is important to realize that all this space is a single compartment for water, but not for gas. Gas can only circulate in the intercellular space system; in embolized fibers and tracheids, gas is isolated. These separate intracellular spaces (fiber and tracheid lumina) are probably mostly water vapor filled, because boiling a piece of wood, vacuum infiltrating, or merely soaking it, fills most of this space. Unless there are 3-μm-wide holes in the cell walls of embolized fibers, it is impossible to remove gas from them by vacuum infiltration (Eq. 4.3).

The fact that xylem water pressures are usually below atmospheric give the land plant two advantages. First, in the case of injury, the system is self-sealing (see Chap. 5). Second, negative pressures provide embolized xylary elements for capillary storage as we have seen. But what happens in plants where pressures are perennially positive as in aquatic angiosperms? Of course the plant needs a separate air duct system and has to seal off the apoplast! We shall come back to this in Chapter 7.4.

5 Hydraulic Architecture of Woody Shoots

5.1 The "Huber Value"

As a painter and naturalist, Leonardo da Vinci was a keen observer of nature. He made the following entry in his notebook about the construction of trees:

> "All the branches of a tree at every stage of its height when put together are equal in thickness to the trunk (below them). All the branches of a water [course] at every stage of its course, if they are of equal rapidity, are equal to the body of the main stream. Every year when the bows of a plant (or tree) have made an end of maturing their growth, they will have made, when put together, a thickness equal to that of the main stem; and at every stage of its ramification you will find the thickness of the said main stem; as i k, g h, e f, c d, a b, will always be equal to each other; unless the tree is pollard – if so the rule does not hold good." (Fig. 5.1; Leonardo's notes 394, 395, Richter 1970).

Although this is no evidence that Leonardo ever did any measurements to confirm this remarkable observation, botanists are well aware of the approximate correctness of this statement; measurements were made around 1900 to investigate the significance of stem dimensions in satisfying both mechanical and hydraulic demands (Metzger 1894, 1895; Jaccard 1913, 1919; Rübel 1919). This concept of even conductance throughout a tree has been called the "pipe model" by Japanese workers (Shinozaki et al. 1964), because a tree can be imagined as consisting of many thin, tall plants, bundled together. It must be emphasized, however, that these 'rules' provide only 'order of magnitude' approximations of the real situation.

It was a considerable conceptual step forward when Huber (1928) measured transverse-sectional xylem areas of stems and branches and expressed them per fresh weight of leaves that were supplied by that part of the axis. We refer to this as the "Huber value". The advantage of relating transverse-sectional area to supplied leaf mass is that measurements taken at different points in a plant, or in different plants, become directly comparable. Huber avoided the complication of heartwood formation by working primarily with young trees or tops of older ones. In some cases, he measured the most recently formed growth ring separately. He was also aware of the fact that transverse-sectional area alone was not sufficient to describe hydraulic properties; he therefore measured, in some cases, hydraulic conductivity separately. This automatically removed the disturbing effect of non-conducting xylem such as heartwood. Today, the Huber value is usually expressed as stem cross section or sapwood cross section per unit area of leaves, since leaf area is now measured much more easily than in Huber's day.

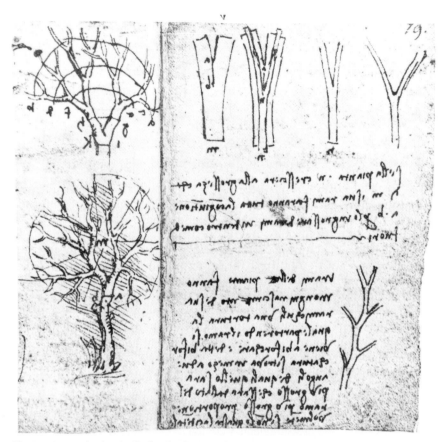

Fig. 5.1. Leonardo da Vinci's sketch of tree architecture showing that the transverse-sectional area of the trunk is equal to the sum of branch transverse-sectional areas. Note that Leonardo wrote in mirror image in Latin. (Richter 1970)

In addition to the innovation of relating the xylem to the supplied leaf mass, Huber (1928) made two major contributions. His measurements permitted a comparison of axes within an individual tree. An example of this is shown in Fig. 5.2, which illustrates the top of a young *Abies concolor*. The average transverse-sectional area in most parts of the tree was around 0.5 mm^2 g^{-1} fresh weight of supplied needles. However, this number increased sharply upward along the main stem to a maximum of 4.26 at the base of the leader. If only the most recent growth was measured, the increase was even more pronounced (numbers in parentheses in Fig. 5.2). The implication of this construction is obvious: the leader is better supplied by xylem transport than the laterals. This was a previously unsuspected expression of apical dominance. The second contribution was a comparison of species. The Huber value of stems and branches of dicotyledons and conifers of the north temperate region is about 0.5, while that of plants of more humid habi-

Fig. 5.2. Diagram of the top of a young *Abies concolor*. The age of each segment is shown by the *number of parallel lines*. *Printed numbers* indicate transverse-sectional areas of xylem in mm^2 g^{-1} fresh weight of supplied leaves. *Numbers in parentheses* concern the most recent growth ring alone. (Huber 1928)

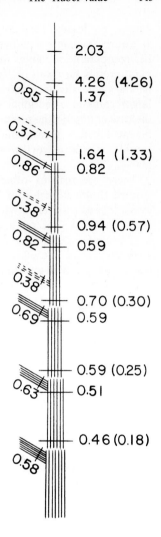

tats is considerably lower. Herbs of the forest floor show an average value of 0.2 (extremes 0.01 and 0.8), and aquatic angiosperms (Nymphaeaceae) 0.02 (Gessner 1951). Plants of dry habitats have much higher Huber values, the average for Egyptian desert plants was found to be 5.95 (extremes 1.4 and 17.5); Stocker 1928; Firbas 1931b). These latter values exclude succulents, which depend heavily on water storage, whose xylem is therefore not directly comparable with that of other plants. Interestingly, plants of raised bogs have Huber values rather comparable to desert plants (Firbas 1931a), probably because such plants experience periodic droughts (see also Chap. 4.1).

Filzner (1948) computed xylem transverse-sectional area per supplied surface area of *Rhynia*, and arrived at a figure of about 1.4 mm^2 dm^{-2}. *Rhynia* has, of course, no leaves. In dicotyledonous trees, 1-g fresh weight corresponds to 1–2 dm^2. The value obtained for *Rhynia* by Filzner is therefore somewhat high, i.e., comparable to values found in desert plants. Differences found in different plant groups may well be based primarily on tracheary diameters. Plants in dry habitats generally have narrow vessels; it takes many more narrow elements (i.e., a larger transverse-sectional area) to move a comparable amount of water. In the case *of Rhynia*, the pits connecting the tracheids may still have been very inefficient. Filzner's measurements were made on a single 2.5-mm-long piece; it would certainly be worthwhile to make more detailed investigations on early land plants.

There are reasons other than hydraulic considerations why investigators studied the relationship of leaf mass and wood production. Burger (1953), for example, was concerned with the amount of foliage that is required in different forest types to produce a certain volume of wood. More recently, ecologists have become interested in finding ways to estimate foliage quantity by non-destructive means. Thus, Grier and Waring (1974) established the ratio of sapwood transverse-sectional area to leaf dry weight of three coniferous species. This information enabled them to estimate the total weight of leaves of a tree from the dimensions of an increment core. Their results range from 1.3-2.1 mm^2 g^{-1} *dry* weight of supplied leaves, values quite comparable to those reported by Huber (1928).

Today, the Huber value, H_v, is reported in m^2 of wood per m^2 of leaf area. In some publications, wood area is reported as sapwood area and in others as total area. Most measurements have been done on branches of <50 mm diameters, so there is generally little pith or non-conductive wood. The range of observed values of H_v for 48 species is shown in Fig. 5.3. Species with large vessels, e.g., tropical lianas, tend to have the lowest Huber values. However, there are some remarkable exceptions. The largest Huber values observed are in rapidly growing, pioneer, tropical trees. For example, *Schefflera morototoni* grows to a height of 20 m in 5 years and has vessels >100 μm in diameter yet the H_v is 2×10^{-3}, perhaps because the tree has very few vessels per unit cross section.

Comparisons of H_v, or other hydraulic parameters below, between species is difficult since these parameters often change with stem diameter (D), hence differences at say D=6 mm may be reversed at say D=20 mm. Furthermore, the stem morphologies may be such that the smallest segments bearing leaves were 20 mm in diameter in some species (*Schefflera morototoni*) but just 3 mm in another (e.g., *Ficus*). It is not clear whether it is more sensible to compare hydraulic parameters between species at the same D or between stems at the same 'morphological diameter', i.e., stems that are, say, three times the diameter of leaf-bearing segments. The values plotted in Fig. 5.3 were computed from regression values of H_v versus diameter and read at D=15 mm except for three tropical, pioneer species with large stems (*Schefflera*, *Ochroma*, and *Pseudobombax*). The range observed at different diameter size classes for *Abies concolor* is shown as '• • • • •' in Fig. 5.3 (recomputed from Huber 1928).

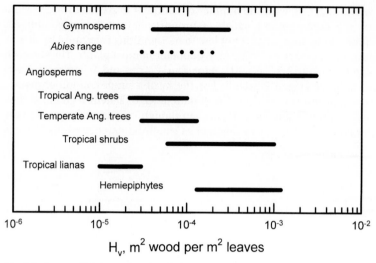

Fig. 5.3. Range of Huber values by phylogeny or growth form. *Horizontal bars* demark the ranges. The ranges are reported for stems 15 mm in diameter except as noted in the text. The original range observed by Huber in *Abies concolor* is shown by *dots* assuming a conversion factor of 0.015 m² of leaf area per g fresh weight

5.2 Measures of Conductivity in Stem Segments

Measurements on stem segments are performed with a conductivity apparatus. Excised stem segments are fitted into water-filled lengths of plastic tubing. One end of a segment of length L (m) is connected via tubing to an upper reservoir of water and the other end to a lower reservoir. The height difference between the reservoirs is usually set at 0.3–1 m to create a pressure drop, ΔP, of 3–10 kPa across the stem segment. The lower reservoir is usually placed on top of a digital balance to measure water flow rate, F, in kg s^{-1}. The fundamental parameter measured is the hydraulic conductivity (K_h) defined as:

$$K_h = F \frac{L}{\Delta P} \tag{5.1}$$

Values of K_h are usually measured for stems of different diameter, D, and regressions are used to obtain allometric relationships of K_h versus diameter of the form:

$$K_h = A\, D^B \tag{5.2}$$

where A and B are regression constants. Log-log plots of K_h versus D are best for evaluation of A and B in Eq. (5.2), as shown in Fig. 5.4.

Stem segments can be viewed as bundles of conduits (vessels or trachieds) with a certain diameter and number of conduits per unit cross section. If all the conduits were the same diameter and number per unit cross section, then $B=2$, because K_h would increase with the number of conduits in parallel which would increase with the cross section which is proportional to diameter squared. Usually, B is found to be more than 2 but less than 5, and this is because the diameter of conduits tends to increase with stem diameter in most plants. The slopes in Fig. 5.4 are approximately 2.5, 2.6, and 2.8 for *Thuja*, *Acer*, and *Schefflera*, respectively. According to the Hagen-Poiseuille law, the K_h of a single conduit of diameter, d, increases in proportion to d^4, so the conductance of a stem segment with N conduits would be proportional to Nd^4. So even though you cannot pack as many large-diameter conduits into a stem segment as small-diameter conduits, there is

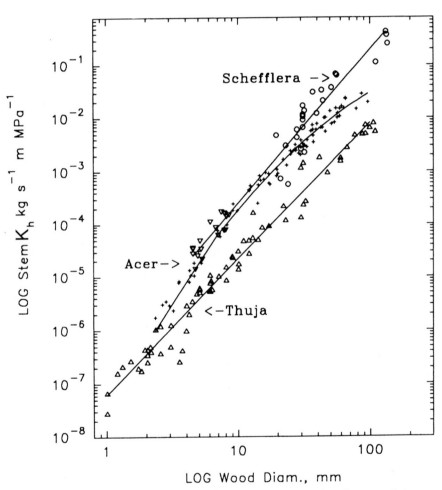

Fig. 5.4. Log hydraulic conductivity (K_h) of stem segments versus log diameter of the stem (excluding bark). *Circles* and *inverted triangles* are values for *Schefflera morototoni* (Aralaceae) stems and petioles, respectively. *Triangles* and *pluses* are for stems of *Thuja occidentalis* and *Acer saccharum*, respectively

a net gain in K_h for a stem segment to have bigger diameter conduits even though fewer would fit into that available space (Fig. 1.11).

Since K_h of stem segments depends on stem cross section, one useful way of scaling K_h is to divide it by stem cross section to yield specific conductivity, K_s. Specific conductivity is a measure of the 'efficiency' of stems to conduct water. The efficiency of stems increases with the number of conduits per unit cross section, and efficiency increases with their diameter to the fourth power (see also Chaps. 1.3 and 2.4). In large woody stems, the central core is often non-conductive heartwood. So it is often better to calculate K_s from K_h/A_{sw}, where A_{sw} is the cross-sectional area of conductive sapwood.

Leaf specific conductivity (K_L), also known as LSC (Tyree and Ewers 1991), is equal to K_h divided by the leaf area distal to the segment (A_L, m^2). This is a measure of the hydraulic 'sufficiency' of the segment to supply water to leaves distal to that segment. If we know the mean evaporative flux density (E, kg s^{-1} m^{-2}) from the leaves supplied by the stem segment and we ignore water storage capacitance, then the pressure gradient through the segment:

$$-\frac{dP}{dx} = \frac{E}{K_L} \qquad (5.3)$$

Therefore, the higher the value of K_L, the lower the $-dP/dx$ required to maintain a particular transpiration rate.

The Huber value, the specific conductivity, and leaf specific conductivity are related to each other. It follows from the aforementioned definitions that:

$$K_L = H_V K_S \qquad (5.4)$$

In making conductance measurements on pieces of stem, one often runs into the problem of more or less sharp conductance drops over time. This was first reported by Huber and Metz (1958) for coniferous wood, and ascribed to aspiration of bordered pits because they used rather large pressure gradients and found that the conductance drop could temporarily be reversed by reversing the flow direction (one of their figures is reprinted in Zimmermann and Brown 1971 as Fig. IV-20). When the conductance drop was also found in dicotyledonous wood whose bordered pits lack tori, it was ascribed to tiny air bubbles (and other suspended particles) that settle on the intervessel pit membranes and thereby obstruct the flow (Kelso et al. 1963). Working with dicotyledonous wood, we found that the conductance drop could be almost entirely avoided by using a dilute salt solution rather than distilled water (Zimmermann 1978). The reason for this is not yet known, but we suspect that intervessel pit membranes are swollen and therefore less permeable when saturated with distilled water (Zwieniecki et al. 2001). With most species (and specimens), 5 or 10 mM KCl gives maximum and stable flow rates. Larger stems often require higher concentrations (in some cases we used 100 mM KCl). The KCl effect is absent in coniferous wood (*Thuja occidentalis*, Chen et al. 1970; *Pinus radiata*, R.E. Booker, pers. comm.; *Tsuga canadensis*, tested in our own laboratory). This is not surprising because the margo of the coniferous bordered pit appears to contain little swellable matrix material and on electron micrographs looks more like an inert screen (Fig. 1.14). Short-term flow-rate measurements are reasonably stable, but long-range flow-rate measurements show the conductance drop one would expect from the accumulation of suspended particles on the pit membranes. The greatest difficulties so far have been encountered with the palm *Rhapis excelsa*, where the conductance drop is very sharp even if all precautionary measures are taken. Membranes of intervessel pits (Chap. 2.2) appear to be particularly easily plugged here. Some plugging has been attributed to growth of bacteria in the measuring apparatus and stems (Sperry et al. 1988a), but stable, long-term measurements can be obtained by using acidic media (pH 2–3) and conducting measurements at 2 °C (Tyree and Yang 1992; see also Fig. 4.20).

5.3 Hydraulic Architecture of Woody Stems

The hydraulic architecture of a tree can be described by the pattern (or distribution) of hydraulic conductances throughout the crown of a tree. Of particular

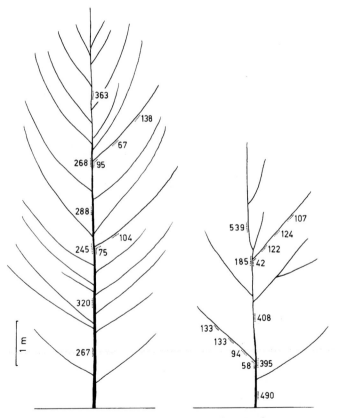

Fig. 5.5. Leaf-specific conductivities, K_L, along the axis of two open-grown large-toothed poplar trees (*Populus grandidentata*). Units are Zimmermann's original as described in the text. (Zimmermann 1978)

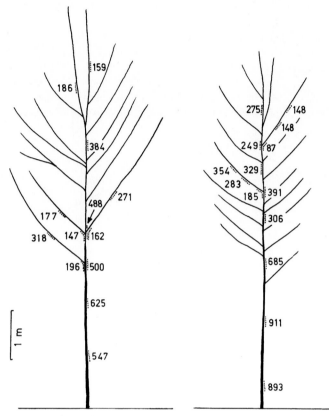

Fig. 5.6. Leaf-specific conductivities, K_L, along the axes of two open-grown birch trees (*Betula papyrifera*). Units are Zimmermann's original as described in the text. (Zimmermann 1978)

interest is the distribution of pressure gradients in a large tree. In this regard leaf-specific conductivity is a useful concept because of the direct relationship between K_L and dP/dx (Eq. 5.3). The concept of leaf-specific conductivity was first advanced by Zimmermann (1978a) and was based on fresh weight of leaves; flow rates were measured in µl h⁻¹ and pressure gradients were measured as multiples of 'gravity gradient' (=9.86 kPa m⁻¹). Zimmermann mapped out the 'hydraulic architecture' of a number of large trees, i.e., how K_L varied with position in the tree. The conversion factor between Zimmermann's original units and modern units is 1 µl h⁻¹ (gravity gradient)⁻¹ (gFW)⁻¹≈2×10⁻⁷ kg s⁻¹ m⁻² MPa⁻¹, depending on the conversion factor between g fresh weight of leaves and m² leaf surface area, which was not reported by Zimmermann. It is useful to look at some of the original patterns of hydraulic architecture reported by Zimmermann before looking at hydraulic architecture in more detail.

Figures 5.5–5.7 show some results. Let us first compare K_L of main stems. In *Populus* (Fig. 5.5) and one of the *Acer* specimens (Fig. 5.7, left) the values fluctuate

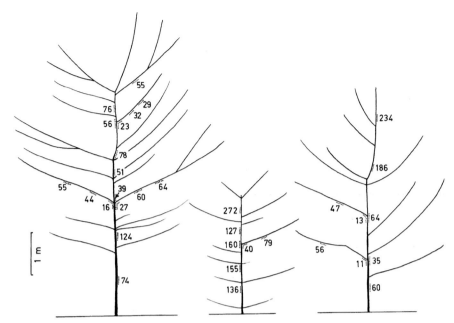

Fig. 5.7. Leaf-specific conductivities, K_L, along the axes of three sugar maples (*Acer saccharum*). Units are Zimmermann's original as described in the text. (Zimmermann 1978)

somewhat along the trunk, but do not seem to show any particular trend. In the two younger *Acer* specimens, however, there is a distinct increase in K_L from the stem base to the top (Fig. 5.7, center and right), and, in *Betula* (Fig. 5.6), there is a distinct decrease in distal direction. Leaf-specific conductivities of lateral branches are consistently smaller than those of the main stem, i.e., about half as large. Finally, junctions often show the lowest K_L values along any specific path of the water; however, this may be of little significance to the overall hydraulic architecture of large trees for reasons discussed in Section 5.6. Similar special patterns for K_L have been noted in conifers (Tyree et al. 1983, Ewers and Zimmermann 1984a,b) except that trees with strong apical dominance tend to have a hydraulic dominance, too, i.e., with large increases in K_L towards the dominant apex compared to primary and secondary branches (Fig. 5.8).

While Figs. 5.5–5.8 provide interesting 'pictographs' of the range and distribution of hydraulic parameters (specifically K_L), the number of measurements in the early studies were insufficient to provide insight into ranges of values versus diameter and statistical variability at any given diameter. Log-log plots of hydraulic parameter versus diameter give a better statistical 'feel' for ranges, as illustrated in Fig. 5.9 for *Ficus glabrata*. At any given diameter size class, the values of K_h, K_L, and K_s can vary by up to a factor of 10 and the statistical trends show values changing by one or two orders of magnitude as diameter changes from 3 to 30 mm. Very few measurements have been made on stem segments up to 250 mm in diameter but the range of parameters continues to follow similar allometric

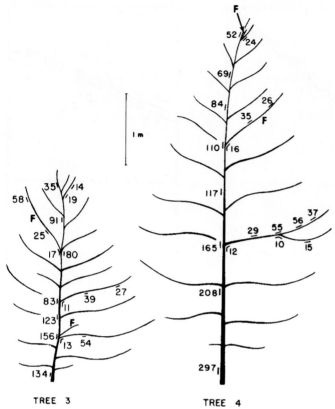

TREE 3   TREE 4

Fig. 5.8. Leaf-specific conductivities, K_L, along the axes of conifers. F is the location of former leaders with aborted apex. K_L values are higher in the trunk than in branches and are particularly low in second-order branches. Units are Zimmermann's original as described in the text. (Ewers and Zimmermann 1984a,b)

trends (Fig. 5.10). It is interesting to note that K_L values of the smallest leaf-bearing branches can differ by more than three orders of magnitude between species (compare smallest stems in *Thuja*, *Acer*, and *Schefflera* in Fig. 5.10), and this has profound consequences for predicted pressure gradients, as will be discussed in the next section.

Trends in hydraulic parameters are now known to vary with growth form between species in a genus and with growth conditions within a species (Ewers et al. 1991; Patiño et al. 1995; Cochard et al. 1997; Heath et al. 1997). The range of K_L and K_S values observed in 15-mm-diameter wood samples summarized over 46 species is given in Fig. 5.11.

Measurements of K_L were also taken in 1-year-old and current-year twigs and in petioles. Not as many measurements were made as with stems and branches, because, at very slow flow rates, conductance measurements require somewhat different techniques. On petioles measurements were only made with *Acer pensyl-*

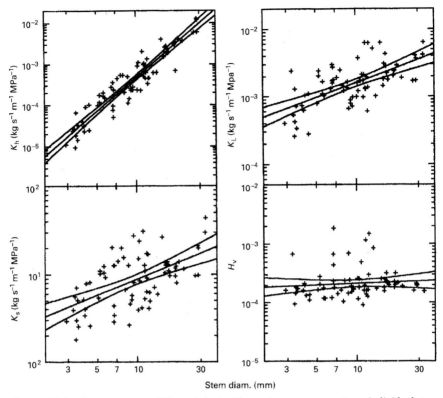

Fig. 5.9. Hydraulic parameters of *Ficus glabrata*. *Pluses* give measurements on individual stem segments. *Straight lines* show the linear regression and *curved lines* show the 95% confidence intervals. K_h Hydraulic conductivity; K_s specific conductivity (=K_h divided by sapwood area); K_L leaf specific conductivity (=K_h divided by leaf area supplied by the stem segment); H_v Huber value. (Patiño et al. 1995)

vanicum and *Populus grandidentata* because these are relatively large. Leaf-specific conductivities of last year's twigs (i.e., those with two growth rings) were between 10 and 30, and those of leaf-bearing shoots and of petioles were between 5 and 10. The lowest values were measured at the petiolar junction, namely 1 to 3. Petiolar junctions will be discussed in more detail in the next section.

It will be of considerable interest to link hydraulic architecture types to the morphological tree-model classification of Hallé and Oldeman (Hallé et al. 1978), although one must realize that the morphology is an expression of all functional adaptations, not only the hydraulic one. It will also be of considerable interest to study developmental events such as the changes in hydraulic properties of a lateral as it assumes the function of the main stem in sympodial branching, etc. To date, very few studies of this type have been attempted.

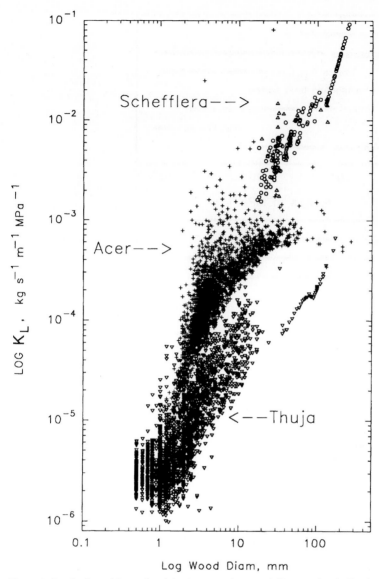

Fig. 5.10. Log leaf-specific conductivity K_L versus log wood diameter (excluding bark). Most values were calculated from hydraulic maps (see Sect. 5.5) and values of K_h versus diameter. *Inverted triangles* are for *Thuja*; *pluses* are for *Acer*; and *circles* are for *Schefflera*. Triangles are directly measured values for *Schefflera*. (Tyree et al. 1991)

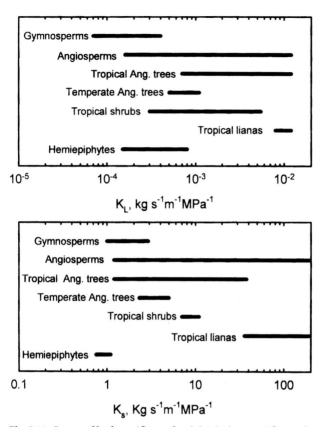

Fig. 5.11. Range of leaf-specific conductivity (K_L) or specific conductivity (K_S) by phylogeny or growth form. The ranges are reported for stems 15 mm in diameter except for *Schefflera*, *Ochroma*, and *Pseudobombax*, which have quite large leaf-bearing branches. So these values were read from regressions for 50-mm branches

5.4 Hydraulic Maps and Pressure Profiles

Let us now consider how leaf-specific conductivity is related to pressure gradients. According to Eq. (5.3), pressure gradient should be inversely proportional to K_L. Equation (5.3) strictly applies only to horizontal branches, i.e., it calculates the hydrodynamic pressure gradients. If branches are vertical or inclined at a gradient relative to the horizon, then a hydrostatic gradient must be added. The hydrostatic gradient is the gradient needed to maintain a static (non-flowing) column of water against the force of gravity. So the full expression should be:

$$-\frac{dP}{dx} = \frac{E}{K_L} + \rho g \frac{dh}{dx} \tag{5.5}$$

Readers should note the similarity between Eqs. (5.5) and (3.10). In Eq. (3.10) the hydrokinetic gradient is given by F/K_h versus E/K_L, in Eq. (5.5), where E is the average evaporative flux density from the leaves supplied by the stem segment in which K_L was measured. Since $E=F/A_L$ (=flow rate through the stem segment divided by leaf area) and since $K_L=K_h/A_L$, we can conclude that the equations are indeed identical. Looking at the range of K_L values in the crowns of trees (Figs. 5.9 and 5.10), we can conclude that the hydrokinetic gradients can vary by more than two orders of magnitude in a large tree! The total gradient (hydrostatic plus hydrokinetic) may not change quite as much because when K_L is very large the hydrokinetic gradients can be a small fraction of the hydrostatic gradient in vertical branches.

Pressure gradient profiles have been computed for quite large trees by combining data of K_h versus stem diameter (as in Figs. 5.4 and 5.9) with a 'hydraulic map' of a tree. A tree is divided up into numbered segments between 'nodes'. The data for a hydraulic map includes (1) the segment number of each segment and the number of the segment connected to its base; (2) segment diameter (without bark); (3) segment length; and (4) the area of leaves attached to the segment. A hydraulic map is very much like a road map, branches along which water flows are like roads along which vehicles move. The hydraulic maps just documents where the 'road junctions are' and the length and diameter of the stems (roads) between the junctions. The hydraulic map also tabulates the amount of leaf area attached to the minor branches (roads) near the distal extremes of the crown.

A hydraulic map can also be extended to include the hydraulic conductance of leaves. A simplified hydraulic map is visualized for a small plant in Fig. 5.12. Stem segment conductances are represented by dark-gray rectangles and leaf conductances by light-gray rectangles. If we assume an equal evaporative flux density for

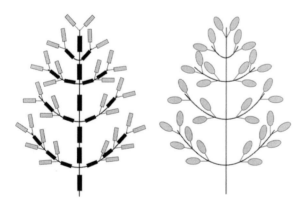

Resistance diagram   Tree

Fig. 5.12. Diagrammatic tree (*left*) and the equivalent 'conductance diagram' or 'hydraulic map'. Stem segments are represented by *black rectangles* with conductance values, K_h, and leaves with *gray rectangles* with leaf conductance values. This small tree is finely divided into short stem segments with two leaves per segment. In larger trees, the divisions would include more leaves per segment

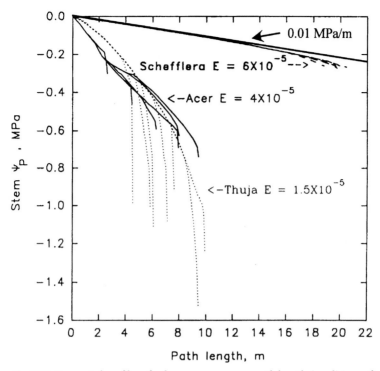

Fig. 5.13. Computed profiles of xylem pressure versus path length, i.e., distance from the base of a large tree to the tip of selected minor twigs. All values include the gravitational potential gradient required to lift water up the tree. These curves are calculated from hydraulic maps and representative evaporative flux densities indicated for each species in the graph. The hydrostatic slope is the *straight line at the top*. (adaPted from Tyree et al. 1991)

each leaf in the crown, we can calculate the flow rate ($F=E\,A_L$) for each leaf-bearing twig. Since the hydraulic map records the pattern of connection between the smaller branches to the larger branches, the flow rate in every stem segment can be calculated. Each stem segment is then assigned an average K_h based on the regression of K_h versus stem diameter from which the pressure gradient can be calculated for each segment using Eq. (3.10). So the whole pressure profile can be computed from the base of a tree to any selected twig in the crown. An example of computed pressure profiles in quite large trees is shown in Fig. 5.13 for *Thuja*, *Acer*, and *Schefflera*. Even though the average value of E from *Thuja* leaves is only one-quarter that of *Schefflera* at midday, the pressure gradients can exceed 5 MPa m^{-1} in the minor twigs of *Thuja*! Of course, these gradients exist for <0.1 m distance.

The computations in Fig. 5.13 show large differences in pressure gradient depending on K_L. *Schefflera* has such high K_L values that the total pressure gradient only slightly exceeds the hydrostatic pressure gradient (=−0.01 MPa m^{-1}). The high K_L values may be correlated with the fast growth rate of the species, which will be discussed later (Chap. 6.4). The figure is based on two assumptions. First, pressures at ground level are assumed to be zero. This does not have to be the case

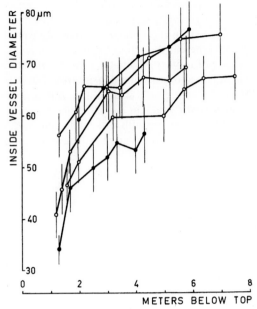

Fig. 5.14. Inside tangential diameters of the largest vessels in the most recent growth rings of the stems in three birches (*open circles*) and two poplar (*closed circles*). Each *point* is the average of 30–150 measurements. *Vertical lines* indicate standard deviations. (Zimmermann 1978)

in reality, but it does not matter; a different pressure at that level would merely shift the graph down with no change in slope. In the absence of transpiration, pressure gradients would be identical to the hydrostatic slope. At maximum transpiration, i.e., all leaves transpiring at the same rate (the second assumption), the gradients would follow the curve drawn below the hydrostatic line. It can be seen that pressure gradients along branches are very much steeper than those of the trunk. Note that some small branches near the base of the tree experience more negative pressure than branches a little further up. This is because of the lower K_L values of some of the lower minor branches. This effect explains in part why the giant redwoods failed to show higher pressure gradients based on pressure-bomb measurements on minor branches (Fig. 3.10).

Measurements were made to investigate the anatomical basis of the characteristic K_L distribution within tree stems. Inside diameters were measured in 30–150 of the largest (i.e., hydraulically most significant) vessels at different points within the tree axes. Figure 5.14 shows the diameter distribution along the trunks of three birches and two poplars. The steady basipetal diameter increase, well known to plant anatomy previously (e.g., Fegel 1941), was thus illustrated. Vessel diameters in branches are distributed as in the main stem: they increase with increasing distance from the leaves.

Let us now compare the actual pressure profiles in Fig. 5.13 with what might be predicted by the unit-pipe model of tree structure (Shinozaki et al. 1964). The unit-pipe model views a tree as a bundle of independent 'plants' or units. Each unit of leaf area is connected to a unit of root area by a unit pipe of uniform diameter, i.e., similar to Leonardo da Vinci's view (Fig. 5.1). In the unit-pipe model, K_L would have the same value for all stem diameters whereas the actual situation is quite different (Figs. 5.9 and 5.10). If the pipes are of uniform diameter then the pressure gradient should be the same everywhere if evaporative flux density from leaves is the same everywhere. In Fig. 5.13 this would appear as a straight line with a slope more negative than the gravitational hydrostatic gradient ($=-0.01$ MPa m^{-1}). The sharply downward-curved pressure profile (consistent with a distal decline in K_L) shows that we must view the unit-pipe model as a very crude first approximation of tree structure.

5.5 Impact of Light Interception on Pressure Profiles

Let us briefly reexamine one of the assumptions used in calculating pressure profiles in Fig. 5.13. We assumed that E (evaporative flux density from leaves) is the same everywhere in the crown of a tree. Is this assumption reasonable? Very different pressure profiles might result if we assumed otherwise.

Light interception will not be uniform within the crown of a tree; some leaves will be shaded by others and all leaves present a different angle of orientation of the leaf surface with respect to the sun's rays. If we measure the angle of the sun's rays with respect to a line perpendicular to the leaf plane (α), the intensity of incident light, I, would be $I=I_o \cos(\alpha)$ where I_o is the intensity of light measured perpendicular to the sun's rays. Light intensity has a dramatic effect on E. As light intensity (measured in PAR units) increases from, say, 0 to 200 μE s^{-1} m^{-2}, stomates go from a fully closed to approximately fully open state. The magnitude of the fully open state depends on the water status of the plant (which will be discussed in Chap. 6). As light intensity increases from 200 to 'full sunlight' (1500–1800 μE s^{-1} m^{-2}), the leaf temperature will be increased by the energy of the absorbed light. This will increase the vapor pressure of water at the evaporative surface of the leaf; hence, E will increase. The dramatic impact of light intensity on E can be seen at the branch level from potometer measurements of water uptake by excised branches during the course of a partly cloudy day (Fig. 5.15). About 80–90% of the variation in average E for the whole branch can be accounted for by variation in incident light measured parallel to the earth's surface without even accounting for variations in stomatal conductance due to changes in stomatal aperture.

While we have been aware of the problems caused by assuming a uniform value of E in shoots, there has been, until recently, no analytical method to estimate differences in E on a leaf-by-leaf basis in the complex geometrical structure of a plant. This analysis can now be performed with modern computer systems by making use of three-dimensional hydraulic maps of plants. With about 8 h of effort it is now possible to map out the three-dimensional position of every branch and leaf in a plant up to 2 m tall with a few hundred leaves. Pearcy and Yang (1996) have worked out the computer code needed to 'reconstruct' a plant on

Fig. 5.15. Branches of *Baccharis pendunculata* (Mill.) Carb. (Asteraceae) with about 2 m² of leaf area were attached to a potometer and water uptake was measured at 5-min intervals and expressed as equivalent evaporation rate per unit leaf area (E). A net radiometer was used to measure net radiation over a grassy surface, measurements were logged every 6 s and averaged every 5 min. *Upper two graphs* show a very close temporal correlation between net radiation and E. The rapid fluctuations in net radiation are due to clouds passing overhead. *Lower graph* shows net daily evaporation versus net daily radiation. Net daily variation varied with the amount of cloud cover on any given day from fully cloudy to nearly fully sunny

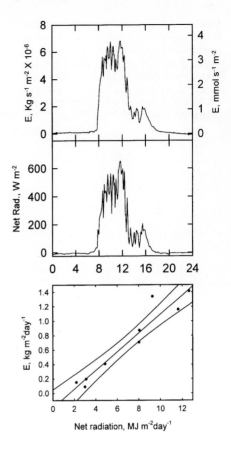

a computer and calculate the light interception of every leaf and how the shadows cast by higher leaves and stems influence light interception below. This model, called Y-Plant, was used to compute photosynthesis and carbon gain. Tyree and Pearcy collaborated (unpubl. results) to extend the model to include estimates of transpiration on a leaf-by-leaf level and to include the hydraulic architecture. This was done by writing an entirely new code (T-Plant), while simultaneously advancing Y-Plant. The reason for duplicating the code was to provide an independent check on Y-Plant (which had been written by several programmers over several years) and to try to improve the speed and accuracy of calculation. T-Plant, which I wrote, used a faster computational algorithm that speeded up computation by a factor of 2 to 3 times and improved the accuracy of the computed results, but both programs gave effectively the same answer.

T-Plant uses the Ball-Berry equation to estimate stomatal conductance, which can be written in the following format:

$$g_{sw} = g_1 \frac{AH_s}{C_s} + g_0 \tag{5.6}$$

Fig. 5.16. A *Psychotria chagrensis* plant has been mapped three-dimensionally and the coordinates have been input into T-Plant. Here, T-Plant displays a two-dimensional projection of the plant onto a plane perpendicular to the sun's rays at 7:15 A.M. on June 5th for a plant located at the latitude of central Panama

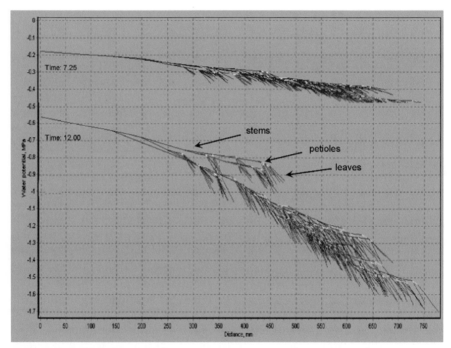

Fig. 5.17. Pressure profiles through roots, stems, petioles and leaves of the *Psychotria chagrensis* plant depicted in Fig. 5.16. See text for more details

where g_{sw} is stomatal conductance to water, A the assimilation rate (µmol s^{-1} m^{-2}), H_s the relative humidity at the surface of the leaf (scale 0 to 1), C_s the concentration of CO_2 at the surface of the leaf (ppm), and g_o and g_1 are constants. Estimates of A come from the original computations in Y-Plant (also incorporated in T-Plant); the assimilation rates are estimated from light-response curves for the species in question. It is beyond the scope of this book to go into the details of the computational methods used, but suffice to say that Eq. (5.6) and energy budget calculations are used to estimate leaf temperature departure from the air temperature and hence the fluxes of water vapor, heat and long-wave (black-body) radiation. T-Plant also incorporates the hydraulic calculations described in Section 5.5. The model also uses estimates of root, petiole and leaf blade hydraulic conductance, which will be discussed in Chapter 6.4.

Figure 5.16 shows a *Psychotria chagrensis* shrub viewed as the sun would see it at 7:15 A.M. on June 5th at the latitude of central Panama. Stems are shown in red, leaves in full sunlight are outlined in white, and shadows cast by leaves are color-coded. T-Plant can 'look' at this plant at different times of the day and compute evaporation from each leaf and each sunny and shaded portion of each leaf. In addition, for each time of the day, T-Plant can calculate the pressure profiles at each time. The pressure profiles at 7:15 and 12:00 are shown in Fig. 5.17. The red lines indicate pressure profiles in the stem, the yellow lines are the pressure profiles in petioles and the green line shows the pressure drop from the leaf base to the center of the leaf. The starting pressure at the base of the plant indicates the pressure drop across the root system, which is assumed to be in soil at zero water potential for the purpose of this calculation. Nevertheless, the graph gives the reader an idea of how T-Plant can be used in more realistic models of hydraulic architecture.

5.6 Branch Junctions and 'Hydraulic Bottlenecks'

The junction of branch and stem xylem is of particular interest. Junctions were perfused with dyes in the basipetal direction. Different dyes were used for the main and lateral in such a way that one could identify that part of the xylem of the main stem which leads into the lateral (Fig. 5.18, inset). This showed that, as one followed vessels from the branch down into the stem, vessel diameters suddenly widen, i.e., they become "stem vessels". This is developmentally rather interesting. Although a branch controls cambial activity in the stem immediately below it, it does not seem to control vessel diameter.

However, the most interesting aspect of a branch junction is its bottleneck. In the trees that were analyzed, the bottleneck was not due to a local drop in the Huber value; in fact, the opposite was often found: a slight bulge in the basal part of the branch just outside the attachment. This may have been due to cambial stimulation at the point of greatest mechanical stress (comparable to the thickening of the trunk base). The constriction can have one or more of several causes. First, we often found a sharp drop in vessel diameter just above the branch junction (Fig. 5.18). This coincided approximately with the slight bulge. A second observation, made often in birch, was that many branch vessels just proximal to the branch attachment were non-conducting, often occluded by gums, which could have been caused by branches swaying in the wind. Vessels may have been torn and become momentarily leaky at the point of branch attachment; the reader should remember that a very minute leak is sufficient to cause cavitation. Vapor blockage is very often followed by gum secretion from paratracheal parenchyma

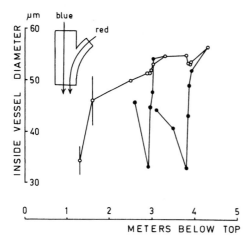

Fig. 5.18. Inside diameters of the largest vessels in the most recent growth rings of a poplar stem (*open circles*; same tree as in Fig. 5.14). In addition, two branches are shown (*closed circles*), attached at 3 and 4 m below the top. Vessels of the branches are distinctly narrower just a few centimeters above their attachment. Branch vessels inside the stem (identified by dye infusions) very quickly reached the diameter of the stem's own xylem. Standard deviations are comparable for all points. They are shown only at two points for the sake of clarity. *Inset drawing* indicates how dye perfusion with different colors identified stem and branch xylem. (Zimmermann 1978)

Fig. 5.19. Diagram showing a proximal basal segment (*A*), a branch junction with two pathways (*B* and *C*) and two proximal distal segments (*D* and *E*). Hydraulic conductance, K_h, in each segment path and all path lengths were equal

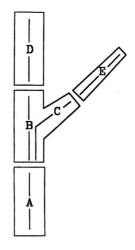

(Chap. 8.3). A third possible cause of the bottleneck was found a few years later when vessel-length distribution was studied. In many species there are more vessel endings at a branch junction than along clear lengths of the axes.

Although hydraulic constrictions or 'bottlenecks' have been observed in the stem insertions on large diameter stems (Figs. 5.5–5.8), it is not clear if they will

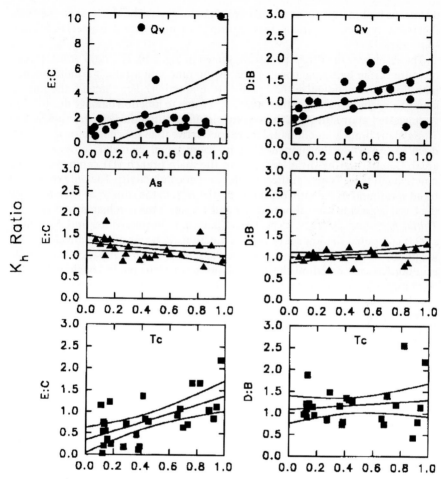

Fig. 5.20. Ratios of conductivity, K_h, observed in segment paths. The *vertical axis* gives conductivities as ratios of K_h measured on segments as labeled in Fig. 5.18. The *horizontal axis* is the ratio of cross-sectional areas of segments E to D. The *solid straight lines* are linear regressions of the data points. The *curved lines* above and below the *straight lines* are 95% confidence intervals of the regression

impact pressure profiles in large trees very much, because K_L values are quite large in large-diameter stems versus small-diameter stems. So Tyree and Alexander (1993) extended the study of branch junction conductivities in smaller diameter stems of *Quercus velutina, Acer saccharum,* and *Tsuga canadensis.* Branch junctions were cut into pieces as indicated in Fig. 5.19. The lengths of the piece were standardized so that the path lengths of A, B, C, D, and E in Fig. 5.19 were equal to 15 cm for *Quercus* and *Acer* and 6 cm for *Tsuga.* The conductances were measured along the five different pathways and compared to each other as ratios. Branch junctions were selected to have ratios of cross-sectional area of 0.1–1.0. A ratio of

0.1 means that the diameter of the stem segment E was about less than one-third the diameter of stem segment D, and a ratio of 1.0 means the diameters of E and D were equal.

The results for the three species are shown in Fig. 5.20. The ratio of E/C gives a measure of the hydraulic constriction of the junction; a ratio >1 indicates that path E is more conductive than path C (the junction). For *Quercus*, the ratio is about 1.5 excluding a few outliers that may be due to junction damage due to wind as suggested above. In *Acer*, there is no significant junction constriction and in *Thuja* many branch junctions are more conductive than the continuing distal segment except when the branches are of about equal diameter (cross-sectional area). Comparing the distal segment of the main branch (D) compared to the junction (B) there is no significant difference in conductivity. The only consistent trend was found in the ratio of conductivities of A to the sum of B+C. In this case the basal segments were about 1.1, 1.3 and 1.5 times the conductance of B+C for *Thuja*, *Acer*, and *Quercus*, respectively (data not shown). Tyree and Alexander (1993) computed the impact of branch junctions in terms of an equivalent extra length the tree would have to be to have the same hydraulic impact without junction bottlenecks. Based on these calculations, it is possible to conclude that the combined effect of all junction constrictions in a tree are likely to impact pressure profiles by less than 1% for trees >10 m in height. Hence, branch junctions are an interesting anatomical anomaly, but are unlikely to have much impact on the overall water relations of trees.

5.7 Leaf Insertions

The vascular connection between stem and leaf petiole is of particular interest, because the leaf is the most distal and most disposable part of the plant's segmented structure. The literature describing the anatomy of stem-leaf connections is very extensive. Howard (1974) gave a detailed survey of the variety of vascular connections between stem and leaves in dicotyledons. What interests us here is the question of how structural features affect the flow of ascending xylem sap. Meyer (1928), reviewing earlier literature and adding his own investigations, concluded that the xylem tracks of stem and petiole are rather separate. Stem vessels never continue into the petiole; leaf traces consist primarily of tracheids. He generalized this concept of "organ-restricted vascular bundles" so much that later investigators made some effort to prove him wrong. When he stated that no vessels are continuous from roots to stem, from stem to branch, and from shoot to petiole, he was probably correct as long as he dealt with the primary plant body, and possibly with very early secondary growth. We certainly know today that in secondary xylem of trees there are many vessels continuous from roots to stem, across branch junctions, etc. Rouschal (1940) studied water movement from stem to leaf with the aid of fluorescent dyes in relatively transparent herbaceous plants. His attention was concentrated on velocity and on transverse-sectional area. As we have seen in the previous section, these two features are not very informative by themselves. However, his method did yield interesting information. The dye moved up the main stem very fast and often entered top leaves before it entered bottom leaves. Rouschal also confirmed earlier reports of a xylem constriction at the petiole base (e.g., Salisbury 1913; Conway 1940).

Dimond (1966) conducted a very detailed study of pressure and flow relations in vascular bundles of the tomato plant. The tomato stem contains six vascular bundles along the stem, three large ones alternating with three small ones. One small bundle emerges from each node entering the petiole as a central bundle at the node above. Lateral petiolar bundles arise as branches from the two large bundles that lie on either side of the petiolar attachment. Flow rates were calculated from transpiration-rate measurements and from diameters of all individual vessels. These flow rates were then compared with experimental flow-rate measurements. What was not taken into account is the difficult question of flow resistance from vessel to vessel through intervessel pits. As we have seen in Chapter 1.3, this can be about as much as longitudinal resistance. Nevertheless, at the flow rates indicated by transpiration measurements, pressure gradients along the stem were of the order of 13 kPa m^{-1}, along petioles of the order of 72 kPa m^{-1}. Conductance of large (i.e., stem-) bundles decreases in the stem in the apical direction, but fewer leaves have to be supplied higher up. The very highest leaves require a lower pressure drop to obtain water than intermediate leaves.

One paper by Begg and Turner (1970) also contains information about the leaf insertion. By making pressure measurements along tobacco stems on bagged and unbagged leaves, they found a maximum difference of 0.55 MPa between stem and leaf. However, we must realize that this measure includes the pressure drop in the leaf as well as the petiole (more about this in Chap. 6.4). Other authors have reported similar findings. However, let us now return to the central problem of this section, namely, the constructional principle which "separates" the leaf from the stem, so that the leaf may become disposable if the plant cannot afford to maintain it.

Larson and coworkers studied the development of the vascular system in shoots of *Populus deltoides* in very considerable detail. The poplar leaf petiole is supplied by three vascular bundles. By measuring individual vessel diameters along the way, Larson and Isebrands (1978) found at the base of the petiole a constricted zone where vessel diameters are narrower. If the relative conductance (the sum of the fourth powers of the vessel diameters) is recorded along the xylem path, there is a sharp drop at the constricted zone to about one-quarter of the rest of the path. This constricted zone is located proximally to the abscission zone of the petiole (Isebrands and Larson 1977).

A constricted zone has also been detected in the form of an LSC drop across the petiole insertion of *Acer pensylvanicum* and *Populus grandidentata*, as mentioned in the previous section.

The significance of the hydraulic construction of the leaf insertion can be illustrated most dramatically with palms. Most palm species have a single perennial stem to which the disposable leaves are attached. This is hydraulically a very sharply defined case: the leaves are disposable because the stem cannot, under any circumstances, be lost. Furthermore, the stem xylem must remain functional for many years because the stem lacks a cambium. As one inspects the mature stem anatomically, one can see that leaf traces, containing narrow xylem elements, are

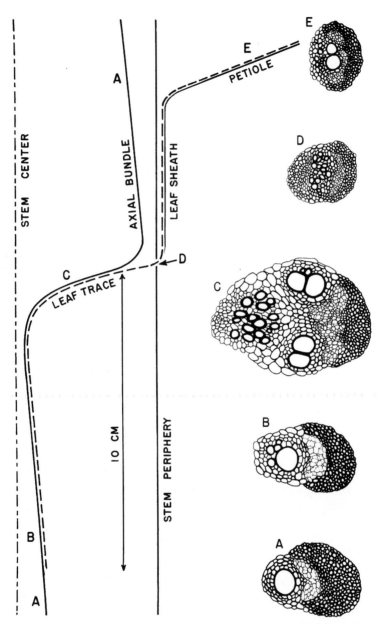

Fig. 5.21. Leaf-trace departure in the palm *Rhapis excelsa*. *Solid lines* indicate presence of metaxylem vessels. These are not necessarily continuous; a more precise plot of the vessel network is shown in Fig. 2.7. *Dashed lines* indicate narrow "protoxylem" tracheids. The letters *A* to *E* show traverse sections of vascular bundles and their location. An axial bundle (*A*) contains only metaxylem. At 10 cm below the leaf-trace departure, the first evidence of protoxylem can be seen. At *B* a few protoxylem elements are present. *C* is a leaf trace just below the point where it breaks up into its branches. *D* is the leaf trace proper, the branch entering the petiole; it contains narrow tracheary elements only. *E* shows a vascular bundle in the petiole, which contains again relatively wide metaxylem. (Zimmermann and Sperry 1983)

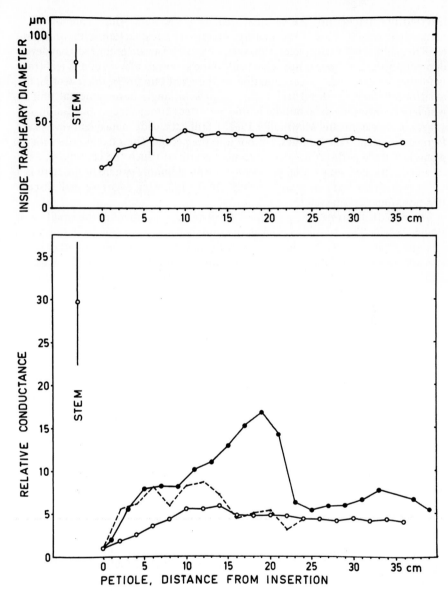

Fig. 5.22. Tracheary diameters and relative conductance along the stem (*point far left*) and the petiole from the insertion (including the leaf sheath) to the blade. *Top* Average diameter of the largest tracheary elements; diameters are most narrow in the region of the leaf attachment (distances zero on the horizontal scale). *Bottom* Relative conductance calculated as the sum of the fourth powers of all tracheary diameters, based on a leaf insertion of 1. Three different petioles are shown. (Zimmermann and Sperry 1983)

inserted in (attached to) the vascular system of the stem, which contains wide metaxylem vessels. This is the mature, functional construction with which we shall deal in the following pages. However, we must be aware of the fact that developmentally the leaf trace is not "attached" to the stem vascular system. A leaf trace is formed in direct distal continuation of the axial bundle. In other words, the axial bundle and its distal leaf trace comprise a single developmental unit to which the continuing axial bundle is attached later (Zimmermann and Tomlinson 1967; Zimmermann and Mattmuller 1982b). Still later, wide metaxylem vessels are formed along the axial bundles, but not into the petiole. The axial bundle with its distal leaf trace is therefore a developmental unit to which the continuing axial bundle is attached. Functionally, however, the axial bundle system of the stem is a unit to which the leaf traces are attached. On the following pages we shall discuss this functional xylem system.

Figure 5.21 illustrates the principle of leaf-trace attachment in the small palm *Rhapis excelsa*. This species is used as an example because we know it in most precise quantitative detail. It is, however, safe to state that the situation in other palm species is the same in principle (Zimmermann and Tomlinson 1974). The diagram on the left shows the course of a vascular bundle. The reader may want to refer back to Figs. 2.4–2.6 for general orientation. The solid line indicates the flow path provided by wide metaxylem vessels. This may not be a single, continuous vessel, but rather a succession of vessels with alternate tracks leading into neighboring axial bundles via bridges (see Fig. 2.7). The dashed line indicates the narrow-vessel path via leaf trace into the petiole. The drawings on the right illustrate transverse sections of a vascular bundle at different levels. It can be seen from this that the axial path along the stem is one of relatively large conductance through wide metaxylem vessels. The path into the leaf is via narrow (protoxylem-like) tracheids. In the petiole itself there are again relatively wide metaxylem vessels. Dimensional measurements for the *Rhapis* leaf insertion are given in Fig. 5.22. Diameters are recorded in the upper graph. Diameters of stem metaxylem vessels are of the order of 85 μm and in the petiole 45 μm. The narrowest diameters are found at the point of insertion itself. If we measure all inside diameters of the tracheary elements of the leaf trace leading into a petiole (about 100) and take the sum of the fourth powers, we obtain a measure of relative conductance. This is shown in Fig. 5.22 (bottom) and is based on the conductance being 1 at the insertion, which represents the bottleneck. The relative conductance of the stem, calculated as one metaxylem vessel for every leaf-trace bundle, is about 30 times greater than that of the insertion. We have done this calculation with 27 other palm species and found ratios in the range 1:5–1:267 (*Rhapis* 1:30).

The constriction indicated by summing the fourth powers of tracheary diameters does not indicate the full extent of the hydraulic constriction. First, the narrow tracheary elements of the insertion are mostly tracheids. The resistance to flow from one compartment into the next is probably greater in tracheids than in vessels. Second, and probably far more important, the flow resistance from the wide metaxylem vessels in the stem to the narrow leaf-trace tracheids is probably very considerable, although the total length of the contact area is ca. 10 cm in *Rhapis*. This is clearly illustrated in Fig. 2.6. The numbers 6, 4, and 3 correspond to the letters D, C, and B in Fig. 5.21. By inspecting these bundle transverse sections carefully, the reader will become aware of the fact that the contact between the wide metaxylem and the narrow leaf-trace tracheids is very poor indeed. There are very few vessel-to-tracheid pit areas; the two xylem strands are even separated by a layer of parenchyma in most places (e.g., in Fig. 2.6 at 4). It is extremely difficult to measure conductance from the stem metaxylem into the leaf-trace xylem (Zimmermann and Sperry 1983). Hydraulically then, the leaf is very sharply separated from the stem. The significance of this will be discussed in the next section.

5.8 The Significance of Plant Segmentation

We are only just beginning to appreciate the significance of the segmented structure of perennial higher plants. An arborescent monocotyledon, such as a palm, is perhaps the organism that demonstrates the situation most clearly. Let us consider a coconut palm that has a crown of approximately 24 leaves. This number is more or less constant, new leaves are added at the top, and old leaves are shed at the base of the crown at a rate of about one leaf per month. This means that the visible crown renews itself within a period of about 2 years. There are an additional dozen or two leaf primordia hidden within the bud, but these do not concern us here. While producing leaves the palm also grows in height, but unlike dicotyledonous and coniferous trees, the vascular system of the stem, once formed, remains the same. There is no cambium that adds new vascular tissue periodically. The primary vascular tissue of the stem is overefficient when the palm is short, and it may become limiting as the palm approaches considerable height. It may well be possible that long-distance transport limits the height palms can ultimately attain, because, as the palm gains height, the same transpiration rate requires an ever-increasing pressure drop to lift the water to the top.

The palm stem thus represents many years of "investment". Leaves, on the other hand, are virtually disposable parts. From roots, via stem, to the leaves, water must move along a pressure gradient. Leaves are the organs from which water is lost. The plant must therefore be designed in such a way that if water loss is excessive, transpiration is shut down. This is, of course, the function of the stomata. But there is always a residual cuticular transpiration. As the pressure drops in the xylem because of continued water loss, water columns must be induced to break before irreparable damage is done to living cells. The plant accomplishes this by providing xylem ducts with pores of such size as to admit an "air seed" at a given negative pressure, thus vapor blocking the compartment that has been "air-seeded" (Figs. 4.1 and 4.6). The problem with this mechanism is that it may be irreversible unless the plant experiences positive xylem pressure soon thereafter, e.g., in herbs that produce positive root pressure (and guttate) at night (Chap. 4.7). One of the most important design requirements is that such vapor blockage does *not* happen in the stem. Leaves are renewable and can be dropped prematurely. The sharp conductance drop at the petiole insertion provides the structural basis for this: it lowers the pressure in the leaf xylem quite drastically as soon as water moves, to a value distinctly lower than that in the stem xylem. Some of the xylem ducts may thus be vapor-blocked, decreasing conductance further. The sharp conductance drop at the petiole insertion therefore provides the mechanism to sacrifice the leaf in order to save the stem, if conditions require such a drastic measure. The mechanism is perhaps comparable to that of a lizard who loses its tail if caught by it: the tail is sacrificed, thereby saving the lizard's body that can, in time, grow another tail.

Sperry (1985, 1986) studied the water relations and embolism in petioles of *Rhapis excelsa*. He found a large P_x drop from stems to leaves of 0.6 MPa during midday transpiration. Embolism was fairly extensive in petioles and this embolism was reversed after rain events in the absence of any root pressure.

Some palms drop their leaves by abscission while they are still green and healthy-looking, e.g., arecoid palms with a crown shaft like *Roystonea* when grown under ideal (humid) conditions. In other palms, the oldest leaves merely dry out and remain attached to the trunk. If we take the conductance into a healthy leaf to be 100%, it may well deteriorate during the course of its life span until it is finally zero when it is completely dry. This gradual loss of conductance must follow a positive feedback mechanism (a vicious circle of runaway embolism which will be discussed in more detail in Chap. 6.2); the more vessels are embolized, the greater the pressure drop, hence still more vessels will fail. If this is true, then we should see very few leaves in the crown of a plant that grows under dry conditions or, for some reason, has lost many of its roots. At the other extreme are palms grown in a greenhouse under extremely favorable conditions of water supply. Old leaves remain green and functional for a long time.

Another feature of monocotyledonous leaf attachment is its meristematic leaf base. When a palm leaf unfolds and becomes functional, its base may still be meristematic and the xylem in the basal part may not yet have reached its full water-conducting capacity. Rouschal (1941) reported this for grasses. Thus, the conductance of the attachment may, during the life span of the leaf, initially increase by maturation of the basal xylem then decrease again by irreversible xylem failure. Within grass leaves, Rouschal (1941) found very rapid long-distance water transport in approximately every other one of the parallel veins, and slow cross transfer of water into the veins located between the fast conductors. The difference between the fast and the slow conductors could not be seen by superficial observation; in transverse section, however, the greater vessel diameter of the fast conductors could be seen and measured. Rouschal also found that the xylem tracks of the leaf blade became fully connected to the stem only after leaf-base maturation. We do not yet understand all aspects of xylem-water supply to monocotyledonous leaves and have here a wide-open field of potentially very interesting future research. Some recent advances are reported in Martre et al. (2000).

Let us now look again at the hydraulic construction of dicotyledonous trees. As far as the leaf insertion is concerned, the situation is similar to that of the monocotyledons. It is perhaps not as extreme, but has the same effect. Excessively low pressures due to drought or loss of part of the root system will wilt leaves before vessels in the stem are lost by cavitation. However, the segmentation of dicotyledonous trees is more extensive than that of palms. The main stem represents the greatest investment of the tree and should not, under any circumstances, be lost. Indeed, one can often find that in trees whose root system has suffered excessive damage, individual lateral branches die. It is possible that this loss is due, in part, to vapor blockage of the branch xylem. Again we are dealing here with a feedback mechanism. If the water supply to a lateral branch becomes poorer, it cannot sup-

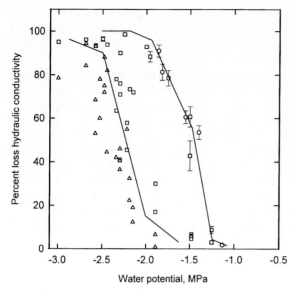

Fig. 5.23. Vulnerability curves of walnut petioles and stems. *Open and closed circles* represent petioles measured by bench-top and pressure-chamber dehydration, respectively. *Error bars* are standard errors of the mean (*N*=20, error bars smaller than symbol size are not shown). *Open and closed squares* represent stem segments of current-year and one-year-old shoots, respectively. Each *point* represents an individual shoot. The *x*-axis gives the minimum value of Ψ reached based on pressure-bomb determination of leaves for petioles and of stem hygrometer readings for stem segments

port as many leaves; as the carbohydrate supply from leaves declines, cambial activity diminishes and decreases xylem development.

We must recognize two possible mechanisms of plant segmentation that could lead to leaf or small stem shedding in response to embolisms. One mechanism is hydraulic segmentation and the other is vulnerability segmentation. The dramatic decline in K_L observed in the distal direction of some trees (Fig. 5.10) will lead to steep declines in pressure profiles distally (Fig. 5.13). However, the efficacy of hydraulic segmentation is progressively lost as water stress progresses and hence the evaporative flux density from leaves decreases. A much more effective mechanism of plant segmentation would be vulnerability segmentation, i.e., if it could be shown that distal portions of a tree are more vulnerable to cavitation than basal portions. Tyree et al. (1993) have found an excellent example of vulnerability segmentation in walnut trees. Walnut petioles are much more vulnerable to cavitation than the stems to which they are attached. The petioles are >95% embolized before stem embolism exceeds 10% (Fig. 5.23). Hence, even without evaporation, petioles will cavitate before stems. Walnut leaves are easily shed during a drought.

It would be interesting to study the hydraulic construction of fossil trees. I assume that a dichotomously branching tree was constructed according to the pipe model. At least successive daughter axes are equivalent. Adverse conditions thus could damage both stems equally. It has been suggested by P.B. Tomlinson (pers. comm.) that the hydraulic construction of the main stem with lateral branches was such a distinct adaptive advantage that it caused dichotomously branching trees to become extinct.

Tomlinson (1978) described the morphology of divaricating shrubs from New Zealand. These are characterized by microphylly, absence of terminal flowers, and interlacing branches of different orders, often so intricately that it is impossible to extricate a severed branch from a shrub. Interlacing can be caused by wide-angle branching or by zigzag growth. In other words, we are dealing here with a very highly segmented construction. The interesting thing is that some divaricates are juvenile forms of species in which the adult form is an upright tree. The description suggests a construction with very low LSC values in the juvenile condition that is maintained until an extensive root system has been established. Once this has been accomplished, a highly conductive main stem is formed. This is, of course, wild speculation, and it will certainly be worthwhile to study the hydraulic construction of these remarkable plants.

A final example of plant segmentation is the peculiar structure of certain desert shrubs whose stems become lobed and finally split longitudinally as a result of the death of strips of the cambium (Fahn 1974; Jones and Lord 1982). If there is not sufficient water available, some of the aerial shoots may die and thereby leave enough water so the remaining ones can survive.

We close this chapter by pointing out another area about which we know little, namely the question as to whether small plants and colonies of lower plants (e.g., mosses) are constructed according to the segmentation principle which is able to "save" certain (perhaps younger) parts at the expense of other (perhaps older) parts. Little is known about this. Furthermore, we do know that there are plants with nodal barriers (e.g., Meyer 1928). It would be very interesting to explore these from the point of view of hydraulic segmentation.

6 Hydraulic Architecture of Whole Plants and Plant Performance

6.1 Where Are the Main Resistances?

Most of the early work on the hydraulic architecture has focused on the stems of the shoots. The tacit assumption has been that because trees are tall, most of the resistance ought to be in the stems. Based on an incomplete data set, I once even suggested that the shoots of large trees might have more hydraulic resistance than the roots in the ratio of 3:1 (Tyree 1988) but this is unlikely to be true generally and may have been wrong in the specific case cited. Roberts (1977) measured the midday xylem pressure, P_x, of large trees and then cut off the trees near ground level and placed them in a bucket of water. Removing the resistance of the roots raised P_x to a less negative value, which we will designate P'_x. If we approximate the root and shoot as simple resistances in series with the same flow rate, F, across them, then we can write for the whole tree:

$$P_x - \Psi_{soil} = F(R_{root} + R_{shoot}) \cong P_x, \qquad (6.1)$$

where the approximation on the right is true in wet soils because $\Psi_{soil} \cong 0$. And for the excised shoot in water we have:

$$P'_x = F R_{shoot} \qquad (6.2)$$

Dividing Eq. (6.2) by (6.1) yields:

$$\frac{P'_x}{P_x} = \frac{R_{shoot}}{R_{shoot} + R_{root}} \qquad (6.3)$$

Experiments on large *Pinus sylvestris* trees revealed that shoot and root resistances were approximately equal. To be precise, gravitational effects (10 kPa m^{-1}) must be subtracted from P_x and P'_x, but this appears not to have been done in the original paper (Roberts 1977). Other examples for smaller plants will be quoted later. However, this observation still ignores an important question. Where in the roots and shoots are the main resistances located? In the woody parts of most shoots, the leaf specific conductivity, K_L, declines rapidly with diameter (Chap. 5.3) so a disproportionate part of the pressure drop and resistance might reside in the small branches. We shall soon see that the same is likely to be true for the vascular part of the roots, i.e., the vascular resistance of small roots may dominate.

There are, however, also non-vascular resistances at the distal and basal extremes of plants. The non-vascular resistances of roots comprise the radial water pathway from the surface of the fine roots (<2 mm diameter) to the vessels.

In shoots the non-vascular resistance comprises the water pathway from minor veins to the evaporative surface in the mesophyll air spaces and in the peristomatal cavities, where most of the water may evaporate (Tyree and Yianoulis 1980, 1984). This distribution of non-vascular and vascular resistance has never been worked out on any one plant let alone on a large tree, although we can piece together some very suggestive facts.

Let us start out by examining the situation of water flow in roots. Axial hydraulic conductivity in roots tends to increase from apex to base, i.e., with root age, because as roots get older the number and/or diameter of vascular conduits (vessels or tracheids) tend to increase. The Hagen-Poiseuille law states that the hydraulic conductance of a cylindrical pipe of uniform diameter increases with the fourth power of the diameter. The Hagen-Poiseuille law does not strictly apply to vascular conduits in roots, because they are neither cylindrical nor of uniform diameter along their lengths, but it does provide a useful first approximation so, for a root segment with n vessels of diameter d:

$$K_{axial} = \sum_{i=1}^{n} \frac{\pi d_i^4}{128 \eta L} \tag{6.4}$$

where η is the viscosity of the solution in the vessels and L is the length. Steudle and Peterson (1998) report that young maize roots have protoxylem vessels near the tip that are 5–10 μm in diameter, then 25 mm from the tip, early metaxylem vessels are 23 μm and finally at distances of 250 mm from the tip the late metaxylem elements are about 100 μm. So Eq. (6.4) would predict that one old metaxylem vessel would be as conductive as 357 early metaxylem vessels. There are typically 14 early metaxylem vessels versus 7 late metaxylem vessels. If we compare the hydraulic conductance of 100 mm of root with early metaxylem versus 100 mm of root with late metaxylem, the old metaxylem part of the root would have an axial conductance that is 179 times that of the early metaxylem part. The theoretical axial conductance of maize roots is much more than the total root conductance and this has been confirmed experimentally as indicated below. Hence the radial (non-vascular) path limits the rate of water flow through roots.

Radial water flow is probably limited by the endodermis and exodermis (Fig. 6.1). Root conductances per unit surface area, $k_{r/a}$, have been measured on maize roots about 1 mm diameter and 200–500 mm long by the root pressure probe method and values tend to be around 2.3×10^{-5} kg s^{-1} m^{-2} MPa^{-1} (Steudle and Peterson 1998) which agree with those measured by more traditional methods (Newman 1973; Miller 1985). A 225-mm length of maize root containing early metaxylem has a surface area of $A = 7 \times 10^{-4}$ m^2 hence k_r for that region is A $k_{r/a} = 1.6 \times 10^{-7}$ kg s^{-1} MPa^{-1} or a resistance of 6.2×10^6 MPa s kg^{-1} ($= 1/1.6 \times 10^{-7}$). The 14 early metaxylem vessels in this same length of root would have an axial conductance of 4.4×10^{-7} (from Eq. 6.4) or a resistance of 2.25×10^6. Hence in young roots the radial resistance is 2 to 3 times the axial. In slightly older maize roots with old metaxylem we have already shown that the axial resistance of the same 225 mm would be 179 times less, hence the radial resistance would be 300 to 500 times more than the axial resistance. Clearly, the ratio of resistance depends rather strongly on the diameter and number of xylem conduits in the roots so the theoretical calculations are rather approximate. These theoretical calculations are,

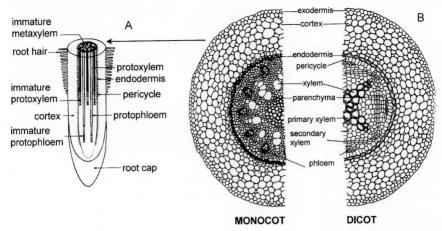

Fig. 6.1. A Enlargement of a dicot root tip with about 0.6 mm basal diameter. **B** Cross section of monocot and dicot roots. The Casparian band is a layer of denser cell wall (not shown) in the endodermis. (Adapted from Tyree 1999b)

however, consistent with experiments on young roots in which hydraulic resistance is measured before and after root tips are excised, which decreases the resistance by a factor of 3–20 (Nobel and Sanderson 1984).

From the above considerations we can conclude that the main resistance to water flow may be non-vascular, i.e., somewhere in the radial pathway. But where in the radial pathway is the barrier located and is it at the same location for water and solutes? Some useful insights result from wounding experiments in which the effect of mechanical damage to the cortex and/or endodermis is measured in terms of the effect of such damage of root pressure, reflection coefficient, hydraulic conductivity and solute permeability.

The Casparian band is thought to reduce the water permeability of the cell walls in the endodermis forcing the pathway of water movement to be transcellular in the vicinity of the endodermis. Hence water would have to pass through at least two plasmalemma membranes. The surface area of this membrane pathway would be at least as great as the surface area of the endodermis. The area could be more if a significant fraction of the water transport is symplastic via plasmodesmata; hence, water could enter through several layers of cortical cells, pass through the symplast, then exit though several layers of cells in the stele before reaching the xylem conduits. This notion is approximately consistent with observed values of root and membrane conductance. Maize root conductance (Steudle 1993) measured on roots 90–180 mm long and 1 mm in diameter is about 2.7×10^{-8} m s^{-1} MPa^{-1} (normalized to the epidermal surface area) and, since the endodermis is about half the diameter of the epidermis, the conductance normalized to the endodermis surface would be about double (5.4×10^{-8}). In contrast, maize-root, cortical-cell membrane conductance is 2.4×10^{-8} m s^{-1} MPa^{-1}, so the conductance of two membranes in series would be 1.2×10^{-8} or about one-quarter that of the root conductance normalized to the endodermal area. From this we

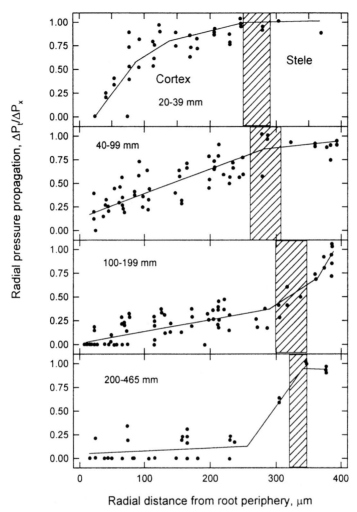

Fig. 6.2. Radial propagation of pressure across the cortex in four different zones of a maize root. *Data points* represent turgor responses (ΔP_t) following either an increase or a decrease in xylem pressure (ΔP_x). Locations of the endodermis within each zone are marked by *cross-hatched areas*, as determined by microscope observations. Pressure gradient trend lines are drawn in by hand; steeper slopes indicate where the hydraulic resistance is most. Each graph is for a different distance back from the root tip as indicated in mm. (Adapted from Frensch et al. 1996)

can conclude that the membrane (transcellular) pathway accounts for at least one-quarter of the root conductance and the rest is via the apoplast. Alternatively, a larger fraction of the root conductance could be transcellular if water enters several cells and travels through the symplast. In an earlier study, where 90 measurements were carried out on shorter maize roots (70–110 mm), $L_r = 1.1 \times 10^{-8} \pm 0.1 \times 10^{-8}$ (mean and SEM, Steudle et al. 1987) and cortical-cell was again $2.4\times10^{-8}\pm0.3\times10^{-8}$ so from these data we might conclude that about half the water transport was transcellular across two membranes in the endodermis.

Wounding experiments provide more insights. Peterson et al. (1993) removed fractions of the cortical tissue of maize roots (15–38%) and this caused an approximately proportional increase in L_p, from which they conclude that the barrier to water flow is diffusely distributed over the entire cortex.

Another approach to the location of the water barrier is to measure the pressure drop across the radius of a root while water is flowing; the principle hydraulic resistance would correspond to the region with the biggest drop in pressure. This has been accomplished fairly directly in experiments done on 500-mm-long maize roots by Frensch et al. (1996). They used a cell pressure probe to measure changes in cortical and endodermal cell turgor pressure resulting from changes in xylem pressure. Although we do not know if the main pathway is transcellular or apoplastic, the changes in cell turgor should reflect changes in apoplastic pressure in adjacent cell walls. Results were obtained for root regions of different ages, i.e., different distances from the root tip (Fig. 6.2). Over the first 100 mm of the root (measured from the tip) there is a more-or-less uniform pressure gradient from the stele to the epidermis, which is consistent with the earlier finding. In the next section (100–200 mm from the tip) about two-thirds of the pressure drop is across the stele and endodermis and the rest across the cortex. For the oldest portions of the root (>300 mm from the tip) almost all the pressure drop is across the endodermis and adjacent cortical cells. Hence the situation is more complex than one might conclude from experiments on young roots 90–180 mm long.

The axial conductivities, K_h, of older woody-root segments have not been measured as often as woody stems of similar diameter but they are generally much more conductive. This is because vessel diameter and vessel length in woody roots exceed that in stems of comparable wood diameter (Fig. 6.3; Kolb and Sperry 1999). From all of the above, it seems probable that well over half of the total hydraulic resistance of large woody-root systems is confined to small roots <2 mm in diameter and that approximately three-quarters of the hydraulic resistance of fine roots is in the non-vascular radial pathway.

We will now turn to the issue of leaf resistance to water flow. The high-pressure flowmeter (HPFM) provides an excellent way of determining the relative magnitude of resistances of leaf blade, petiole, and stem segments of different ages on large branches. The HPFM is an instrument that can be used to perfuse water into the base (stump) of a root or shoot system at a known pressure, P_2, while measuring the flow rate of perfusion. The objective is to perfuse water into a shoot faster than water evaporates from the leaves. This causes infiltration of leaf air spaces and insures that the leaf water potential is zero during the measurement. Readers interested in how the HPFM is designed and its application to the measurement of both shoot and root hydraulic conductance should refer to Tyree et al. (1993, 1994, 1995).

The hydraulic architecture of whole shoots is a complex mixture of parallel and series resistance components. However, for the sake of analysis, it is often easiest to assume that 'components' of shoots can be treated as components in series. Examples of components are leaf blades, petioles, current-year stems, 1-year-old stems, 2-year-old stems, etc. Tyree et al. (1993) harvested 1.5-m-long shoots of oak with 1.1–2.1 m^2 of leaf area and basal diameters of 16–19 mm. A whole shoot

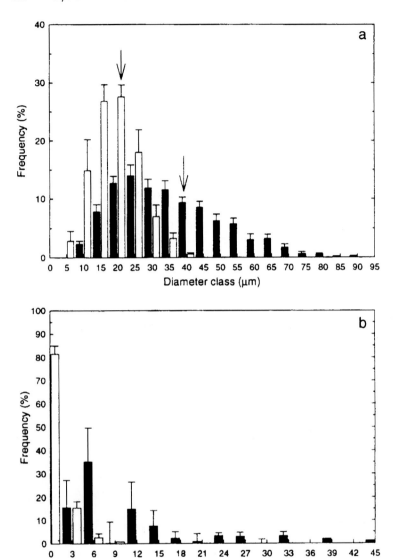

Fig. 6.3. *Top* Percentage of vessels versus diameter class for stems (*open bars*) and roots (*filled bars*). *Arrows* indicate mean diameters, which differed between roots and stems (*P*<0.0001). *Error bars* are standard errors for *n*=6 segments. *Bottom* Percentage total vessel versus length class for stems (*open bars*) and roots (*solid bars*). *Error bars* are standard errors for *n*=4 segments. (Kolb and Sperry 1999)

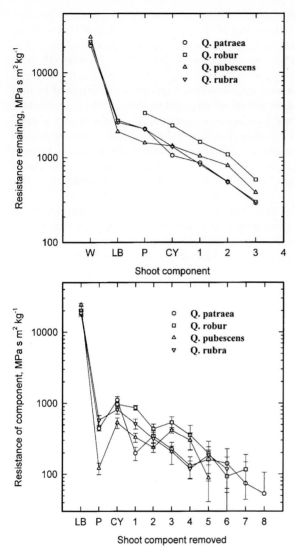

Fig. 6.4. Shoot resistances of *Quercus* species. All points are means of five branches per species. *Top* Shoot resistance values plotted as resistance remaining after removal of components. On the x-axis: *W* whole shoot resistance with nothing removed; *LB* resistance remaining after removal of leaf blades; *P* resistance remaining after removal of leaf blades and petioles; *CY* resistance remaining after removal of leaf blades, petioles and current-year shoots, *1, 2* and *3* refer to resistance remaining after removal of everything including 1-, 2o-, and 3-year-old shoots, respectively. *Bottom* Resistance of each component of the shoot. On the x-axis: *LB* leaf blade; *P* petiole; *CY* current-year shoots, *1, 2, 3*, etc., refer to 1-, 2-, and 3-year-old shoots, respectively. *Error bars* are SEMs and are shown only when errors are larger than the symbol size. In the *upper graph* representative error bars are shown only for *Q. petraea* and *Q. robur* (*Q. robur* had petioles too short to measure). All resistance values have been normalized to leaf area (resistance×leaf area)

resistance, R_w, was measured with an HPFM. Then the leaf blades were removed. The resistance of the shoot without leaf blades, R_{-LB}, was less (Fig. 6.4). Assuming the resistance of the leaf blade, R_{LB}, is in series with the remainder of the shoot, the leaf blade resistance can be computed from $R_{LB}=R_W-R_{-LB}$. Other 'serial' components removed were petioles, current-year shoots (the shoots bearing leaves), and then all the other shoots were removed based on age (as determined by bud scars). The age of the stem segments removed could be determined from the bud scar count assuming only one leaf flush per year, which is not always a valid assumption since some trees can have multiple flushes per year. The resistance of the shoot remaining after removal of each component is plotted on a log scale in Fig. 6.4. The leaf blade resistance accounts for more than 90% of the resistance of the whole shoot! The resistance of all the stem components decreases approximately log-linearly with age.

In Fig. 6.4 (bottom) the resistance of each component is calculated from the difference in resistance of the shoot measured with the component present minus the resistance with the component removed. This computational approach should not in theory be generally valid for a branched catena network of hydraulic resistances. However, the approximation works surprisingly well. Zotz et al. (1998) measured the hydraulic conductance, K_h, of stem segments of several tropical trees by two methods. A HPFM was used to measure the resistance of stem segments, R_{seg}, of different diameter size classes, which was equated to the resistance change of the whole shoot before and after removal of the stems of the particular size class. This measurement was repeated for increasing diameter size classes until only the base of the shoot remained. K_h was then computed from:

$$K_h = \frac{L_R}{nR_{seg}}$$
(6.5)

where L_R is a 'resistance-averaged length' of the stem segments removed., i.e.,

$$L_R = \frac{n}{\dfrac{1}{L_1}+\dfrac{1}{L_2}+..+\dfrac{1}{L_n}}$$
(6.6)

where L_1 to L_n are the lengths of the n stem segments removed. Equation (6.6) is identical to the equation that would describe the equivalent resistance of n resistors in parallel and we have assumed that, in a diameter size class, the resistance of a stem segment would be proportional to the length of the segment. Once K_h has been calculated then the leaf-specific conductance can be estimated from:

$$K_L = \frac{K_h}{A_T/n}$$
(6.7)

where A_T is the total leaf area attached to the n stem segments removed. Zotz et al. (1998) compared K_h and K_L computed from Eqs. (6.5)–(6.7) to values measured in the conventional way (Chap. 5.2) and the values were found to agree quite well in most cases (Fig. 6.5).

The resistance of a leaf decreases as leaf area increases (larger leaves are more conductive). Hence, to compare hydraulic parameters of leaves between species, it

Fig. 6.5. Comparison of hydraulic parameters measured by conventional means (*open circles*) and by an HPFM using Eqs. (6.5)–(6.7) (*solid circles*) in *Baccharis pendunculata* and *Croton draco*. See text for more details. (Adapted from Zotz et al. 1998)

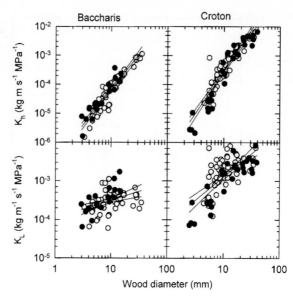

is necessary to divide conductance by leaf area or to multiply resistances by leaf area. Leaf resistances have been measured on more than 35 species including the four *Quercus* species mentioned above. Leaf resistances do not correlate with specific leaf area (SLA), which is the leaf area per gram dry weight and often taken as a measure of sclerophylly. *Quercus* spp. tend to be in the upper range of the 35 species measured but there are several others in the same range (Fig. 6.6). *Acer saccharum* and *Acer rubrum* (not shown in Fig. 6.6) also have very large resistances of 1.3×10^4 and 2.4×10^4 MPa s m^2 kg^{-1}, respectively (Yang and Tyree 1994). Leaf resistance as a percentage of branch resistance of large branches with 1–4 m^2 of leaf area has not been measured very often, but *Quercus* appears to be in the upper range. This probably is because *Quercus* spp. tend to have high stem conductances (low stem resistance) compared to many other species and because leaf resistance is high (Fig. 6.6). Very large leaf resistances have also been reported in *Tsuga* needles (Tyree et al. 1975).

The relative contribution of trunk, crown and leaves was measured in *Acer rubrum* and *A. saccharum* as the trees grew larger. Yang and Tyree (1994) measured this partitioning of resistances in trees from 3–20 m in height corresponding to trunk diameters at ground level of 0.02–0.2 m. Leaf resistance remained a constant value of 45 or 55% for *A. rubrum* and *A. saccharum*, respectively, regardless of tree size (trunk diameter; Fig. 6.7). As the trunk diameter increases the percentage of the woody part of the crown resistance that is trunk increases perhaps because of the increased trunk length, but the crown resistance plus trunk resistance remains more or less constant. Hence the crown resistance must be decreasing perhaps because of increased 'branchiness' resulting in more parallel paths. However, the percentage resistance due to leaves does not change.

The contribution of vascular versus non-vascular tissue to whole leaf resistance has been measured less frequently; however, there were early indications

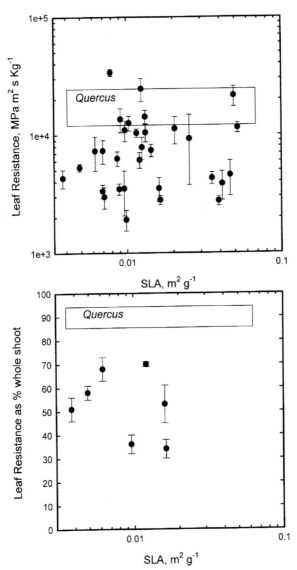

Fig. 6.6. *Top* Leaf resistance (normalized to leaf area) as a function of specific leaf area (SLA) were known. *Quercus* values will be somewhere in the *rectangular box* shown since SLA was not measured. *Bottom* Leaf resistance to water flow as a percentage of stem resistance in large shoots of 1–4 m^2 leaf surface area. The seven points with standard error of the mean are seven tropical species from Zotz et al. (1998), but the leaf resistance data was referred to in the methods section but omitted by error in results. The four *Quercus* values from Fig. 6.4 are somewhere in the box drawn. All error bars are standard error of the mean. (Data reproduced from Zotz et al. 1998; Tyree et al. 1999b).

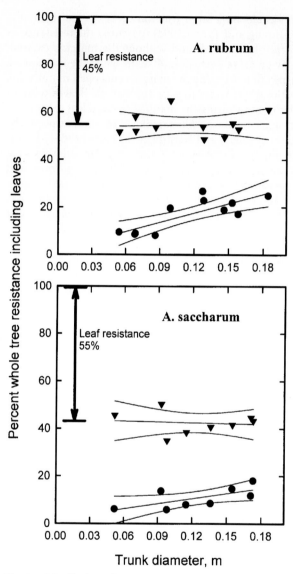

Fig. 6.7. Distribution of resistance to water flow in whole shoots versus basal wood diameter. Resistances were normalized to a leaf area basis and expressed as a percentage. 100%=Resistance of trunk+crown+leaves. *Inverted triangles* are trunk+crown resistance and *circles* are trunk resistance. Linear regressions and 95% confidence intervals are shown. (Adapted from Yang and Tyree 1994)

that non-vascular resistances could be a large fraction of whole leaf hydraulic resistances (Tyree et al. 1975). Yang and Tyree (1994) measured the resistance of *Acer saccharum* and *A. rubrum* leaves with and without the margins excised. Resistances were measured by pressure perfusion of stems while the leaves were under water to prevent transpiration. Excising the margins cuts open minor veins and opens an alternative pathway for water to pass through the leaves without passing through non-vascular tissue. Cutting the margins reduced leaf resistances to 17 and 25% of the intact values for *A. saccharum* and *A. rubrum*, respectively. Similar experiments by Tyree et al. (2001) and Nardini et al. (2001) produced similar magnitude changes when minor veins were cut or non-vascular pathways were destroyed by freezing and thawing in *Prunus laurocerasus*, *Quercus patraea*, and *Viburnum tinus*. From this we can conclude that >75% of the resistance to water flow in leaves is non-vascular in the species studied so far. Recently, attention has been focused on the hydraulic architecture of single leaves and models have been developed to describe the hydraulic architecture of leaves. These models are beyond the scope of this chapter and will likely improve in the coming years. Readers interested may consult Wei et al. (1999) and Martre et al. (2000) for models of monocot leaves and Nardini et al. (2001) for models of dicot leaves.

Unfortunately, no studies have been undertaken to partition the resistances in all parts (roots, stems, and leaves) of any one plant and we await a further resolution of the fascinating question of 'where are the main resistances to water flow in large trees'? Tentatively, however, I would venture to say that at least 50% of the hydraulic resistance for liquid water to flow from the root surface to the leaf evaporative surface is non-vascular. And, if we include the high resistance of the small vessels in the basal and distal extremes of large trees, we might even suggest that only 20–25% of the total resistance is in the main part of the woody roots and stems comprising >95% of the total transport path.

Before moving on from this topic we should include the resistance of the vapor phase, which largely has been ignored so far in this book. We can imagine the soil-plant-atmosphere is a linear catena of resistances all of which sustain the same water flux, F, during steady-state transport of water from the soil to somewhere outside the plant. During steady-state flow, the biggest water potential drop, $\Delta\Psi$, will occur across the biggest resistance since for each resistance in series in a linear catena: $\Delta\Psi = F R$. Water potential of water in air compared to pure liquid water at the same temperature is a log function of the relative humidity (see Nobel 1983). Hence, if water evaporates from a leaf at the same temperature as the air to air at 50% relative humidity, the $\Delta\Psi$ across the gaseous resistance is 97 MPa at 30 °C. Since $\Delta\Psi$ from soil to leaf is generally <2 MPa, we can conclude that >98% of the resistance to water flow is in the gaseous phase. Much of this resistance can be ascribed to the stomates, but, under some circumstances, it is a more diffuse 'resistance' limited by the convective mixing over many meters of air space above the canopy. As discussed in Chapters 3.1 and 5.6, the driving force on water evaporation is determined by leaf temperature, light interception patterns and the total energy budget of leaves. In order to figure out patterns of evaporation from leaves in the canopy of trees, we have to use models such as T-Plant (Chap. 5.6) to compute the pattern of evaporation from leaves in a shoot. Once we know this pattern we can use hydraulic architecture models to compute patterns of pressure profiles in the roots, stems, and leaves.

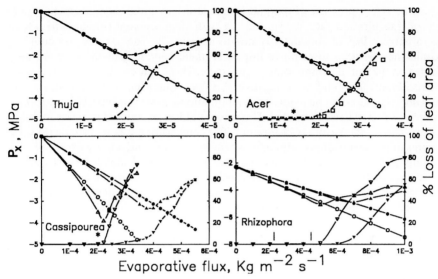

Fig. 6.8. Results of a model predicting runaway embolism in various trees. A hydraulic map of large branches (for *Cassipourea elliptica* and *Rhizophora mangle*) or of whole trees (for *Thuja occidentalis* and *Acer saccharum*) were measured and solved for xylem pressure (P_x) versus evaporative flux from the leaves. The model predicted loss of conductance of stem segments as described in the text and loss of an increasing percentage of leaf area as evaporative flux increased. The * near the x-axis represents the extreme values of evaporative flux observed in field measurements. For *Cassipourea* and *Rhizophora* solutions were made for two different functional dependencies of K_h on stem diameter (Eq. 5.2 with *B*=2 *open symbols* and *B*=2.7 *closed symbols*). (Adapted from Tyree and Sperry 1988)

6.2 Runaway Embolism: Limits Imposed on Maximum Transpiration

We have already reviewed the concept of a hydraulic map (Chap. 5.4) and how a hydraulic map can be used to estimate pressure profiles in trees at any given transpiration rate. The steepness of the pressure profile increases with increasing transpiration. We have also introduced the concept of the vulnerability curve (Chap. 4.1), i.e., a plot of percent loss of hydraulic conductance versus xylem pressure caused by cavitation events. In this section we want to examine how these processes combine to place limits on the maximum transpiration that can be sustained by the vascular system and, by analogy, by a soil system.

In 1987, when we were working out methods to measure vulnerability curves in maple and documenting the hydraulic architecture of maple, Dr. John Donnelly (pers. comm.) asked me the simple question: 'Given that xylem pressure declines as transpiration increases, how much transpiration would you need to cause enough xylem cavitation and embolism such that water transport becomes impossible?' It turned out that we had all the information necessary to answer this question for a few species of trees. We had hydraulic maps from which computations of xylem pressure (P_x) could be made in any given stem segment at any given transpiration rate. We started out with stem segment conductivities with zero percent loss of conductivity due to negative P_x. The initially computed P_x could then be used to estimate loss of segment conductance in every segment based on the initial computation of pressures. These new conductances were used to recompute new (and more negative) P_x values from which new estimates of loss of conductance could be made. The computational cycle was repeated until stable values of P_x had been computed. The amount of loss of conductance was found to increase with increasing transpiration. The computation was repeated with ever-increasing transpirations rates and, whenever loss of conductivity exceeded 95% in any given segment, the segment was 'declared' dead and deleted from the hydraulic map and the area of leaves 'declared' dead was noted and expressed as a percentage of the total leaf area. The average P_x in 'living' segments bearing leaves was also computed. The results of these computations are shown in Fig. 6.8.

Let us look at the results for *Thuja occidentalis*, which are similar to those of the other three species. The open circles in Fig. 6.8 give the theoretical relationship between average P_x in leaf-bearing segments versus evaporative flux assuming no cavitations. The model predicted that certain branches would start cavitating and indeed suffer runaway embolism. The negative P_x generated would cause a loss of conductance and hence a more negative P_x would be needed to maintain transpiration, which, in turn, would cause more loss of conductance in a vicious cycle. The solid circles show the average P_x in living branches after the 'death' of the more vulnerable segments by runaway embolism. The triangles show the percent loss of leaf area at any given evaporative flux when stable pressures had been reached. The open circles show the average P_x if no embolisms had occurred. The '*' near the x-axis indicates the maximum evaporative flux observed in *Thuja* on a sunny day with wet soil. Figure 6.8 suggests that these trees operate near their theoretical maximum transpiration rate at which runaway embolism might start.

Of course, trees rarely ever suffer runaway embolism. The reason for this is that the physiology of the stomates is designed to limit transpiration by closing down when trees reach the limit of runaway embolism. Cochard et al. (1997) studied the hydraulic architecture and vulnerability of two hybrids of *Populus*. Both hybrids had similar vulnerability curves and hydraulic architectures, but one hybrid, cv. 'Peace', could not close stomates in mature leaves and the other, cv. 'Robusta', had normal stomatal function. When these two varieties were subjected to similar periods without irrigation, the Peace variety had much more extensive loss of hydraulic conductance. In another study of 30-year-old *Quercus petraea* trees, some of which were subjected to a prolonged drought, stomatal conductance was found to fall to very low values at a midday water potential that would cause 20% loss of hydraulic conductance in small diameter stems. Midday sap flow densities were also monitored continuously for 3 years and it was found that midday sap flow densities were generally less than or equal to the critical flow rate needed to start runaway embolism; however, when subjected to drought, the midday flow rate was quite close to the critical flow rate causing embolism (Fig. 6.9).

Stomatal closure to limit cavitation events seems to be a very rapid response. When excised branches of *Prunus laurocerasus* (Nardini et al. 2001), *Laurus nobilis* (Salleo et al. 2000) and *Laurus nobilis* and *Ceratonia siliqua* (Salleo et al. 2001) are dehydrated in air they reach the critical water potential for initiation of cavita-

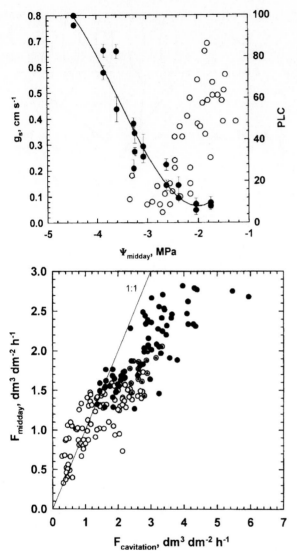

Fig. 6.9. *Top* Percentage loss of hydraulic conductance (PLC) due to xylem embolism (*closed circles*) and midday stomatal conductance (g_s) (*open circles*) versus leaf water potential at midday (Ψ_{midday}). Embolism significantly increased in petioles and twigs when Ψ_{leaf} became lower than −2.7 MPa, the point where the value of g_s was reduced to about 10% of its maximal value. *Bottom* Actual midday sap flow density (F_{midday}) versus theoretical sap flow density ($F_{cavitation}$) inducing xylem embolism. *Closed circles* represent irrigated trees and *open circles* represent drought-stressed trees. (Adapted from Cochard et al. 1996a)

Fig. 6.10. Soil-plant hydraulic continuum model used to evaluate runaway embolism. See text for more details. (Sperry et al. 1998)

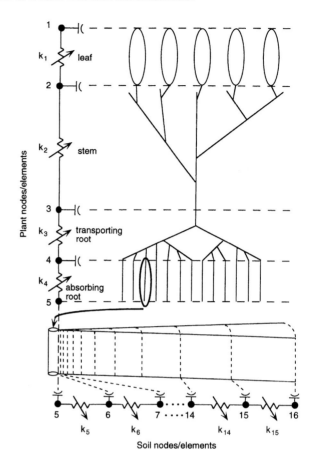

tion events within 90 min and stomatal closure is always coincident. Sperry and Pockman (1993) manipulated woody-shoot hydraulic conductance of small *Betula occidentalis* trees (2–3 cm basal trunk diameter). The total shoot conductance per unit leaf area was decreased by an air-injection method and the effects of reduced woody-shoot conductance on midday transpiration rate and stomatal conductance were measured. They demonstrated that stomatal conductance adjusted to keep the midday water potential near the cavitation threshold (–1.43 MPa) even though woody-shoot conductance per unit leaf area varied over a factor of 10.

Sperry et al. (1998) have extended the runaway embolism model to include breakdown in water transport in the rhizosphere (roots and soil). This model was derived in order to determine if the maximum rate of water extraction by a plant is limited by runaway embolism in the plant (stems or roots) or in localized dehydration in the soil. It turns out that both limitations are possible! The basic model is shown diagrammatically in Fig. 6.10.

In this model the plant was approximated by a linear catena of just four conductance elements and the soil is described by a linear catena of 11 resistance elements. The model takes into account the loss of conduction of soils by using standard expressions for how soil conductance changes with soil water potential, Ψ. Loss of conductivity comes from empirical curve fits of the vulnerability curve, i.e., K equals some function of Ψ, and it is assumed that all plant parts have the same vulnerability curve. Sperry et al. (1998) use the Weibull function, which gives a good empirical fit, but a much superior function is now available to fit vulnerability curves, e.g., Pammenter and vander Willigen (1998). Since the model treats the plant as a linear catena, it is incapable of predicting the gradual dieback of branches predicted by a branched catena model (Fig. 6.8), but this need not concern us here since the fascinating aspect of this model (and its main strength) lies in its ability to predict when water transport fails in soil and when it fails in the plant. There is, of course, the trivial case of dry soils being incapable of transporting the remaining water because the hydraulic conductivity of dry soils is very low. This model predicts that even in wet soil there will be a breakdown in water transport when water flux through the soil to the root surface grows to a critical value. While this prediction is not new, it is the first model I have seen to predict when and where breakdown might occur in the whole soil-plant continuum. As the water flux density near the root surface increases because of increased transpirational demand, the gradient in water potential from the 'bulk' soil to the soil surface becomes steeper and the hydraulic conductance decreases. Hence there will be a theoretical maximum flux density at which transport breaks down. The flux density near the root surface depends on the ratio of root surface area to leaf surface area. The smaller the root surface per unit leaf surface, the larger the flux density needed to maintain the water flow rate to the leaf.

To explain what is going on here let us write a very simplified linear transport model of the soil-plant continuum, which quantifies transport of water across a plant conductance, k_{plant}, and a soil conductance, k_{soil}.

$$\Psi_{leaf} = \Psi_{soil} - \frac{E}{k_{plant}} - \frac{E}{k_{soil}} \tag{6.8}$$

In summary, when Ψ-dependent k in each part of the soil-plant continuum is taken into account, there is no longer a directly proportional relationship between E (evaporative flux density from leaves) and Ψ. Instead, increases in E are associated with progressively disproportionate decreases in Ψ because of declining k everywhere in the continuum. The value of E reaches a maximum (E_{crit}) at a corresponding minimum (Ψ_{crit}) at which water conduction breaks down somewhere in the plant or soil because of runaway embolism in either the plant or the breakdown of water continuity in the soil. So where does this breakdown occur? The model predicts that for most situations the breakdown probably occurs because of embolism in the xylem, but for plants with low xylem vulnerability, i.e., where embolism occurs only when xylem pressures are very negative, then a breakdown in water continuity is more likely to occur in the soil (especially if the soil is dry, i.e., at very negative Ψ_{soil}). The soil texture also plays an important role in determining E_{crit}, since course soils, e.g., sands, fill with air at much less negative water potential than fine soils, e.g., silt or clay.

This model has been tested by comparing the midday values of E and Ψ_{leaf} to the theoretical critical values that would cause breakdown of transport either in

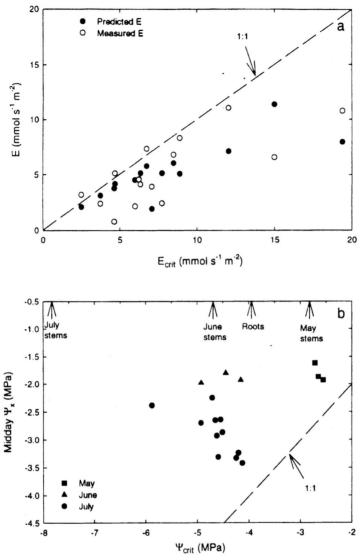

Fig. 6.11. Comparisons of midday values of evaporative flux density (*E*) from leaves and leaf water potential (Ψ_x) versus the corresponding critical values (as defined in the text) for *Artemisia tridentata*. Safety margins ($E_{crit}-E$) correspond to the *y*-axis distance between *E* and the *dashed 1:1 line* in the *upper graph*. Safety margins ($\Psi_x-\Psi_{crit}$) correspond to the y-axis distance between midday Ψ_x and *the dashed 1:1 line* in the *lower graph*. The Ψ_x values corresponding to 99% loss of hydraulic conductance in May stems, roots, June stems, and July stems is shown by *arrows* at the top. (Kolb and Sperry 1999)

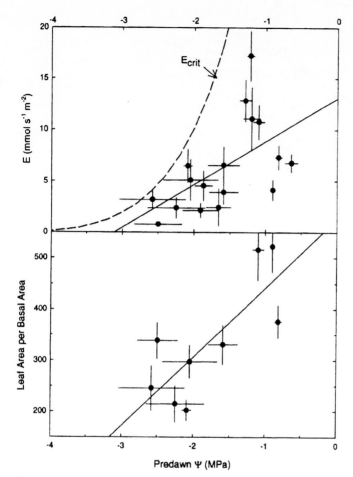

Fig. 6.12. *Top* Evaporative flux density (E) versus pre-dawn leaf water potential Ψ, which is generally taken as a measure of soil water poten-tial. *Bottom* Leaf area per stem basal areas versus pre-dawn Ψ. The maximum E_{crit} that would cause runaway embolism in the soil-plant continuum is shown as the *dashed line* in the *upper* graph. *Bars* show the standard errors of the means. (Kolb and Sperry 1999)

the soil or plant. Plants seem to 'push' these limits with a relatively narrow margin of safety (Sperry et al. 1998; Kolb and Sperry 1999; see Fig. 6.11).

An interesting strategy taken by the arid zone plant *Artemisia tridentata* is its gradual loss of leaf area as the soil dries to keep the transpiration rate within the critical boundary (Fig. 6.12). We can presume that decreases in stomatal conductance may not have been enough to reduce E below E_{crit} if leaf area had not declined.

Relationships similar to those shown in Fig. 6.12 (top) have been found for *Pinus taeda* stands growing on two different soil types – loam and sand. Again, as the soils dried, the value of E approached E_{crit} in both soil types (Hacke et al.

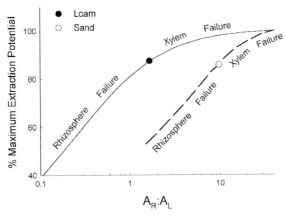

Fig. 6.13. Extraction potential versus root/leaf area ratio (A_R/A_L) for loam (*solid line*) and sand (*dashed line*). Extraction potential is defined as the area under the E_{crit} versus soil water potential. Extraction potential is expressed as a percentage of the value at A_R/A_L=40. This value was chosen as a liberal estimate of the maximum likely physiological value for a woody plant. *Symbols* correspond to measured values. (Hacke et al. 2000)

2000). However, there were major shifts in the root/leaf area ratio (A_R/A_L). In loam the ratio was 1.68 versus 9.75 in sand! These huge differences in A_R/A_L are driven by an interaction of the plant with the soil type and represents a compromise ratio to avoid breakdown of transport in either the soil or the plant. E values were comparable for both stands throughout the growth season, but, in order to avoid transport breakdown in the soil, the *P. taeda* in sandy soil had to maintain much higher root surface areas. Sperry's model (Sperry et al. 1998) would have predicted breakdown of transport in the soil had the A_R/A_L ratio been 1.68 in both the sand and the loam soil. Hacke et al. (2000) defined the extraction potential as the area under the theoretical curve of E_{crit} versus soil water potential for the physiological range of soil water potential observed in the field. When extraction potential is plotted versus A_R/A_L for the loam- and sand-soil types it appears that the observed A_R/A_L falls at the intersection between the ratios that would cause rhizosphere failure versus xylem failure, i.e., breakdown of transport in the soil versus the plant (Fig. 6.13).

6.3 Hydraulic Limits on Transpiration and Carbon Gain

In the previous section we examined the notion that there is a maximum potential water flow rate through the soil-plant continuum that is determined by the hydraulic architecture and the vulnerability to cavitation in either the plant or the rhizosphere. When the water flow rate exceeds a critical value, runaway embolism causes a breakdown in the water pathway either in the soil or in the plant. In this

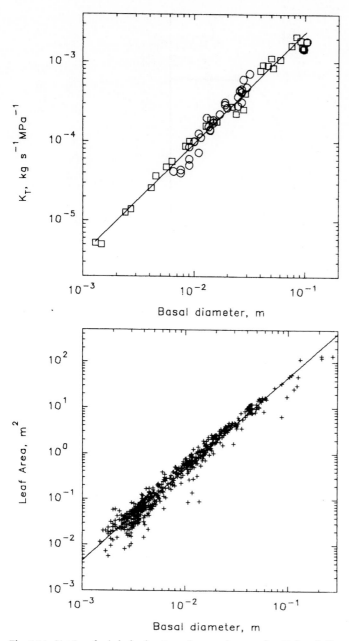

Fig. 6.14. *Top* Log k_T (whole shoot conductance) versus log D (basal diameter of the wood) of *Acer saccharum* branches. *Bottom* Log A_L (area of leaves attached to a shoot) versus log D. (Yang and Tyree 1993)

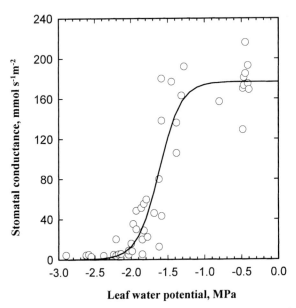

Fig. 6.15. Stomatal conductance of detached *Acer saccharum* leaves versus leaf water potential. The leaves were slowly dehydrated while exposed to saturating light intensity. Each *point* is a different leaf. The *smooth curve* is the data fitted to a three-parameter sigmoid curve of the form stomatal conductance=$a/(1-\exp(b(c-\Psi))$, where a=maximum conductance=176.8, b=6.66 and c=water potential at half maximum conductance=-1.616 MPa. (Adapted from Yang and Tyree 1993)

section we will examine the possibility that the hydraulic architecture of plants might indeed limit transpiration below these critical levels. Since transpiration is a proxy for rate of gas exchange of both water and carbon dioxide, it also follows that hydraulic limits on transpiration may also impose hydraulic limits on carbon gain.

I first became aware that shoot hydraulic architecture might limit gas exchange through stomatal regulation when Yang and Tyree (1993) examined how whole shoot conductance and leaf area scaled with shoot basal diameter, D, in *Acer saccharum*. Whole shoot conductance was given by k_T=0.06 $D^{1.402}$ and leaf area A_L=4667 $D^{2.007}$ (Fig. 6.14). The drop in xylem pressure across the shoot, ΔP_x, should equal EA_L/k_T, hence it follows that:

$$\Delta P_x=(7.781\times10^4 D^{0.605})E \tag{6.9}$$

Hence, we have to conclude that as branches grow larger ΔP_x grows larger too. We can actually turn Eq. (6.9) into an approximate predictor of leaf water potential because in a wide variety of species root and shoot conductances are approximately equal (Tyree et al. 1998; Becker et al. 1999), hence the water potential drop across the whole plant will be double that across the shoot. So if the soil is wet and the soil water potential is nearly zero, we have:

$$\Psi_{leaf}\approx-2(7.781\times10^4 D^{0.605})E \tag{6.10}$$

Fig. 6.16. *Top* Theoretical relationship between maximum possible stomatal conductance, g_s, and whole plant hydraulic conductance is plotted using the measured relationships for *Acer saccharum* in Figs. 6.14 and 6.15. The whole plant conductance values on the *x*-axis correspond to basal stem diameters of 0.022–1.22 m and stem diameters (in m) corresponding to specific plant conductances are indicated on the *x*-axis. See text for computational details. *Bottom* Theoretical relationship between g_s is plotted versus whole plant conductance per unit leaf area: k_{plant}/A_L

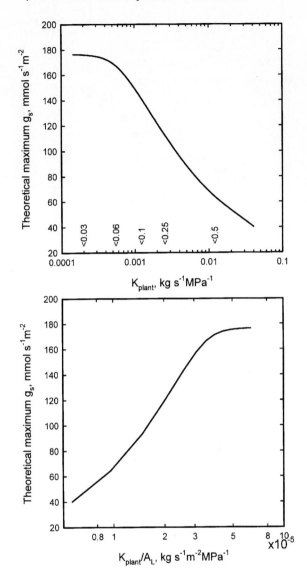

Yang and Tyree (1993) compared predicted values of ΔP_x or Ψ_{leaf} to the response of stomatal conductance to leaf water potential (Fig. 6.15) and concluded that as *A. saccharum* grows larger the change in Ψ_{leaf} should start limiting stomatal conductance.

Midday leaf water potential, Ψ_{leaf}, of *Acer saccharum* leaves are typically –1.2 to –1.5 MPa in wet soil at the base of Mt Mansfield, Vermont, where the data for Figs. 6.14 and 6.15 were collected. So clearly, Ψ_{leaf} limits stomatal conductance. Because Ψ_{leaf} decreases with increasing basal diameter, a proxy for tree size, it seems likely that stomatal conductance will be restricted increasingly as trees

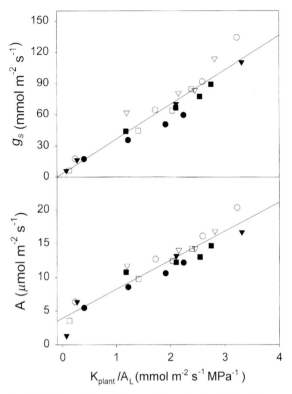

Fig. 6.17. Stomatal conductance (g_s) and assimilation (A) versus whole plant leaf-specific conductance (k_{plant}/A_L). The leaf-specific conductance of *Pinus ponderosa* seedlings was changed by air injection. Each *symbol* is for a different seedling ($n=6$) and multiple points represent multiple air injection pressures. In order to compare conductances here with others in this volume, note that 1 mmol s^{-1}=1.8×10^{-5} kg s^{-1}. (Hubbard et al. 2001)

grow larger. Although Yang and Tyree (1993) did not go on to compute a theoretical limiting stomatal conductance (g_s) versus whole plant conductance (k_{plant}), it can easily be done. The approach you take is to substitute for E in Eq. (6.10) the approximate value equal to $g_s \Delta X$, where ΔX is the appropriate driving force giving a typical midday transpiration rate. You then pick a value of g_s from Fig. 6.15 and look up the corresponding value for Ψ_{leaf}; you then find the stem diameter that yields the same Ψ_{leaf} in Eq. (6.10). This value of D is then used to compute the whole plant conductance from $k_{plant}=0.03D^{1.402}=k_T/2$. When this exercise is carried out for a range of g_s values, you obtain the results in Fig. 6.16 (top). A more typical way of expressing the relationship today is to plot maximum g_s versus hydraulic conductance per unit leaf area. This relationship is shown in Fig. 6.16 (bottom).

What is the mechanism connecting changes in g_s and whole plant conductance? One hypothesis is rather indirect. Meinzer et al. (1995) suggest that as hydraulic

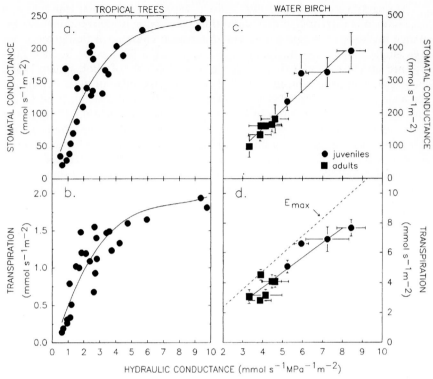

Fig. 6.18. Stomatal conductance (**a, c**) and transpiration rate (**b, d**) versus hydraulic conductance from bulk soil to leaf (k_{plant}/A_L). **a, b** Various tropical gap-species (Meinzer et al. 1995); **c, d** *Betula occidentalis* juveniles and adults. (Saliendra et al. 1995)

conductance changes during plant development, associated changes in xylem sap composition and concentration are sensed in the leaf and result in corresponding changes in g_s. However, this cannot explain all instances since many authors have noted very rapid changes (<15 min) in g_s in response to experimental changes in k_{plant}/A_L (Sperry and Pockman 1993; Saliendra et al. 1995; Fuchs and Livingston 1996). Another explanation is that stomata respond to changes in Ψ_{leaf} caused by changes (short-term or long-term) in k_{plant}/A_L. The link to stomatal response could be a turgor-mediated release of abscisic acid (Raschke 1975).

Two different approaches have been taken to establish a relationship between whole plant hydraulic conductance per unit leaf area, k_{plant}/A_L and stomatal conductance. One way is to induce rapid changes in k_{plant} and look at immediate responses in g_s and carbon assimilation, A. This has been done in *Pinus ponderosa* seedlings, where k_{plant} was rapidly altered by injecting stems with air to induce extra embolism (Hubbard et al. 2001). A strong linear relationship was found between k_{plant}/A_L and g_s and A (Fig. 6.17). Another way (Sperry 2000) is to look for correlations between k_{plant}/A_L in the 'native state' and g_s (Fig. 6.18).

The quantification of the photosynthetic capacity of the total leaf area of a large branch is extremely difficult and time consuming by conventional gas-exchange methods on single leaves. For this reason chlorophyll fluorescence has been employed to determine photosynthetic potential. This works because fluorescence provides information about the reduction state of photosystem II and good relationships have been found between CO_2 assimilation measured by gas exchange and the quantum yield of photosystem II. Brodribb and Field (2000) have used chlorophyll fluorescence to estimate quantum yield on 22 species of woody plants in New Caledonian and Tasmanian rainforests. They found a strong correlation between quantum yield and hydraulic conductance of whole shoots. Since leaf-specific whole plant conductance increases the instantaneous gas exchange (including net assimilation and quantum yield), we might suppose that long-term growth rates might also correlate with k_{plant}/A_L and this will be the subject of the next section.

6.4 Hydraulic Adaptation and Limits on Growth

A number of studies have focused on the hydraulic architecture of large woody tropical plants, e.g., trees, shrubs and vines (Ewers et al. 1991; Patiño et al. 1995; Tyree and Ewers 1996). Most of this work has focused on stem segment parameters of specific conductivity, K_s, and leaf specific conductivity, K_L (see also Chap. 5). Rapidly growing pioneer species have larger K_L and/or K_s values than slower-growing species (Tyree et al. 1991; Machado and Tyree 1994). This has been interpreted in terms of the need of plants to have high levels of water potential to achieve high relative growth rates in meristems and, as we have seen in the previous section, this also correlates with high net assimilation. However, in Section 6.1, we saw that woody shoots might account for a rather small fraction of the total hydraulic conductance from fine roots to leaves. The recent development of the high-pressure flowmeter (HPFM) allows the rapid measurement of root and shoot hydraulic conductances of seedlings. Thus it is now possible to study the growth dynamics of seedlings while monitoring the dynamic changes in hydraulic conductance of roots and shoot. Before proceeding to the data we should review some issues of scaling of hydraulic conductances to aid our understanding of growth dynamics.

Root conductance (k_r) can be defined as water flow rate (kg s^{-1}) per unit pressure drop (MPa) driving flow though the entire root system. Values of k_r could be scaled by dividing by some measure of root size (root surface area, total root length, or mass) or by dividing k_r by leaf surface area. Division by root surface area (A_r) is justified by an analysis of axial versus radial resistances to water flow in roots. In the radial pathway, water flows from the root surface to the xylem vessels through non-vascular tissue. In the axial pathway, water flow is predominately through vessels. The resistance of the radial path is usually more than the axial path (Frensch and Steudle 1989; North et al. 1992). Most water uptake is presumed to occur in fine roots (<2 mm diameter) and fine root surface area is usually >90% of the total root surface area (personal observation). So root uptake of water would appear to be limited by root surface area and hence it is reasonable to divide k_r by A_r yielding a measure of root efficiency. Some roots are more efficient than others. Division of k_r by total root length (L) is not as desirable, but is justified because A_r and L are correlated approximately and L can be estimated by a low-cost, line-intersection technique rather than a high-cost, image-analysis technique.

Scaling by root mass is justified by consideration of the cost of resource allocation. Plants must invest a lot of carbon into roots to grow and to maintain them. The benefit derived from this carbon investment is enhanced scavenging for water and mineral nutrient resources. Total root dry weight (TRDW) is a measure of carbon investment into roots. Thus the carbon efficiency of roots might be measured in terms of k_r/TRDW, A_r/TRDW, or L/TRDW. Scaling by TRDW provides information of ecological rather than physiological importance.

Scaling of k_r by leaf surface area (A_L) provides an estimate of the 'sufficiency' of the roots to provide water to leaves. The physiological justification of scaling k_r to the leaf surface area (A_L) comes from an analysis of the Ohm's law analogue for water flow from soil to leaf (van den Honert 1948). The Ohm's law analogue describes water flow rate (W, kg s^{-1}) in terms of the difference in water potential between the soil (Ψ_{soil}) and the leaf (Ψ_L):

$$\Psi_{soil} - \Psi_L = (1/k_{soil} + 1/k_r + 1/k_{sh})W \tag{6.11}$$

where k_{soil} is the hydraulic conductance of the soil. It is usually assumed that $k_{soil} > k_r$ and k_{sh} except in dry soils, so $1/k_{soil}$ can be ignored. Leaf water potential is approximated by:

$$\Psi_L \cong \Psi_{soil} - (1/k_r + 1/k_{sh})W \tag{6.12}$$

Or, if we wish to express Eq. (6.12) in terms of leaf area and average evaporative flux density (E), we have:

$$\Psi_L \cong \Psi_{soil} - (1/k_r + 1/k_{sh})A_L E \tag{6.13}$$

This equation can also be rewritten so that root and shoot conductances are scaled to leaf surface areas, i.e., to give leaf-specific shoot and root conductances, k_{sh}/A_L and k_r/A_L, respectively:

$$\Psi_L \cong \Psi_{soil} - (1/[k_r/A_L] + 1/[k_{sh}/A_L])E \tag{6.14}$$

Meristem growth and gas exchange are maximal when water stress is small, i.e., when Ψ_L is near zero. From Eq. (6.14) it can be seen that the advantage of high k_r/A_L and k_{sh}/A_L is that Ψ_L will be closer to Ψ_{soil}. Leaf-specific stem-segment conductivities, k_L, are high in adult pioneer trees, so the water potential drop from soil to leaf is much smaller than in old-forest species (Machado and Tyree 1994). This may promote rapid extension growth of meristems in pioneers compared with old-forest species. Also, total conductance (g_s) and therefore net assimilation rate are reduced when Ψ_L is too low. During the first 60 days of growth of *Quercus rubra* seedlings, there was a strong correlation between midday g_s and leaf-specific plant conductance, k_p/A_L, where $k_p = k_r k_{sh}/(k_r + k_{sh})$; Ren and Sucoff 1995). This suggests that whole-seedling hydraulic conductance is limiting g_s through its effect on Ψ_L. Thus, high values of k_r/A_L and k_{sh}/A_L may promote both rapid extension growth and high net assimilation rates in pioneers.

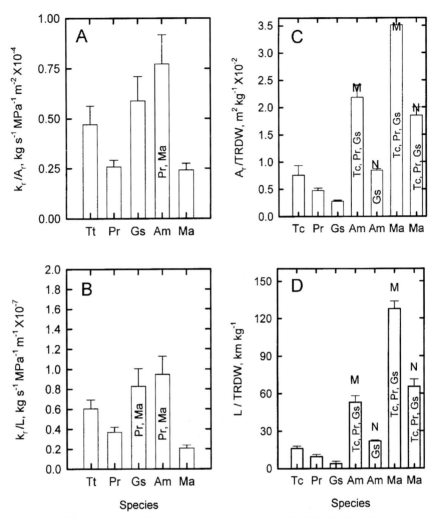

Fig. 6.19. Different ways of scaling growth parameters. **A** k_r scaled (divided) by fine root surface area (A_r). **B** k_r scaled to fine root length. **C** A_r scaled to total root dry weight (TRDW=dry weight of fine plus coarse roots). **D** Fine root length (L) scaled to TRDW. *Error bars* are SEM, n=5–6 for May or November and 10–12 for pooled data. Data were measured in May and November; when means were not significantly different, values were pooled. Letters M and N indicate non-pooled data for May and November, respectively, which were significantly different (Tukey test, $P \leq 0.05$). Tt = Trichilia Tuberculata, Pr = Povteria reticulata, Gs = Gustavia superba, Am = Apeiba membranacea, Ma = Miconia argentea. *Species abbreviations within a bar* indicate which species means are significantly different (Tukey test, $P \leq 0.05$) from the bar bearing the abbreviations. In the original publication there was an error in computation of A_L made by the Delta-T root scanner software that failed to multiply projected area by PI (3.14) to get surface area. This error is corrected in Fig. 6.19A. (Tyree et al. 1998)

Fig. 6.20. Hydraulic conductances of shoots and roots scaled to dry weight or leaf area. *Top* k_r per unit TRDW and k_{sh} per unit shoot dry weight. *Bottom* k_r and k_{sh} both scaled to leaf area (A_L). *Error bars are SEM, n=23–36.* Data from all collection dates combined. Species abbreviations as in Fig. 6.19. Root and shoot means for Am and Ma were significantly different from corresponding root and shoot means for Tt, Pr, and Gs in both A and B (Tukey test, $P \leq 0.05$). (Tyree et al. 1998)

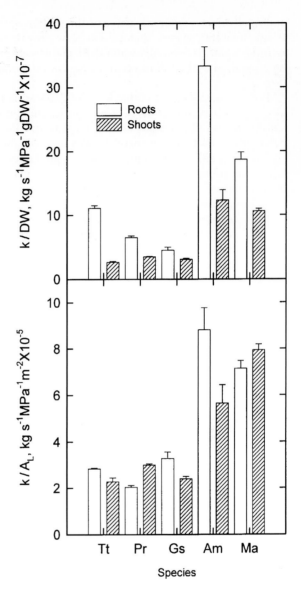

Scaling is always necessary to normalize for plant size. As seedlings grow exponentially in size we would expect an approximately proportional increase in k_r and k_{sh}. Since roots and shoot both supply water to leaves, and since an increase in leaf area means an increase in rate of water loss per plant, we would expect k_r and k_{sh} to be approximately proportional to A_L.

Tyree et al. (1998) studied the root and shoot conductance of five species of tropical tree seedlings over the first 16 months of growth. Two of the species were pioneer (light-demanding) species that exhibit rapid growth (*Miconia argentea* and *Apeiba membranacea*). Three of the species were intermediate to old-forest (shade-tolerant) species that exhibit slow growth (*Gustavia superba*, *Pouteria reticulata*, and *Trichilia tuberculata*). All plants were grown under an intermediate light regime, i.e., they all received the equivalent of about 25% of full sunlight per day.

Tyree et al. (1998) found that k_r and k_{sh} values increased exponentially during growth; hence comparisons between species could not be made with raw data. A suitable means of scaling k_r and k_{sh} to plant size is needed to reveal differences between species that may correlate with light adaptation. The scaling parameter selected may differ depending on the biological question being asked, i.e., physiological questions concerning the mechanism or pathway of transport versus ecological questions related to adaptations to microsites.

A common way of scaling k_r to plant size is to divide k_r values by the surface area of fine roots; in this case we are dividing by the surface area perpendicular to the pathway of water movement to produce a specific conductivity. Implicit in this type of scaling is the assumption that the hydraulic resistance to water flow radially into the root exceeds the hydraulic resistance to water flow along the axis of the root. There is reason to believe this assumption is valid (Nobel and Sanderson 1984; Frensch and Steudle 1989; North et al. 1992), otherwise we would need to divide by the cross-sectional area of the roots. Figure 6.19A shows values of k_r/A_r for the five species in this study. Data were pooled for the two harvest months since there was no significant difference for the two harvest dates. Values of k_r/A_r differed significantly between a few species, though no pattern with light adaptation (successional stage) was evident, i.e., the two light-demanding species (*M. argentea* and *A. membranacea*) were not distinct from the shade-tolerant species (*P. reticulata*, *G. superba*, and *T. tuberculata*). Another common way of scaling to plant size is to divide k_r by total fine-root length (L). From Fig. 6.19B it can be seen that the relative ranking of k_r/L is nearly the same as the ranking of k_r/A_r.

The light adaptations (successional stages) more clearly separated when we used $A_r/$TRDW or $L/$TRDW to measure the cost of producing roots. The advantage of additional root surface or length per gram is better access to water and mineral nutrient resources in pioneer versus old-forest species. By contrast, the roots of shade-tolerant species are more robust, and presumably less susceptible to damage and predation (cf. Kitajima 1994). The advantage gained by more favorable values of $A_r/$TRDW and $L/$TRDW in pioneer species may be gradually lost as they age, but more work is needed to confirm this trend.

In Fig. 6.20 (top) the root and shoot conductances are scaled by root and shoot dry weights, respectively, and in Fig. 6.20 (bottom) the root and shoot conductances are scaled by dividing both by leaf area. In both cases the adaptive advantages of the pioneer species become evident. All pioneer species differed significantly from the non-pioneer species even though all species were growing in the same light regime. Figure 6.20 illustrates two advantages of pioneers versus other species in this study. The higher values of k_r/A_L and k_{sh}/A_L mean that the pioneer species can maintain less negative leaf water potentials than the other species at any given transpiration rate. This could lead to higher rates of extension growth and net assimilation. The higher values of k_r/DW and k_{sh}/DW in pioneers means that pioneers spend less carbon to provide efficient hydraulic pathways than do the other species. Both of these advantages (Fig. 6.20) mean that pioneers can be more competitive in gap environments than old-forest species.

Hydraulic architecture appears to be an adaptation that correlates the adaptation to light environment with manifest differences in growth rate. However, even in high light environments, there can be differences in growth rate correlated with differences in water availability. Other factors that might affect growth rate are temperature (in an altitudinal or latitude gradient) or mineral nutrition. Do differences in hydraulic architecture correlate with these environmental differences that are known to influence growth rate? More work will answer aspects of this fascinating question. Whole plant conductances per unit leaf area, measured with an HPFM, ranged over a factor of 10 in one survey of 12 tropical angiosperms and conifers from Borneo (Becker et al. 1999). A wide diversity of conductances increases the potential for a correlation with growth rate. One relevant survey of seven European oak species supports the notion that drought adaptation correlates with hydraulic conductance in the expected way. Nardini and Tyree (1999) found that the three drought-adapted species (*Quercus suber, Q. pubescens*, and *Q. petraea*) had root and shoot conductances per unit leaf area that were significantly lower than those of the water-demanding species (*Q. alba, Q. cerris, Q. robur*, and *Q. rubra*).

What has been lacking so far is a common garden study in which relative growth rate is quantified for each species and related to whole plant conductance. In future experiments, one might actually control differences in light level and water input to induce differences in relative growth rate within species and between species and see how this correlates with whole plant hydraulic conductance.

6.5 Variable Hydraulic Conductance: Temperature, Salts, and Direct Plant Control

So far we have talked about the hydraulic architecture of plants as though the resistance or conductance components are absolute constants, except for the effects of cavitation on loss of hydraulic conductance. We cannot leave the discussion of hydraulic architecture without considering how the hydraulic parameters will change with respect to temperature, sap composition, endogenous rhythms (internal plant effects) and implicitly with how resistance is defined.

The physical basis for the temperature dependence of stem segment or whole plant conductance is perhaps the easiest to understand. If we imagine that water flowing through vessels obeys the Hagen-Poiseuille equation (Eq. 1.3) we see that conductivity is inversely proportional to viscosity ($1/\eta$), which is called the fluidity. An inspection of the table of water viscosity in the Handbook of Physics and Chemistry reveals that the fluidity of water is temperature-dependent (Fig. 6.21A). The temperature dependence of conductance in vessels ought to be about 2.42% per °C. Even if we treat water transport in the most general sense of water movement through an irregular porous medium (Darcy's law), we would expect the same dependence on viscosity. Another way of treating the temperature

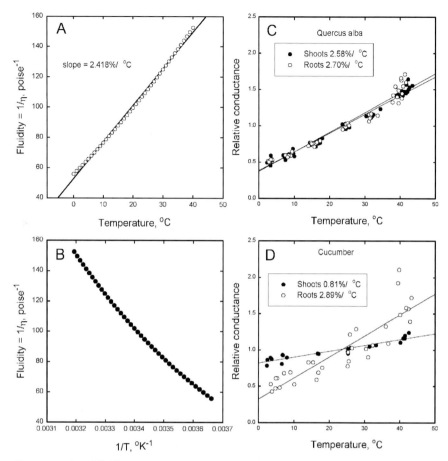

Fig. 6.21. A Plot of fluidity versus temperature. **B** Plot of fluidity versus inverse absolute temperature. **C** and **D** Plots of whole shoot and whole root relative conductance measured with a high-pressure flowmeter versus temperature in *Quercus alba* and cucumber. Relative conductance=*k* at temperature *T* divided by the mean of all values. (**C** and **D** from S. Zhang, M. Tyree, unpubl. data)

dependence of water flow is by plotting fluidity (or conductance) versus the inverse absolute temperature, which is called an Arrhenius plot (Fig. 6.21B). Arrhenius figured out that the slope of this plot equals an activation energy divided by the gas constant. In the case of laminar water flow the activation energy in a vessel would be the energy needed to break hydrogen bonds as water molecules flow past each other. For water flow through membranes the activation energy would be more because water in a membrane is at a higher energy state so the temperature dependence for water flow across a membrane ought to be more than 2.42% per °C. See Tyree et al. (1973) for an example of how Arrhenius plots are used.

During a visit to my laboratory in 1998, Prof. Shuoxin Zhang, Northwest Forestry University, P.R. China, undertook a study of the temperature dependence of

Table 6.1. Data collected by Prof. Zhang in 1998 concerning the temperature dependence of whole shoot and root conductances of seedlings of various plants. Plots like that in Fig. 6.21C,D were used to compute mean slope (expressed as percent change) and standard error of the mean of $n=7$ or 8 roots and shoots. Means marked with * were significantly different from 2.418 ($P<0.01$)

Species	Root %/°C	SEM	Shoot%/°C	SEM
Acer saccharum	2.74*	0.07	3.27*	0.08
Quercus alba	2.70*	0.03	2.58*	0.06
Picea rubens	2.79*	0.25	2.85*	0.20
Pinus strobus	2.30	0.18	2.44	0.14
Pinus taeda	2.68	0.30	2.50	0.39
Tsuga canadensis	-	-	2.97	0.14
Cucumber	2.89*	0.34	0.81*	0.07
Squash	2.35	0.05	1.50*	0.18

whole shoot and whole root conductance of the seedlings of various plants. Representative curves are shown in Fig. 6.21C,D. Root and shoots were immersed in water of various temperatures between 2 and 42 °C and the conductance was measured within 3 to 5 min of the change in temperature. Since we are interested in the relative change of conductance for each plant the relative conductance is computed relative to the mean conductance at all temperatures. In many cases the temperature dependence was significantly more than the temperature dependence of the fluidity of water (Table 6.1). This is in contrast to similar measurements of water flow through stem segments which are not significantly different from the fluidity of water (Tyree, unpubl. data).

On a typical day the air temperature will change by 10 °C from dawn to early afternoon. So if stem and leaf temperatures tracked air temperature changes then whole shoot conductances are likely to increase by about 24–30%. However, the temperature variation of stems and leaves in sunlight are likely to be even more. Cylindrical objects about 1 cm in diameter in direct sunlight with low transpiration can generally warm up 5 or 10 °C above air temperature (Nobel 1983; Patiño et al 1995), and leaf temperatures of many trees can also be 5–8 °C warmer than air when in full sunlight (Tyree and Wilmot 1990). Root temperature, on the other hand, is likely to change very little with time of day; the roots <10 cm deep might change a couple of degrees and roots >50 cm will be at approximately constant temperature although there will be gradual seasonal trends. Considering the distribution of resistances discussed in Section 6.1, it seems likely that hydraulic conductance of whole plants could change by 10–15% diurnally.

An interesting anomaly, first reported by Zimmermann (1978a), was that the hydraulic conductance of stem segments varies with the concentration of salts in the xylem. This salt-concentration dependence is not always present (pers. observ.) but it has been reported from time to time (van Leperen et al. 2000; Zwieniecki et al. 2001). Xylem sap concentration in trees tends to vary inversely with transpiration rate, being higher at dawn than at midday. The reason for this is that salt and water uptake by roots are decoupled, i.e., occur through independent means. The salt uptake rate, J_s mol s^{-1}, tends to be more or less constant with time of day, whereas water uptake, J_w kg s^{-1}, varies. So the concentration in the vessels of minor roots will equal J_s/J_w molal at any instant. But, as the sap ascends through the roots and stems, there is a dynamic exchange of salts with living cells so the composition and concentration is dynamically variable. The typical range of concentration change is 10–20 mmolal at dawn falling to 1–2 mmolal at midday. This range of concentration change is likely to cause a 15% decrease in stem-segment conductivity from morning to midday. However, since most of the hydraulic resistance is in non-vascular portions of root and leaves, the whole plant conductance is likely to change by just 3–6% (Sect. 6.1). So the concentration-dependent and the temperature-dependent changes tend to cancel each other out, with the temperature dependence tending to be bigger most of the time. Most authors think that the salts influence the porosity of the pit membranes between xylem conduits and hence effect the overall xylem conductance (Zimmermann 1978; Zwieniecki et al. 2001), hence the effect is likely to be most pronounced with species with short vessels because the intervessel resistances are likely to be a larger faction of the vessel lumen resistance.

A question that has fascinated plant physiologists for many years is whether plants have some direct physiological control over their hydraulic conductance. The most likely place where such control could be exerted is in membranes, because transmembrane water flow might control the resistance of the non-vascular pathway, which, as we have seen, is quite substantial in plants. Recently, some evidence for diurnal changes in root resistance has emerged. Using a root pressure probe, Henzler et al (1999) have observed diurnal changes in root conductance in *Lotus japonicus* roots that are 4–10 times more conductive at midday than at midnight. Using an HPFM, Tsuda and Tyree (2000) observed a twofold diurnal change in root conductance of sunflower plants. In both publications, the authors tentatively attributed changes in whole root conductance to assumed diurnal expression of aquaporin activity. Aquaporins are membrane-bound proteins that are known to enhance membrane permeability to water at the cellular level (Tyerman et al. 1999). However, this is only speculation and may not be true.

Recently, I have measured periodic changes in tobacco root conductances measured in wild-type and anti-sense strains. In the anti-sense strains the aquaporin activity is greatly reduced (Franka Siefritz and Ralf Kaldenhoff, pers. comm.). Plants of both strains were grown in pots until they reached the flowering stage. The shoots were excised and the pots were enclosed in a root pressure bomb similar to that shown in Fig. 3.8. The rootstalk protruded through the rubber seal to the outside air. The rate of water efflux was continuously measured while a constant pressure of 300 kPa was applied to the soil mass, and water was pumped into the pot at a rate that equaled the rate of flow out of the pot in order to prevent soil dehydration. A strong periodicity of hydraulic conductance (flow rate divided by applied gas pressure) was observed in both wild-type and anti-sense tobacco roots (Fig. 6.22). This means that the reduced expression of the NtAQP1 aquaporin (Siefritz et al. unpublished) was not totally responsible for the periodicity. More work needs to be done to determine if other aquaporin proteins are responsible for some of the periodicity or if other unknown factors are contributing to the phenomenon.

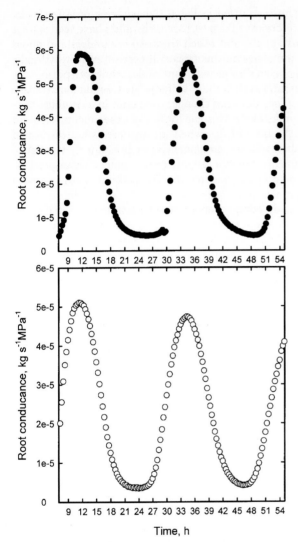

Fig. 6.22. Periodicity of the hydraulic conductance of tobacco root systems is shown. Shoots were excised from pot-grown plants and the roots were placed in a pressure bomb with the root-stock protruding to the outside air. A constant gas pressure of 0.32 MPa on the soil mass caused a periodic flow rate of water out of the soil-root system. The root conductance is the flow rate divided by 0.32 MPa. A pump added water back to the soil every 20 min to replace the exact quantity of water lost through the root system; hence the soil did not dehydrate. Over 2 days >1 kg of water passed through the root system. *Top* Roots with reduced NtAQP1 aquaporin expression. *Bottom* Wild-type roots with normal NtAQP1 expression. The roots were of different size hence conductances cannot be compared. (Tyree, Siefritz and Kaldenhoff, unpubl. data)

Observations similar to those reported in Fig. 6.22 have been repeated many times in tobacco in my laboratory over the past 18 months, and I have seen similar periodicity in roots of locust, apple, and peach trees. All we understand about hydraulic architecture and the cohesion theory is still correct even if hydraulic conductances of plant components change with time of day. However, periodicity needs to be taken into account, because these shifts in root conductance alone (assuming the shoot conductance does not change) can result in a twofold, diurnal change in whole plant conductance! Designs for future experiments will have to be rethought since now the time of day when measurements are taken will greatly influence the value observed, and no longer can we presume that a change in leaf water potential necessarily implies a proportionate change in transpiration rate, because roots can no longer be considered passive hydraulic pathways with constant conductance in every species.

The last issue for our consideration is represented by a suite of related questions: How do you define whole shoot or whole root conductance? If you define it by a method of measurement do all methods give the same result? Could the methodological definition make the whole shoot or root conductance non-constant in theory? We will confine the discussion to shoots or leaves, but most of what is said about shoots will also apply to roots.

In reality a shoot is a complex, branched catena of resistance elements in series and in parallel. In some hydraulic architecture models we have used more than 4000 resistance elements to describe a large tree (Tyree 1988). Is there such a thing as an equivalent resistance value that will measure the equivalent resistance of several hundreds to several thousands of resistance elements? In an electrical circuit there is an equivalent resistance that correctly describes a resistance network, however complicated, provided certain criteria are met and provided each resistance element has a resistance that is constant, i.e., independent of the flow of electric current. The most important criterion is that a complex circuit must begin and end at two discrete points. If that structural criterion is met then there is an equivalent resistance, regardless of how complex the branches in between these two points. If the electrical circuits have non-resistive components that store charge (capacitors) then we have to speak of an equivalent impedance. The equivalent of an electrical capacitor in a plant would be cells that store water. The equivalent impedance changes with the speed of change of current flow (e.g., the frequency of an AC current), but to make things simple let us confine ourselves to situations where there is steady state flow (DC current), which still allows us to speak of an equivalent resistance.

A shoot alone does not meet the structural criterion above, because the water flow does not begin and end at single points. Water flow enters at a point near the base but liquid water flow ends at millions or billions of points distributed throughout all the leaves in the shoot. In order to meet the structural criterion we would have to include the resistance elements of the air space inside the leaf and beyond to some imaginary point very distant from the shoot. This has two fatal disadvantages, because if we include the gas phase in the equivalent circuit: (1) The resulting equivalent resistance would be 'uninteresting' because the equivalent value would be dominated by the gas-phase resistance, and (2) the gas-phase resistances are not independent of the flow rate. In theory the resistance is inversely proportional to the water vapor concentration in the gas phase (Tyree 1999b), so the constant-resistance criterion is not met. This is not a problem in the liquid phase because water concentration of a liquid is much more constant. So we are left wondering: (1) How much does the value of an equivalent conductance depend on the methodological definition? (2) How much does the value changes with the method used to define it? (3) How 'constant' is the equivalent value in theory? If all the component resistances in a branched catena are constant then ideally the equivalent resistance, which we measure or define, should also be constant. And if the component resistances do all change by a certain percent then we would hope the equivalent resistance would change by a 'comparable' amount.

Very little theoretical work has been done to date to answer these questions. One approach to getting answers is to use a branched catena model to compute how the equivalent resistance depends on the methodological definition. I will answer these questions with just two preliminary examples. We could consider two possible methodological definitions of equivalent resistance, e.g., a HPFM method and an evaporative flux (EF) method (Tsuda and Tyree 2000). In the EF method the average evaporative flux is estimated by some means (e.g., gas-exchange cuvette or potometer) and the average leaf water potential is estimated by another (e.g., pressure bomb or psychrometer) and the equivalent resistance is equal to $R=$(average water potential drop)/(average evaporative flux).

In theory, these two methods should not produce the same resistance value. This is because the pattern of water flow in a single leaf or whole shoot is not the same during pressure perfusion and free evaporation. During pressure perfusion with the HPFM the perfusion rate exceeds the rate of evaporation in every leaf, so leaf air spaces fill up with micro-pools of water. Hence, the pressure of water at the evaporative surface in each leaf is zero, i.e., relative to atmospheric pressure. However, resistance from the base of the shoot to leaves near the base is less than the resistance from the base to more distant points and rather more water will flow over the shorter path-length than over the longer, so the HPFM will disproportionately weigh the lower (shorter) resistance pathway. During free evaporation the rate of evaporation differs from leaf to leaf; leaves in sun evaporate faster than leaves in shade and in sun-exposed leaves the rate of evaporation differs with the angle of the leaf surface relative to the sun's rays. Hence the pressure drop from the base of the plant to each leaf differs. If the EF method is applied to a single leaf and if the light intensity is uniform over the leaf surface, then the flow to each region of the leaf will be approximately the same at each point, in contrast to more flow near the base in the HPFM method.

As an example we can consider evaporation from a maize leaf, which is an extreme case because it is so long and narrow. The hydraulic architecture of maize has been worked out (Wei et al. 1999) and from the resistance network it is possible to compute the profiles of P_x in the xylem from base to tip in both the EF and HPFM methods. In the HPFM method the pressure goes from high positive values near the base to low positive values near the tip. In the EF method, the pressure is negative everywhere, but still goes from high (less negative) to low (more negative) values from base to apex, respectively. So if we plot the values as percent change from the base (100% at the base to 0% at the tip), we have different theoretical profiles and equivalent resistances that differ by about 10% (Fig. 6.23).

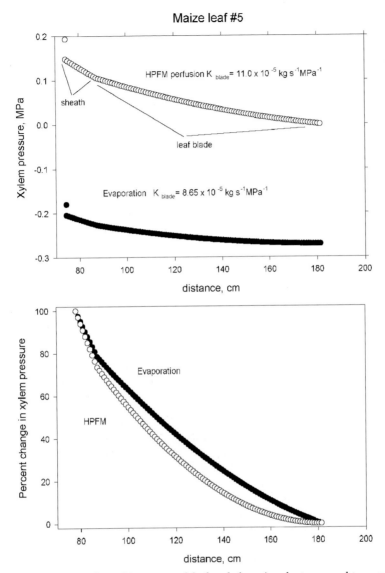

Fig. 6.23. Hydraulic architecture model of a whole maize plant was used to compute pressure profiles in the fifth leaf from the base of a large maize plant under two conditions: (1) during perfusion with a HPFM and (2) during free evaporation from the leaf surfaces. The figure at the *top* shows the absolute pressure profiles in the xylem of the leaves. The *lower curve* shows the percent change in pressure from the base to the tip of the fifth leaf. (Reproduced from Tyree et al. 2001)

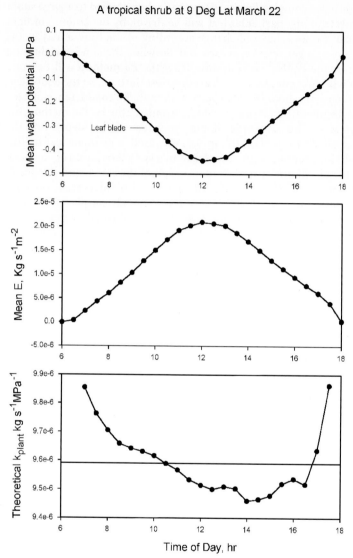

Fig. 6.24. Theoretical output of T-Plant, which is a hydraulic architecture model: T-Plant inputs information on the three-dimensional structure of a plant and then computes the light interception, photosynthesis, leaf energy budget, and evaporation of 2-mm-wide rectangular grid areas on every leaf. This information is used to compute water potential profiles in the stems, petioles and leaves of the plant at 30-min intervals based on the position of the sun with respect to the location of each leaf in the canopy and shading by leaves within the canopy. The model has been used to estimate the average evaporation rate from each leaf, average water potential of each leaf, and the theoretical conductance of the plant. Plants with symmetrical crowns, with leaves approximately equidistant from the base of the plant and with little mutual shading have theoretical conductances independent of the time of day (data not shown). Plants with asymmetrical crowns and with leaves of unequal distance from the base have theoretical stem conductances that vary a few percent with time of day. In the lower graph, $k_{plant}=-0.202E/($mean water potential$)$ where 0.202 is the leaf area in m^2 of the plant

The situation is more complex for the EF method when applied to a large shoot because the flow rate to any particular leaf will be driven by the amount of light intercepted by that leaf and that will change according to the time of day as the position of the sun changes relative to each leaf. So when the sun is brighter on more distant leaves the equivalent resistance from the EF method will be more than when the sun is most intense on the leaves nearest the base. So the equivalent conductance of the whole shoot can change even though the conductance of each component (stem segment, petiole and leaf) is constant. In the HPFM method, the shorter pathways are still over estimated, but the value will not depend on the time of day (solar position). T-Plant (Chap. 5.6) was used to estimate leaf-by-leaf differences in light interception with time of day. In a symmetrical shrub in which all leaves are evenly distributed over the outer surface of the crown and all equally distant from the base, the theoretical value of equivalent conductance is constant with time of day to within 1%. However, in a less symmetrical plant or one in which leaf distances from the base to any given leaf are unequal, the equivalent conductance is somewhat more of a function of time of day for the EF method (Fig. 6.24). However, if *every* leaf is assigned the same transpiration rate (the average value in Fig. 6.24, upper graph), then the equivalent conductance is constant regardless of time of day (horizontal line in lower graph of Fig. 6.24). The major advantage of computing equivalent conductance is the simplicity of the single value and the ease of comparing values between plants. The main disadvantage, previously unrecognized, is that the value measured might be a function of the methodology used. In most cases I have examined the 'errors' are less than 5%. Tyree et al. (2001) have presented an example where the theoretical conductance by the EF method changed by a factor of 2 for the special case of leafless stems.

7 Other Functional Adaptations

The subject matter of this book is multi-dimensionally related in so many different ways that it is difficult to accommodate it logically in a one-dimensional text representation that proceeds linearly. The organization of the subject material into chapters and sections has therefore been somewhat arbitrary. Frequent cross-references help to restore some of the multi-dimensionality. There are nevertheless topics which do not fit easily into the linear text sequence and so some of these have been gathered in this chapter under the heading, *Other Functional Adaptations*. The chapter has thus become rather a collection of odds and ends. Functional adaptation is a very attractive topic for botanists who are interested in xylem evolution, but it tempts many to walk on thin ice. Arber (1920) wrote: "One of the unfortunate results, which followed the publication of *The Origin of the Species*, was the acutely teleological turn thus given to the thoughts of biologists. On the theory that every existing organ and structure either has, or has had in the past, a special adaptive purpose and "survival value", it readily becomes a recognized habit to draw deductions as to function from structure, without checking such deductions experimentally." Too many structural features have more than a single function, and too many functions depend on several factors. Correlations become therefore easily unreliable and one begins to speculate wildly. However, there is more to it: random mutations may produce structural features that are functionally rather unimportant. It is rather futile to try to interpret such features for their adaptive value (van Steenis 1969; Baas 1976). Nevertheless, we must learn a great deal more about wood function before we can begin to speculate about adaptation. I suppose we are all entitled to a certain amount of speculation, although I normally prefer to reduce questions to basic simplicity, thus making them accessible to experimental tests. This chapter is the place where I have most often disregarded my principle.

7.1 Radial Water Movement in the Stem

Leaves of deciduous trees are attached to shoots that have been formed during the current growing season. In the case where we are dealing with low-efficiency, low-risk trees, axial water conduction in the stem goes through several growth rings. As water moves from the small absorbing roots into older roots, it must cross over into older growth rings in order to reach the leaves. Radial water movement is also required during the course of secondary growth. The onset of cambial activity proceeds quite slowly basipetally in the main stem of diffuse-porous trees and conifers. The water supply to the distal young growth layers must therefore be initially supplied through the basal older layers.

We have relatively little experimental evidence of this radial transport, although there are two structural features which obviously serve the purpose of radial water transport, the vessel (and tracheid) network and the rays. If one analyzes the network carefully, one finds primarily a tangential spread of the water path within a growth ring; this is experimentally evidenced by dye ascents (e.g., Fig. 2.4). However, there is also a very slight radial contact between successive layers of tracheids and vessels, so that over long distances water can move axially in a centrifugal and centripetal direction within the growth ring. In conifers, bordered pits can occasionally be found on tangential tracheid walls throughout the growth ring. They are particularly conspicuous when they are situated at the growth-ring border. These have been known for well over 100 years and have been discussed repeatedly (e.g., Laming and ter Welle 1971). Bosshard (1976) called them growth-ring bridges. Similarly, one often finds intervessel pit fields on the growth-ring border of secondary dicotyledonous wood, whereby the latest latewood vessels connect directly with the earliest earlywood vessels of the next growth ring (e.g., MacDougal et al. 1929, pp. 56–65). Braun (1959) measured this contact quantitatively in a poplar species and found that about 30% of the earlywood vessels at the growth-ring border have contact with the latewood of the older ring.

In addition to these axial water paths, which connect inner and outer xylem layers over long distances, the rays provide direct radial contacts. Certain conifers are well known for their ray tracheids, and we can assume that in this case radial movement of water is relatively efficiently provided for. Ray vessels are very rare in dicotyledons; they have been described for a few species only (e.g., Chattaway 1948; van Vliet 1976; Botosso and Gomes 1982). Nevertheless, ray transport of water seems to be easily possible, even in the absence of special conducting cells. This becomes obvious as one observes dye movement in the xylem: water appears to spread radially very easily along rays, probably through the walls (Fig. 5.1, below). Volume flow must be very small through any one ray, but sufficient if one considers the large number of rays available along the length of the tree stem.

In the terminal 2 years of growth of a shoot one can experimentally show the path of water from last year's xylem of the 2-year-old twigs into this year's xylem of this year's twig. One merely has to inject a dye in the backward direction, i.e., basipetally from the current shoot. The dye will then move back into both growth rings of the 2-year-old shoot, and the path can be studied by cinematographic analysis. Somewhat surprisingly, but indeed quite logically, one finds only a very small amount of 2-year-old xylem at the tip of the 2-year-old shoot. This is quite logical because this xylem had to supply only the terminal leaves of last year's shoot. The path of water from the tip end of this 2-year-old xylem leads axially directly into this year's xylem of this year's shoot. This is very difficult to describe in words, although easy to see in a film. Figure 7.1 (top) may help to clarify the situation.

Although we know where the water flows, we have little quantitative information about volume flow of water through individual rings and from ring to ring. Radial conductivity measurements have been made many times (e.g., Huber 1956), but these do not take into account movement resulting from the radial deviation of the axial path, which is probably very important.

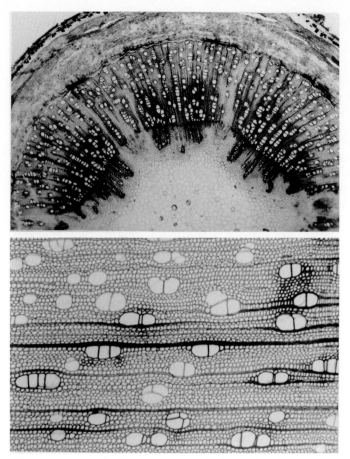

Fig. 7.1. *Top* A 2-year-old twig was perfused with dye (gentian violet) in the basipetal direction from the current-year distal end to show the path of water from last summer's growth ring of the 2-year-old twig into this year's ring of the current shoot to which the leaves are attached. This transverse section was made of the distal end of the 2-year-old axis. Heavy staining of the primary and early secondary xylem of last year's growth ring shows the direct axial path across the growth-ring border. *Bottom* A transverse section through the stem of *Acer rubrum*. A dye (gentian violet) has been pulled up into the wood through a few vessels (which can be recognized by their dark-appearing walls) and thus has been forced to move into other vessels. Dye has also moved radially through rays as indicated by their dark appearance

In arborescent monocotyledons without secondary growth, radial channels are provided by the metaxylem vessels of the leaf traces that run more-or-less radially within the stem, and their continuation in the continuing axial bundle and via bridges (Fig. 2.5). However, in those monocotyledons that have a secondary xylem, radial paths are unknown and will have to be found experimentally, if they exist at all (Tomlinson and Zimmermann 1969; Zimmermann and Tomlinson 1970).

7.2 Xylem Structure in Different Parts of the Tree

There is a voluminous literature about the dimensions of xylem elements in different parts of trees. It begins with Nehemiah Grew (1641–1712) who noted that root wood vessels are generally wider than those of the trunk (Baas 1982b). Systematic investigations of this topic probably began with Sanio's (1872) classic work, describing his "five laws". This is not the place for a comprehensive review; our interest concerns specifically the relation of structure and function. However, it is very important that variations within an individual tree are respected. Variations of structure, not only within a species, but also those within an individual plant, have made, and continue to make, wood identifications difficult. For example, every wood anatomist knows that root wood cannot be identified with a standard key: even branchwood may cause considerable difficulties. Paleobotanists had, and undoubtedly still have, the same problem. Taxa of extinct plants are usually established on the basis of relatively small plant parts. It therefore happened that different genera were described on the basis of wood fragments from different parts of the same species (Bailey 1953).

Variations in anatomical characteristics from roots to twigs in an individual tree make it extremely dangerous to set up correlations of certain xylem features (such as vessel diameter) with habitat and draw conclusions about functional adaptations when one has only a random sample of wood from each species. Wood anatomy varies so much throughout a specimen that ecological trends based on vessel diameters, densities, etc., make sense only if the anatomy of the compared species is known precisely throughout the individual plants.

The distribution of tracheid dimensions in coniferous trees has been summarized briefly, but thoroughly, by I.W. Bailey (1958). The situation can be illustrated with two of his graphs. Tracheids become longer and wider with increasing age (diameter) of a stem (Fig. 7.2). There is also a distinct increase in length and width in the basipetal direction from twigs to branches, to stem and finally to roots (Fig. 7.3). The two statements are of course related, except that the large tracheid

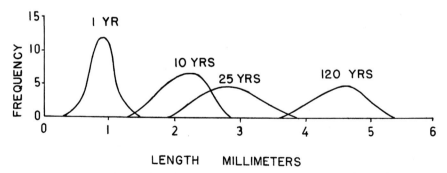

Fig. 7.2. Variations in length of tracheids with increasing age or diameter of a coniferous stem. In such a stem the fusiform initials of the cambium are of nearly equivalent length. Frequency distributions based on measurement of 100 tracheids. (Redrawn from Bailey 1958)

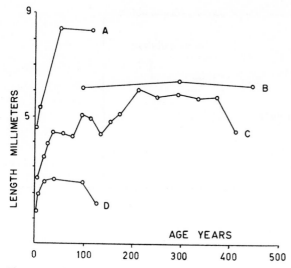

Fig. 7.3. Variations in average length of tracheids in stems and roots of *Sequoia sempervirens*. *A* Root 6.4 cm in diameter and 120 years old. *B* Root 30 cm in diameter and 450 years old. *C* Stem 1.5 m in diameter and 420 years old, at stump height. *D* Branch 4.55 cm in diameter and 130 years old, from the crown of a huge, old tree. Averages based on measurements of 100 tracheids. (Redrawn from Bailey 1958)

size in roots is not strictly related to age, and certainly not to root diameter. Rundel and Stecker (1977) measured tracheid dimensions and xylem pressures in young branches at different heights in a 90-m-tall *Sequoiadendron giganteum*. Pressures followed the hydrostatic slope quite closely, and tracheid diameters were linearly correlated with pressure. Tracheid length, on the other hand, correlated very poorly with either diameter or pressure.

In woody dicotyledons, vessel diameters increase the basipetal direction from the top down (Fig. 5.14); they are usually greatest in the roots. Although this was known to the earliest botanists, there are not nearly as many systematic investigations of dicotyledons as there are of conifers. The earliest systematic report known to me is a study of comparative anatomy of branch, trunk, and root wood of tree species of northeastern North America by Fegel (1941). Not only do vessel diameters increase basipetally from branches and down along the trunk, they often continue to increase in the roots with increasing distance from the trunk. This observation was reported by von Mohl (Riedl 1937), confirmed many times, and quantitatively represented by Fahn (1964, Fig. 3). However, roots cannot always be included in such a statement; for example it is important that one distinguishes between the horizontal, rope-like conducting roots with their wide vessels, and the vertical taproots which serve primarily as storage organs (Riedl 1937).

Vessel diameters as well as vessel lengths increase in the basipetal direction. Figure 7.4 shows the basipetal increase in vessel size of red maple. Even within the trunk there is a very distinct basipetal gradient of increasing vessel diameter and length (Zimmermann and Potter 1982, Fig. 2). At 11 m height, the longest length

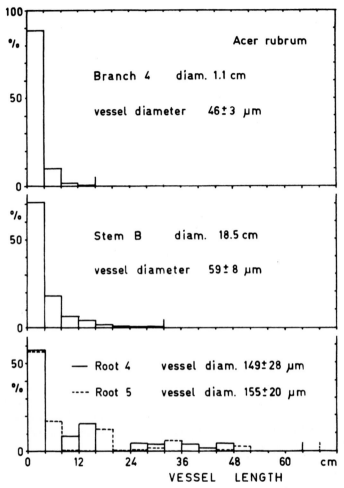

Fig. 7.4. Vessel dimensions in a branch, the main stem and two roots of red maple (*Acer rubrum*). Vessel lengths and diameters (average and SD given) increase in the basipetal direction. (Zimmermann and Potter 1982)

was 20 cm and about 93% of the vessels were in the shortest length class (0–4 cm), and, at the base of the trunk, the longest length had increased to 30 cm and only 54% of the vessels were in the 0–4 cm length class.

Plant anatomists often measure vessel-element length in order to study "Baileyan trends" (Fig. 1.1). Tracheids of vessel-less woods are very long; length is obviously an hydraulic advantage. However, as soon as vessels evolved, vessel elements shortened, because the hydraulic length requirement, previously filled by a single cell, was now filled by a cell series. In plants with vessels, cambial initial length is independent of hydraulic requirements.

I cannot hold with Carlquist's (1975) argument that shorter vessel elements are an adaptation to greater tensions because they make vessels mechanically stronger and thus prevent collapse. A little added wall thickness does this too well, and there are too many other factors involved. Vessel-element length, it seems to me, is merely a by-product, e.g., the result of cambial initial length, which might be under some other control, perhaps fiber length, although fibers can also elongate by intrusive growth. The variation in vessel-element length seems to me a perfect example of what Baas (1982b) calls "functionless trends imposed by correlative restraints".

Hydraulically then, vessel-less trees like conifers, as well as vessel-containing trees, are constructed in the same way: the conductance of individual compartments decreases acropetally from roots to leaves. The compartments are large and fewer in number at the base of the tree; they become smaller and more numerous in the acropetal direction, i.e., they become safer (Chap. 1.4). This construction appears to be a perfect adaptation to prevailing pressure gradients. Pressures are always higher at the base of the tree and always decrease acropetally; the danger of embolism therefore increases with height. Exceptions to this rule are probably very rare and restricted to brief periods after heavy rains. It is not surprising that the very largest vessel diameters (0.6–0.7 mm) found anywhere in the plant kingdom have been found in the conducting roots of wide-vessel species (Jenik 1978). Klotz (1978) found that in wide metaxylem elements of palms, the bars of the scalariform perforation plates are most closely spaced in leaves, intermediate in the stem, and farthest apart in the roots. The plates are also least oblique in the roots. From a functional point of view it is reasonable to propose that vessels originated in roots, stems, and leaves in succession during evolution (Cheadle 1953).

Vessels are narrow and short in seedlings. This has been shown dramatically in ring-porous trees. In American elm, e.g., 1-m-tall seedlings had a longest vessel length of only 4 cm, and 91% of the vessels were in the shortest (0–1 cm) length class. In a 1.5-m-tall seedling, longest vessels length had increased to 26 cm and the percentage of vessels in the shortest length class (0–2 cm) had dropped to 70% (Newbanks et al. 1983). In mature elms, on the other hand, vessel lengths range over many meters. As a dicotyledonous tree grows in height, its water-conducting system becomes more efficient from the base up. This is the same trend as the one found in conifers (Fig. 5.2).

Palms do not have secondary growth; in this case, the xylem of the basal part of the stem must be made efficient enough for the mature plant from the beginning. The xylem of a short (young) palm stem is therefore over-efficient (Chap. 5.8). But the distribution of efficiency and safety features is shown here in another way. As we have seen in Chapter 2.2, the longest vessels are in the stem center, because this is where the long (major) axial bundles are found. Vessels at the stem periphery are shorter, because this is where the shorter (minor) axial bundles and the vast network of bridges are. The peripheral xylem near the stem surface is therefore a safe network of vessels with many alternate pathways that can lead around possible injuries.

Finally, let us remember vessel distribution within individual growth rings at any one height. We have already seen that vessel width and length in dicotyledonous stems of temperate regions show a consistent decrease from early- to late-wood (Fig. 1.10). This can perhaps be regarded as a built-in safety feature in temperate angiosperms, just as in present-day temperate conifers there is a certain specialization of early- and latewood. Earlywood is primarily conducting xylem, and latewood serves a more mechanical function (e.g., Bailey 1958; Bosshard 1976).

7.3 Wall Sculptures and Scalariform Perforation Plates

We have seen at the very beginning of this volume (Chap. 1.1) that the presence of scalariform perforation plates indicates an intermediate stage of vessel evolution. In many cases, the intermediate character is quite obvious. In palms one finds all stages of transition from protoxylem tracheids to the wide metaxylem vessels that still have scalariform perforation plates (Klotz 1978; Zimmermann and Sperry 1983). There are other cases, e.g., species of the genus *Betula*, where the xylem has consistently maintained the scalariform perforation plates (Fig. 7.5, lower right). It looks then as though the adaptive advantage of keeping them is greater than the advantage (namely decrease of flow resistance) of getting rid of them. What could be the nature of this advantage?

The function of scalariform perforation plates may be to catch bubbles when the ice in the vessels thaws at the end of the winter. It may have this function in common with other wall irregularities such as spiral thickenings (illustrated, e.g., by Meylan and Butterfield 1978b). When Bailey (1933) investigated vestured pits (Chap. 1.5), he occasionally found vestures extending from the inside of the pit border to the inside of the vessel wall. These wall outgrowths have been studied in some detail by electron microscopy and are commonly called warts (Frey-Wyssling et al. 1955/1956; Côté and Day 1962; Liese and Ledbetter 1963). As will be discussed in the next chapter (Chap. 8.2), freezing displaces dissolved air and produces bubbles. These bubbles are present in the freshly thawed xylem water and present a grave danger. They must be dissolved before tensions appear in the xylem. The smaller the bubbles, the more easily and therefore faster they will dissolve. Plants that experience winters with freezing temperatures are in danger of suffering vapor blockage when such bubbles expand because of dropping xylem pressure when transpiration begins. One of the greatest dangers during the time of thawing is the coalescence of bubbles. Simple arithmetic tells us that eight coalescing spherical bubbles will form a single one twice the diameter. This not only drops the tensile strength of water to one-half (Chap. 3.4), but it also will take a very much longer time for the bubble to dissolve. Not only does this single bubble have only half the surface area of the eight small ones, more importantly, the air is concentrated in one spot. It is therefore a very great advantage to keep bubbles separated upon thawing.

A few years ago I observed, with a horizontally mounted microscope, the melting of ice in thick radial sections of birch. As soon as bubbles were liberated from ice they began to rise. However, the slightest obstacle in their path caught them; the scalariform perforation plate was a particularly effective trap. These experiments have since been expanded by Ewers (1985). In model experiments he froze air-saturated water in glass capillaries of different diameter. He then let them thaw

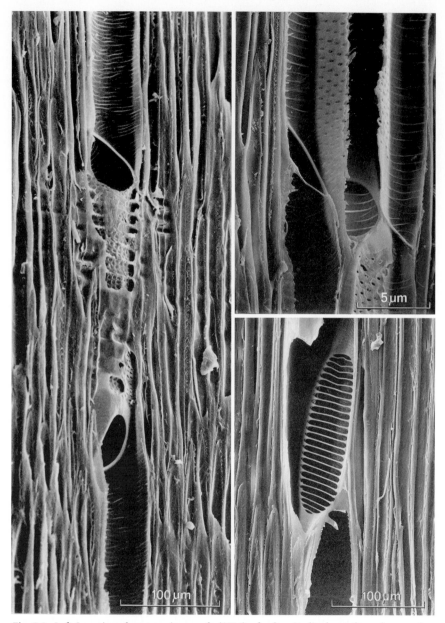

Fig. 7.5. *Left* Scanning electron micrograph (SEM) of a longitudinal cut through secondary wood of red maple (*Acer rubrum*). A vessel has been cut in such a way that the central element remained intact but the elements above and below have been cut open. *Upper right* SEM of a longitudinal cut of red maple wood, showing the spiral thickenings inside the vessel walls. *Lower right* SEM showing the scalariform perforation plate between two vessel elements in birch (*Betula* sp.)

in the horizontal position and examined them periodically with a microscope. Even when the capillaries remained in the horizontal position, the bubbles in larger capillaries fused and did not redissolve after several days when the smaller bubbles in the narrow capillaries had already dissolved. In additional experiments, Ewers (pers. comm.) froze pieces of wood that had been perfused with air-saturated water. He observed the upper transverse-sectional surface of the wood during thawing with the piece in an upright position. Observation was carried out with water immersion objectives and epi-illumination. In our local diffuse-porous species (of northern New England), no bubbles could be seen to emerge at the cut surface, while, in ring-porous (i.e., large-vessel) species like ash, large, obviously coalesced, bubbles appeared at the upper surface.

It makes no sense to argue that certain habitats require higher flow rates than others and thereby exert a selection pressure that eliminates scalariform perforation plates. Flow rates depend on too many factors other than perforation plates (vessel width and length, effectiveness of intervessel pit, number of functioning vessels per leaf surface area, transpiration rates, etc.) to permit such a simplistic assumption. However, Baas (1976) reported that the percentage of genera with scalariform perforation plates increases with altitude and latitude. This would support the notion that scalariform perforation plates have a function (such as bubble catching) in cold climates. A similar trend was found for vessels with spiral thickenings (Fig. 7.5, left and upper right; van der Graaff and Baas 1974; Carlquist 1975) which might well have a similar function. The thin spiral thickenings in some secondary woods are also very effective in strengthening vessel walls without increasing flow resistance excessively (Jeje and Zimmermann 1979).

7.4 Aquatic Angiosperms

Aquatic angiosperms are perhaps somewhat comparable to whales: they returned to the water, taking with them some features of terrestrial organisms. In whales, it appears to have been easier to modify the breathing system for a periodic return to the surface than to redevelop gills. In angiosperms, the root system had developed into such an important secretary organ in the mind of Kursanov (1957) that it was easier to maintain the xylem transport system than to have the shoot system regain functions that had been taken over by the roots. There is an enormous amount of literature about aquatic plants; however, there are fortunately two quite thorough summaries, books which are not only informative, but also make delightful reading (Arber 1920; Sculthorpe 1967). Furthermore, a reference volume on the anatomy of the Helobiae has been published (Tomlinson 1982). We are here concerned only with questions related to xylem transport.

By 1861, Unger had already shown water transport from roots to shoots in two aquatic species (*Potamogeton crispus* and *Ranunculus fluitans*). The upper and lower parts of the plants were placed in different containers while the central part was protected from drying by a bent glass tube. Unger recorded an increase of water volume in the upper container, but not when the roots were removed from the plant. The relatively high flow rates recorded by Unger could not be obtained by Wieler (1893), but Wieler himself demonstrated root pressure in many aquatics by observing bleeding after cutting off a shoot, raising the cut end just above the water level and enclosing the space above the cut end to maintain a humid atmosphere. von Minden (1899) observed guttation from leaves of aquatics that were slightly raised above the water surface. When he wiped the water droplets from the water pores of the leaves, they quickly reappeared. Gardiner (1883) confirms Sachs' finding that root pressure depends on soil temperature. The phenomenon of guttation was already known at that time. Duchartre (1858) had discovered earlier that land plants, whose aerial parts were enclosed in a humid atmosphere, guttated water from leaves. By 1900 there was still a lively discussion in progress as to whether or not there was water transport in aquatic angiosperms; many prominent botanists were involved. As a further milestone one can perhaps regard the paper by Snell (1908), a student of Goebel, who, in many careful experiments, established the roots as essential for nutrient uptake in many aquatics. When the roots were permitted to penetrate soil or sand, they developed root hairs. Even the floating plant *Pistia stratioides* (Araceae) needs roots for proper growth. In two genera of the Lemnaceae, however, Snell (1908) found that roots were merely balancing organs, preventing the plants from flipping over. Here, water and mineral uptake was taking place primarily through the underside of the leaves. Snell (1908) had plants take up potassium ferrocyanide, which he could later detect in the vascular tissue by the Prussian blue reaction. The reported advance of ferrocyanide was of the order of 50 cm/day. It had thus been quite firmly established at the beginning of the twentieth century that roots and transport from roots to shoots are essential to growth of some aquatic angiosperms. The modern literature confirms this. Mantai and Newton (1982) found that the roots are very important for growth in *Myriophyllum spicatum*. They even reported that root growth was stimulated by a low mineral nutrient content of the water. Waisel et al. (1982), on the other hand, found other species that are less dependent on the roots for mineral uptake. However, there are also certain hormonal factors, such as cytokinins, which originate in the roots and must move via the xylem to the shoot (see Bristow (1975 and the literature cited therein).

The question of the driving force of the xylem stream in aquatic plants is rather interesting; transpiration is out of the question in totally submerged plants. Root pressure is obviously the most likely candidate: it is suggested by the observation of bleeding from cut tops by Wieler (1893) and others. For example, Thut (1932) tested a number of species and showed water movement of around 10 μl h^{-1} per plant with roots; this figure dropped to one-tenth when the roots were removed. A driving force that can operate even in the absence of roots is a side effect of phloem transport. Münch's (1930) pressure-flow hypothesis, which is still quite widely accepted today as the best explanation of phloem transport, requires that water be taken up osmotically into sieve tubes at the "source" and where photosynthesis takes place (i.e., in the leaves), and water be lost in "sinks", where translocated solutes are removed from the sieve tubes for storage or growth. This water returns to leaves via the xylem. Münch (1930) estimated it to be about 5% of the transpiration stream in land plants (for a discussion of this, see Zimmermann and Brown 1971, p. 263). This component of xylem transport is independent of transpiration and could provide a significant portion of xylem water in aquatics. Thoday and Sykes (1909), working with submerged shoots of *Potamogeton lucens* in situ, carefully cut shoots and inserted them into vials containing eosin. The eosin solution was taken up and moved rapidly (up to 9.5 cm min^{-1}). When leaves were removed, movement of eosin dropped to 6 cm h^{-1}. As they were dealing with cut shoots, root pressure is out of question as a driving force, but ascent of water to the top due to sieve-tube loading would be a quite reasonable explanation of xylem transport in this case.

Thus, the literature leaves no doubt that xylem transport from roots to shoots is significant and essential in many aquatic angiosperms, although individual species differ in this. What concerns us most here is the structure of the xylem. The xylem of aquatic angiosperms is commonly considered to be reduced. However, to call it "degenerate" is certainly misleading, because it is obvious that the structural requirements of a xylem that operates exclusively with positive pressures are fundamentally different from those of a xylem in which pressures are consistently, or periodically, negative. Xylem in which pressures are always positive should be constructed somewhat like resin ducts: it should consist of canals lined by living cells. Indeed, in perennially submerged plants, we often find such xylem ducts (Fig. 7.6, right), while, in plants whose tops are able to emerge from the water, "conventional", thick-walled xylem cells are present (Fig. 7.6, left).

While some of the fundamental questions concerning xylem transport have been reasonably answered, others have remained open and represent a fascinating area for further research: First, what is the three-dimensional structure of the xylem? Have all compartment barriers disappeared, i.e., is the entire xylem system a single, complex cavity? This sounds very unreasonable, but we have little information about this point. There is some indication that these ducts are not continuous across nodal areas.

Second, and related to the first question, is the problem of safety. If xylem translocates are of vital importance to the plant, how does the plant cope with injuries?

Xylem under positive pressure leaks when injured. The problem of scaling presents itself even in the absence of injury. There are plants that grow actively in front while they die away behind (Arber 1920). The xylem of the rear end must therefore be sealed off. What is this sealing mechanism? Rodger (1933) investigated wound healing in submerged plants. She found cell divisions (wound callus formation), cells growing into air ducts and thickening of cell walls. Sealing has also been observed when a gummy substance plugs the hydathodes of older leaves, thus directing the xylem stream to younger leaves (Wilson 1947). We must assume that cell walls are sealed off against injuries so that the xylem leak is plugged.

Third, intercellular spaces of land plants provide not only some water-storage space, but also air ducts (Chap. 4.8). In the case of frequent or perennial positive xylem pressures these ducts get flooded unless there is a special seal such as an endodermis or a bundle sheath that divides the apoplast into a water and an air compartment. In the case of aquatic plants, the question then arises how the air-duct system, which is very well developed, is separated from the xylem. Where is this seal? How are the cells of the air-duct system supplied with root nutrients? In analogy to the situation in leaves and roots of land plants, one would expect xylem contact via living cells and a seal (a Casparian strip) between the vascular tissue and the air-duct system. On the other hand, we could assume that air pressure in

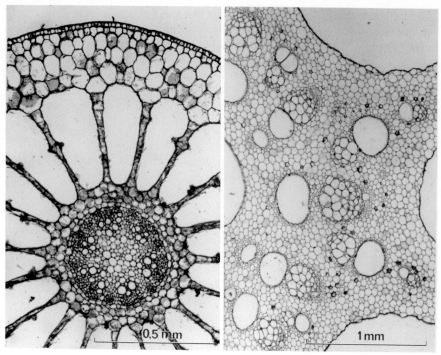

Fig. 7.6. *Left: Myriophyllum brasiliense,* transverse section of a submerged stem, showing large air ducts around the stele. The vascular tissue contains small, thick-walled tracheary elements. Emerging stems, when present, have a very similar structure. *Right: Nelumbo lutea,* transverse section of a stem. The large spaces at the *upper* and *lower right* and in the *center* on both sides, all partly cut off, are air ducts. Within the stem tissue are large vascular bundles with xylem and phloem. Most of these xylem ducts appear wall-less, but some have thin walls. (e.g., the one on the *upper left*)

the air-duct system exceeds water pressure in the xylem apoplast. This could work as long as the plant is uninjured. In the case of injury to the air system, however, the air would escape and the air system would fill with water. Many aquatics subdivide their air-duct system by water-repellent diaphragms that contain small pores that permit passage of air but not water (see also Frey-Wyssling 1982).

Finally, there is the interesting case of similar xylem ducts in land plants; the "carinal" canals in *Equisetum* (e.g., Bierhorst 1958) and the protoxylem lacunae in monocotyledons. Dye movement experiments have shown that these canals can transport water (e.g., Buchholz 1921). In land plants, the questions about function are as follows: if these canals are continuous with the xylem, and there is hardly doubt that this is the case, they could at least function as water-storage areas. The first question which then arises is by what size pores are they connected to the intercellular air system. If the lacunae are tightly sealed off, the size of any air pocket in them will have to follow the gas equation. In the case of positive pressure the air pocket is compressed and water-storage space is thus provided. If the lacunae are connected to the air system by larger pores, they should behave like (injured) vessels. It is unlikely that lacunae can stand much tension without collapse, although in the case of *Equisetum* dye transport at below-atmospheric pressures has been shown to occur in carinal canals (Bierhorst 1958). However, the connections to the air system could be dimensioned in such a way that air seeding takes place at pressures that do not yet cause collapse (Chap. 4.2).

8 Failure and "Senescence" of Xylem Function

8.1 Embolism

In his review of xylem conduction, Huber (1956) wrote: "we are virtually certain that the pressures of the moving water columns in the xylem are negative most of the time, but we are equally certain that such negative pressures cannot be maintained forever. Sooner or later water columns will break. Tracheary elements are thus filled with gas, probably irreversibly. This loss of water from the conducting channels is the beginning of a chain of events which we call heartwood formation" (free translation from the German).

In conifers, water loss from the xylem is relatively easily recorded as water-content measurements. Figure 8.1 shows how the total wood volume of different growth rings is made up of cell wall material, bound water, free water, and "air". "Air" is written in quotation marks here because its quantitative composition may differ from atmospheric air, i.e., it is likely to be much richer in CO_2 and correspondingly lower in O_2 because of respiration. Free water, i.e., water in tracheids, decreases in successively older layers of wood as the number of embolized tracheids increases. Figure 8.1 represents a textbook example of water content, the heartwood is here relatively dry, i.e., most tracheids are embolized. Dry zones, sometimes described as pathological heartwood, can also appear in the sapwood as a result of fungal infection (Coutts 1976). This is undoubtedly due to premature embolism: fungal action makes tracheids leaky so that they embolize. On the other hand, heartwood occasionally has a rather high water content, and is then called wetwood. This special case is discussed in Chapter 9.1.

Although vapor-blocked tracheary compartments are very difficult to demonstrate, our concept (Fig. 4.10) has not changed from that illustrated by Dixon (1914, his Figs. 17 and 18). Water transport through vessels of relatively transparent herbs has been observed directly with fluorescent dyes (Strugger 1940; Rouschal 1941). Direct observation of water transport is more difficult in woody plants. One normally uses dyes to mark the track, but, in doing this, the dye can also be drawn in by water-vapor-filled vessels (Preston 1952). This objection holds only if the dye is injected at atmospheric pressure, because the "vacuum"-filled vessels pull in dye that is administered at atmospheric pressure. Embolized vessels can be kept free of dye if dye is applied under vacuum. We shall have to come back to this problem in Chapter 9.4. The vacuum method is difficult to use on entire trees. Dye ascent experiments stain cell walls only, not the tracheary lumen. One can often distinguish water-filled from vapor-blocked vessels by quick freezing. A cardboard box is mounted around a tree stem and filled with dry-ice. When the stem is entirely frozen, it is cut with a bow saw right across the dry-ice-containing

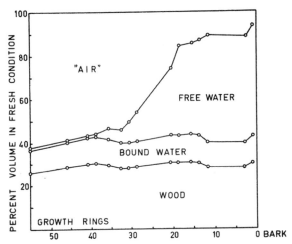

Fig. 8.1. Distribution of wall substance and water in growth rings of a 55-year-old spruce. The portion labeled *air* may be air-enriched in CO_2 and somewhat depleted in O_2 because of respiration. Tracheids have been gradually embolized over a period of 40 years following their formation. Note that in this representation *distances between curves* (not from abscissa to curve) show amounts. (Trendelenburg 1939)

box, thus exposing a frozen transverse section, which is kept cold, cleaned with a razor blade, and observed with a stereo-microscope. Ice-filled vessels can be recognized, especially if they are large (Zimmermann 1965). More recently, cryo-scanning electron microscopes have come into common usage and it is possible to obtain dramatic photographs of air- or water(ice)-filled vessels or to observe vessels with mixtures of water and air (Canny 1997; Tyree et al. 1999a).

Embolism has been discussed a number of times in other sections of this volume; we need not elaborate further upon it. However, it is perhaps good to emphasize that there are many different gas compartments in the xylem of a tree stem. Intercellular spaces provide the regular air-duct system (Chaps. 4.8 and 7.4). Parts of these may be tightly sealed from each other as discussed before. Tracheary compartments may be in contact with the air-duct system if they are badly injured. If they are merely air-seeded, they contain gas under a different, probably much lower, pressure immediately after air-seeding, but would later be filled with gases at atmospheric pressure. Such gas spaces are isolated. There has been relatively little interest in the relation of water and gas space in woody stems, and this may well be a very fruitful area for research. For example, in many large palms, the ground parenchyma of the (primary) stem expands considerably causing the stem to increase in diameter after the vascular tissue has matured (Schoute 1912; Fig. 8.2). The parenchyma then forms a rather loose network of cells, which criss-crosses what looks like an enormous volume of air space. It would be interesting to see if this space is required for respiration of the ground parenchyma, which, even in a very thick palm stem, must remain functional even in the center. Does it, under conditions of positive xylem pressure, get water-filled, or is it sealed from vascular tissues by bundle sheaths to prevent "suffocation"?

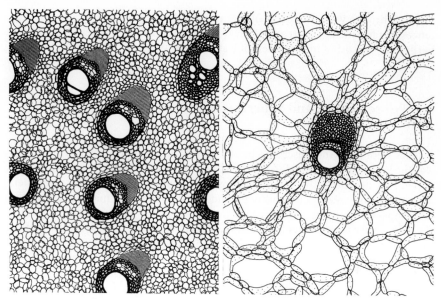

Fig. 8.2. Expanding tissue ("secondary thickening") in the stem of the royal palm. Suitable material from a single specimen was unfortunately not available, but would show about the same thing. *Left* Stem transverse section immediately below the crown. Ground parenchyma of the stem is still quite closely spaced, and the fibers of the vascular bundles are still extremely thin walled (*Roystonea elata*). *Right* Stem transverse section in the older and wider basal part of the stem. Ground parenchyma lacunose, fibers of vascular bundles thick-walled (*Royystonea regia*)

Our north-temperature ring-porous tree species (*Quercus, Fraxinus, Castanea, Ulmus*, etc.) are peculiar in that their earlywood vessels are so large that cavitation takes place at least by the end of the winter. In *Quercus alba* and *Q. rubra*, we know that the embolism occurs on the first freezing event (Cochard and Tyree 1990) and we presume this is the case for most species with wide vessels. The result is that most water is conducted in the wide earlywood vessels of the most recently formed growth ring. This means that these trees have to produce the earlywood vessels in the spring before the leaves emerge (Coster 1927; Priestley et al. 1933, 1935; Huber 1935). The wide vessels are so efficient that those of a single growth ring can supply the entire crown with water (Chap. 1.4). Ascent of sap during early spring while the new set of earlywood vessels is formed is via the small latewood vessels, which remain functional for several years. Their capacity is sufficient to supply the leafless crown, but transpiration from leaves must await the formation of the wide earlywood vessels. However, these generalizations may be subject to change with further research. For example, sap flow measurements have revealed water flow in multiple annual rings of some ring-porous species (Granier et al. 1994) and removal of the outer annual rings of *Quercus macrocarpa* (bur oak) by saw cuts does not immediately interrupt the normal water balance of leaves for 3 years (Tyree, pers. observ.), until the roots have died because of the interruption of carbohydrate transport through the bark. The answer might be that water conduction through wood tracheids in *Quercus* spp. might be quite efficient over distances of a few meters so trees can survive without the large diameter vessels, but perhaps at the cost of higher levels of water stress.

8.2 Tyloses, Gums, and Suberization

Tyloses were seen and described by the very earliest microscopists. However, it was the Viennese baroness Hermine von Reichenbach (1845) who, in a remarkable paper, first described the origin of tyloses and who gave these structures their name. Her study was such a thorough and careful one that it makes many of the later papers about tyloses look redundant. Although she published anonymously, she was remembered for many decades. Nevertheless, she finally was forgotten. The rediscovery of her identity and her splendid work involved some interesting detective work (Zimmermann 1979).

Reichenbach (1845) described tyloses in many species, including some tropical plants. An English summary of her paper is given in Zimmermann (1979) where her 24 original drawings are also reproduced. She discovered that tyloses grow into the vessel lumen from neighboring parenchyma cells (Fig. 8.3A,B). They are real cells, their nuclei can often be seen suspended by the cytoplasm; when neighboring cells touch each other, they form pit pairs (Fig. 8.3C). Their cell walls often thicken secondarily, which makes the pit pairs more conspicuous. They often contain starch (Fig. 8.3D). They behave in other ways like ordinary living cells, e.g., they can be plasmolyzed (Fig. 8.3E) and the cytoplasm often shows streaming. Reichenbach's developmental studies are equally remarkable. In *Robinia pseudoacacia* she observed that all of the current year's vessels were free of tyloses and all older ones were filled with them. Tylosis formation in the current year's vessels began in October (in Austria) and was more or less completed in December. The exact same timing was reported recently for the same species from Kyoto, Japan, by Fujita et al. (1978). Reichenbach (1845) concluded that tyloses form when the vessels are gas-filled, because they appeared in the fall after cessation of water conduction! They remained always in contact with their mother cells, because they could not get water or nutrients from the vessels.

Böhm (1867) showed that tyloses were produced regularly in connection with wounding. They always appeared in the living wood bordering dead wood, either near injuries or heartwood. He found that tyloses represent a very effective seal; they did not let water or air pass through vessels at his experimental pressures of 0.1–0.3 MPa. Gum production had the same sealing effect in smaller cells. He therefore concluded that tyloses and gums provide the plant with a mechanism to seal off living tissue against injuries and other dead parts. Even Reichenbach (1845) had recognized the significance of gums: when she traced them in serial sections, she always found them to originate from injuries.

The sealing function of tyloses and gums was confirmed in many subsequent papers. When both these mechanisms were present in a single plant, tyloses were usually restricted to the larger, gums to the smaller cells (Praël 1888). There is much literature about tylosis formation in wood, in connection with leaf abscis-

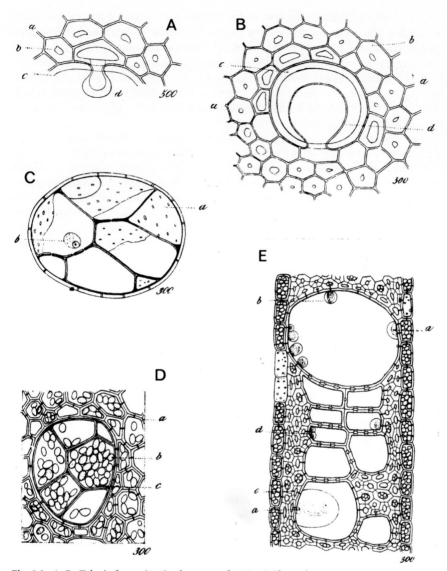

Fig. 8.3. A, B Tylosis formation in the stem of *Vitis vinifera*. The sections were treated with KOH: *a* primary cell wall; *b* secondary cell wall of wood parenchyma; *c* secondary vessel wall; *d* young tyloses with their mother cells. **C** Transverse section through a vessel of *Robinia pseudoacacia*: *a* walls of tyloses with pits in surface view; *b* nuclei. **D** Tyloses in *Vitis vinifera*, stained with iodine: *a* wood parenchyma cells; *b* tyloses containing starch; *c* vessel. **E** Vessels of *Robinia pseudoacacia* with young tyloses; *a* plasmolyzed cytoplasm. (Reichenbach 1845)

sion, etc., which need not concern us any further here. Many of these papers indicated that tyloses and gums are the means by which vessels that had lost their water-conducting capacity were sealed off from functioning tissues. Tyloses are the only means in some plants, in others it is gums only, and in still others it is both. Chattaway (1949) found that in the many Australian species she investigated, parenchyma-to-vessel pits had to be at least 10 μm wide to permit tylosis formation. Furthermore, in her samples (from 1100 genera!), tyloses and gums originated almost exclusively from ray cells.

The question what triggers tylosis formation was specifically addressed in a very detailed paper by Klein (1923). He repeated many of the earlier experiments of Wieler (1888) and others, and found that tyloses always appeared near wounds like drying cracks or cuts. His experiments were carried out with woody and herbaceous species, with stems, roots and aerial roots. Tyloses did not form as a response to the wound stimulus alone, or as a response to the higher oxygen concentration of atmospheric air, or to the cessation of water movement, but always as a response to loss of vessel water. Rarely, he found a few wide earlywood vessels in *Robinia* to be free of tyloses for a second summer, while the narrow latewood vessels were usually free of tyloses for several summers. Klein's (1923) findings remained available for many years while generations of plant physiology students repeated one of his experiments as a routine laboratory exercise. A Y-shaped piece of stem was turned upside down, one of its legs dipped into water, the other left in air, and suction applied at the morphological lower end. Tylosis formation could be observed in the half of the piece through which air had been drawn. The half through which water was drawn remained free of tyloses. Much of this older work was forgotten, and, in the most recent literature, we find the notion that tyloses may be plugging the vessels, thus preventing water transport, particularly as a result of infection. Such statements have no experimental support as far as I know (see Chap. 9.3).

Another misconception has crept into the recent literature, namely that tyloses are "ballooning" into vessels, implying pressure differences as a causal mechanism. Nothing could be farther from the truth. The pressure difference between the inside of a living parenchyma cell and the adjacent vessel lumen is greatest when the vessel is conducting, because the living cell is turgescent (i.e., has an internal positive pressure) while the vessel lumen may be under considerable negative pressure. When the vessel is embolized, the pressure in its lumen is between zero and 100 kPa, while the cell wall surrounding the living cell still contains water under tension, i.e., the turgor remains the same. In other words, the pressure difference between the living parenchyma cell and the vessel lumen is very considerably greater while the vessel is functioning and tyloses do not form. We do not know how the absence of water on the vessel side stimulates the living parenchyma to grow into the vessel.

Electron microscopy made the study of ultrastructural details of tylosis formation possible. Readers interested in this are referred to the papers by Kórán and Côté (1965), Meyer and Côté (1968), and those of Shibata et al. (1981, earlier papers cited therein).

The gums mentioned above are produced by paratracheal parenchyma cells, primarily ray cells (Chattaway 1949), following stimulation such as cavitation or injury of vessels. However, there are many species, coniferous as well as dicotyledons that have special resin or gum ducts (Esau 1965; Fahn 1974). When stem tissue is injured, these resins or gums are released from the ducts, thereby soaking neighboring tissues and sealing them off effectively.

There is a growing literature on the physiological stimulation of gummosis. Readers interested in this topic may refer to one of these papers and trace the literature from there (e.g., Olien and Bukovac 1982). The problem that is hardly ever addressed, however, is the question of how these stimuli are linked to cavitation. If cavitation comes first, then conductance measurements are in vain (Chap. 9.4).

Tyloses and gums are means by which the plant seals off injured xylem that has become useless and could serve as entryways for infections. One or both of these mechanisms exist in many plants, but some plants have neither of them. However, the injured tissue is not entirely isolated from the functioning tissue until the entire apoplast, including the cell walls, is sealed off. This is accomplished by suberization, lignification or other forms of plugging of all cell walls that lead to injury (Chap. 9.5). This seal is effective, like those of the endodermis in the root and leaf, but often broader in extent (not only one cell layer thick). It has been known for a long time and studied particularly in connection with abscission (Fahn 1974, p 276; Addicott 1978). For plants that do not form tyloses or gums, it is the only seal. Some of these topics will need to be taken up again in the next chapter.

8.3 Heartwood Formation

The light outer sapwood is the living xylem, most recently produced by the cambium; it functions in water conduction and storage of starch and other substances. Tree trunks often have a dark core, the heartwood. This is the oldest, dead part of the wood. There are trees in which heartwood is always darker than sapwood and clearly recognizable; there are others in which there is a functional change, but no color change. In still others, the color change is facultative. Finally there are species that show no color change and a very slow decline of physiological activity over time (Bosshard 1974). In some representatives of the first group with regular, darkly colored heartwood, there are tropical members in which the living xylem parenchyma cells survive for a very long time, so that the heartwood occupies a relatively thin core. Some very desirable types of wood, such as ebony (*Diospyros ebenum*) and letterwood (*Piratinera guianensis*), have this characteristic. This makes their desirable dark heartwood rather precious, because even a large tree yields relatively little. The opposite is usually the case in ring-porous species of the north-temperate regions where the sapwood layer may be rather thin. This again indicates that loss of vessel water is probably the first step in heartwood formation.

Heartwood formation is the normal consequence of xylem senescence. Respiration and other measurable parameters of living processes decline in the sapwood (Ziegler 1967). Between sap- and heartwood a transition zone can usually be recognized (Chattaway 1952; Shain and Mackay 1973). The dark coloration of "heartwood" is due to the deposition of lignins and polyphenols (Bauch et al. 1974; Hillis 1977) and may or may not involve centripetal transport in the rays, depending on species (Frey-Wyssling 1959; Höll 1975). Readers interested in heartwood formation are referred to the above references. What interests us here is its relationship to water transport.

The percentage of xylem that remains functional over the years can be seen most easily in conifers (Fig. 8.1). The free water here normally represents mostly water in the lumen of the tracheids, i.e., movable water. Water content measurements of course also include extracellular water; this has been discussed in Chapter 4.8.

In species which have vessels the matter is more complex. This can be illustrated best with the example of the north-temperate ring-porous species. As previously discussed, these conduct water primarily in the large earlywood vessels of the most recent growth ring. The volume of this movable water is relatively small, because flow velocities are very high (Chap. 3.8). When water content measurements are made with pieces from successively deeper layers within ring-porous wood, the movable water is likely to be missed. The outermost layer, containing the most recent growth ring, should have the highest water content, but it usually does not. The water is very likely to be lost in sampling if the xylem is under stress when the tree is felled, and the water withdraws immediately from the cut surface. What is then recorded is (partly) the content of the latewood vessels, tracheids (if present), fibers, etc. The problem is, of course, the same in all vessel-containing wood types; it is less severe if there is little stress at the time of sampling, if the vessels are small, and if the samples are collected fast.

It has been said that the loss of water is the first step in heartwood formation. This is probably the reason why roots have in general very little heartwood and why proportionally more heartwood is often present higher up in the tree (Bosshard 1955; Ziegler 1968). Stresses are always least in the roots, greater higher up in the tree, and water loss by cavitation must therefore be greater at the top. In some species a process called 'cavitation fatigue' contributes to permanent xylem dysfunction that can be considered a function of heatwood formation. In *Populus tremuloides*, xylem >1 year old is pathologically vulnerable, i.e., it is always embolized in nature and much more vulnerable than the more recent xylem. This has been attributed to degradation in the pit membranes. In a number of species, this degradation can be attributed to repeated cycles of cavitation (Hacke et al. 2001).

A final feature of interest in connection with heartwood formation is the status of the coniferous bordered pits. The discussion in Chapter 4.4 indicates that the coniferous bordered pits close when the pressure drop across them becomes excessive, i.e., when the water in one of the two contiguous tracheids cavitates. What we do not know is whether the pit membrane is flexible enough to regain its original position when the other tracheid also becomes vapor-filled, or if the membrane remains stuck permanently to the pit border (Comstock and Côté 1968). In other words, when we look at bordered pits with the microscope, we are not certain whether pit membranes in their normal central position (such as shown in Fig. 4.10, left) have always been in that position, i.e., that both tracheids have been water filled at the time of sampling. One-sided embolism may have displaced the membrane, and later embolism of the neighboring tracheid may have permitted the membrane to regain its original central position. We might speculate that recovery is possible if the displacement is not maintained for a long time. In the sapwood-heartwood transition zone, displaced membranes are common.

Unilateral embolism may have existed for a long time and the membrane may have finally been "glued" into position by incrustation with heartwood substances. The presence of incrusted membranes in their original, central position may suggest that this is the case. Krahmer and Côté (1963) reviewed much of the older literature about this matter and described and illustrated their own observations. Pit closure and incrustation of the pit membranes are of considerable practical importance, because they decrease the longitudinal permeability of coniferous heartwood quite drastically, thus making permeation of wood with preservative liquids difficult. The permeability of heartwood in dicotyledons is, of course, also much lower than that of the sapwood, because of tyloses and gums.

9 Pathology of the Xylem

This chapter is probably the most difficult one for me to write, because, in doing so, I have to step outside my field of expertise more often than I like. Yet, I feel that it is necessary to build a bridge between those of us who are concerned with xylem structure and the ascent of sap, and those plant pathologists who are interested in xylem dysfunction. I hope that my colleagues in plant pathology will not take offense for my "meddling in their affairs", but that instead this will be the beginning of a fruitful collaboration. The two groups can certainly learn a good deal from each other. I also hope that I shall hear from plant pathologists, in order to learn where my concepts are wrong and my knowledge is incomplete. This chapter is not a review of pathological disturbance of water relations in general. Such reviews are available elsewhere (e.g., Ayres 1978). There are also many aspects of xylem dysfunction that are important after conduction has ceased; these are also outside the area of interest of this book. We are concerned here only with the ascent of sap and its disturbance.

9.1 Wetwood Formation

Figure 8.1 shows the "normal" distribution of the water content of coniferous wood of increasing age within a standing tree trunk. The water content is highest in the most recent growth ring and decreases toward the stem center. Such graphs are often somewhat irregular due to what we might call biological scatter. In addition, as we discussed in Chapter 8.1, in ring-porous trees we very likely fail to record the full water content of the most recent, conducting growth ring. The result is that graphs show only the scatter of non-functional wood and therefore look somewhat confusing.

In a normal tree, the heartwood is the driest part of the xylem because the lumen of most of the tracheary elements is vapor-blocked. However, there is a condition, common in some species, rare in others, whereby the water content of the heartwood is as great, or even greater, than that of the sapwood. The heartwood is then referred to as wetwood. It may contain liquid under positive pressure while in the sapwood the transpiration stream moves along a gradient of negative pressures. Why is the water of the central wet core not drawn into the sapwood? In order to develop positive pressure, the heartwood must be sealed within the tree trunk. If this seal, whatever its origin, is semipermeable, and if the solute content of the central core is sufficiently great, the heartwood space could develop a positive pressure by drawing water by osmosis from the sapwood, until the osmotic equation is balanced. Another possibility is an impermeable seal, effective as a Casparian strip. In this case, the solute content of the wetwood would be unimportant, but there would have to be an internal source of water and of pressure. Water could be derived by microbial breakdown of wall substance and pressure could be produced by internal (microbial) gas production. Let us now turn from these theoretical speculations to reality. This will not be a thorough review of wetwood formation; readers interested in this are referred to the review by Hartley et al. (1961) and Murdoch's (1979) bibliographic list of publications about wetwood formation. Our only concern here will be the physiological problem outlined above.

Wetwood has been known to forestry for a long time; it occurs in a number of coniferous and dicotyledonous species and the mechanism of its formation does not have to be the same in every case. Wetwood-water relations are not easy to understand, because most of the papers that have been published about it concern pathology, wood quality, effects on wood processing, etc. and are relatively unconcerned about physiology. Furthermore, I can certainly not claim to have read or even seen all the literature on this topic. Nevertheless, this is an attempt to explain how the high water content (and pressure) in the heartwood can be maintained next to the negative pressures in the sapwood.

Let us look first at American elm (*Ulmus americana*), where wetwood is very common. Wetwood becomes evident when it is exposed by a broken-off branch, frost crack, or by deep holes made for purposes such as fungicide injection. A liquid then emerges from the hole and may flow out for a long period of time, running down along the trunk. This liquid is often secondarily infected and may produce unsightly streaks along the trunk. If excess liquid is taken up by the sapwood, the liquid moves up into the crown and may cause damage or even kill the tree.

The osmotic potential of wetwood liquid is quite considerable. Murdoch (1981) reported an average value of –1.47 MPa (a range of –0.66 to –2.32 MPa) for American elm, while xylem sap has an osmotic potential of only a fraction of an MPa. He obtained an average value of –0.05 MPa for sapwood "expressate"; the value of "expressate" must be greater than that of xylem sap, because it includes cell sap from parenchyma. At any rate, the osmotic potential difference between sap- and heartwood would permit a pressure buildup in the heartwood core, at least during periods of lesser xylem tensions, provided there is a semipermeable "membrane" between the two.

Wetwood often contains gas under pressure. Murdoch (1981) recorded pressures from 0 (i.e., atmospheric) during the winter, to about 0.1 MPa during the summer. Other authors have reported pressures up to 0.4 MPa (Carter 1945). While the gas content of sapwood is similar to that of the atmosphere, with possibly a somewhat lower oxygen and a much higher carbon dioxide content, wetwood may contain up to 50% methane (Hartley et al. 1961; Murdoch 1981). If the high water content of wetwood is maintained osmotically, gas pressure would simply have a cushioning effect. As osmosis in this system may work rather slowly, the gas pressure may reflect the daily average of the wetwood sap pressure. In fact, the entire system is so extensive that it may never reach real equilibrium.

The wetwood area of elm appears to act like a single, giant osmotic cell that is separated by a semipermeable "membrane" from the sapwood area. This can be visualized something like a Traube membrane, as early plant physiologists called it.

In 1867, Traube described how semipermeable membranes can be prepared by precipitating a semipermeable substance, such as copper ferrocyanide. This can be done within the walls of a porous shell such as a clay container, thus providing a rigid osmotic cell that does not fail when pressure builds up. Much of the early work on osmosis was conducted with such cells. We can assume that the microorganisms of the wetwood cause the deposition of semipermeable material within the wall and intercellular spaces, thus surrounding the wetwood area with a Traube membrane. In American elm, Murdoch (1981) isolated and identified 14 species of bacteria and two species of yeasts from wetwood tissue. We can speculate that one or more of these produce the seal. Its chemical nature is not known, but the presence of a pressure difference between wetwood and sapwood makes it a necessity.

Wetwood in fir (*Abies*) appears to be at least superficially different. The liquid content of the heartwood core is not much greater than that of the sapwood, and the two are separated by a dry zone (Bauch 1973). However, the water content curve from sapwood to the adjacent dry zone looks quite normal, i.e., is comparable to that of sapwood and a normal transition zone, as shown in Fig. 8.1. However, instead of remaining dry, the heartwood regains a high water content. The seal we are looking for must therefore be between the dry zone and the wetwood. Bauch (1973) made measurements of radial hydraulic conductance and found a decrease from sapwood to the dry zone, and a drop to zero in the wetwood. Coutts and Rishbeth (1977) reported osmotic potentials of wetwood of –0.3 to –0.5 MPa, and Worrall and Parmeter (1982) –0.53 to –0.76 MPa. This is enough to keep the moisture content relatively high, but perhaps not always enough to have the trunk bleed when punctured. These authors also found that wetwood formation in fir did not have to involve microorganisms as it could be induced aseptically. The "seal" seems to be a product of the plant itself, it did not form below a girdle. Another rather peculiar finding of Coutts and Rishbeth (1977) was that latewood areas sealed off as wet areas first.

In conclusion, we can say that the evidence supporting the osmotic concept is very strong. There must be a Traube membrane around the wetwood area and it must be permeable to water diffusion, otherwise wetwood could not keep bleeding for days or weeks. Solute content does make osmotic pressure possible and the system often appears to be cushioned by gas pressure. Permeability must be very low otherwise diurnal changes of wetwood sap (or gas) pressures would have to be recorded. It is rather ironic that a wound in the wetwood area, which bleeds liquid for a long period of time, thus appears to have the transpiration stream as a source of water, in spite of the fact that the pressure of the transpiration stream is negative most of the time!

9.2 Movement of Pathogens

In order to move in the xylem, a pathogen first has to enter. This usually happens through an injury. A rather unusual situation exists in the case of viruses. Although normally phloem-mobile, they can move freely through the xylem from vessel to vessel if their diameter is smaller than the intervessel pit membrane pores, ca. 25 nm depending on plant species. If it is small enough, it can move through all cell walls with the apoplast water and thus reach the surface (plasma-lemma) of any living cell; we shall come back to this later. Larger organisms have to dissolve enough of the cell wall to get in. The problem of entering a vessel containing water under tension is comparable to the problem of introducing an object into a container which is under positive pressure. Any injury of the container wall carries with it the danger of pressure loss. If a bacterium thus enters a vessel containing tensile water (water at negative pressure), it almost necessarily has to cause air seeding, whereby it is flushed either distally or basipetally to the end of the vessel with the retreating water. Subsequent entries are, of course, also possible; the conditions for these can be varied. In the case of large enough multiple injuries, rainwater can enter broken vessels by capillarity.

Before further discussing entry of fungal hyphae into functioning vessels, let us consider how far a pathogen could proceed under the most "favorable" circumstances. The vessel-length calculations, explained in Chapter 1.2 (Figs. 1.3), are relevant here; the distance to which vessels are incapacitated upon injury is discussed also in Chapter 2.5. The number of injured vessels decreases very sharply as one moves away from an injury (Fig. 2.12). If a spore or bacterial suspension is inoculated via a fresh wound, the inoculant concentration diminishes with distance from the point of entry in a very similar way (e.g., Figs. 1-3 in Suhayda and Goodman 1981). Inoculant concentration also reflects three-dimensional xylem distribution in other ways, e.g., shows constrictions at nodes (Chap. 5.7; Pomerleau and Mehran 1966). The maximum distance a pathogen moves as a result of inoculation depends on vessel length. One has to be fully aware of the vast differences in distribution patterns in small shoots, seedlings or, at the other extreme, trunks of ring-porous trees (Chap. 7.2). It is interesting that elm varieties with narrower vessels are more resistant to Dutch elm disease (e.g., Elgersma 1970). We know now that narrower vessels are also shorter (Chap. 1.2). Smaller vessels limit the extent of inoculation.

The initial inoculation may be bidirectional from the point of injury. For example, fungal spores introduced into the vessels of a stem by the feeding activity of a beetle may be flushed up and down in the injured vessels. Downward movement, i.e., movement against the normal direction of the transpiration stream, must always be considered "inoculation movement" in damaged vessels, not movement in the undisturbed transpiration stream. When a vessel is opened up by an injury, the xylem water at the injury is exposed to ambient pressure (+101 kPa). This means that there is now a pressure gradient away from the injury in both directions, up and down. This can flush the spores either up or down in the broken vessel until they reach the vessel end. The broken vessel is now permanently out of function and upward movement of water resumes around it.

The most important barriers to xylem movement of any particle after the initial inoculation are the intervessel pit membranes. These have been described in detail in Chapter 1.5 and their effectiveness is illustrated in Fig. 1.4, which shows paint-particle penetration. The pores in intervessel pits are of the order of 25 nm; this is very small indeed (ca. 1/20 of the wavelength of visible light!). Once a pathogen has been flushed into an injured vessel, it can move passively in the transpiration stream only if it is small enough to get through intervessel pit membranes. Such movement must be strictly with the transpiration stream, i.e., under normal conditions in the general direction from roots to leaves, or into other transpiring organs such as flowers. Pores in intervessel pit membranes are so small that virus particles are the only candidates for unrestricted movement in the transpiration stream, and only the smaller types. Schneider and Worley (1959) published a report about the southern bean mosaic virus that can be interpreted in terms of movement across intervessel pits. They estimated the size of the virus particle to be of the order of 30 nm, i.e., very close to what we think is the size of the pores of intervessel pit membranes. They demonstrated xylem mobility by its movement across steam girdles (which blocks the phloem path). The interesting aspect of Schneider and Worley's paper in this context is the fact that they refer to a systemic host (*Phaseolus vulgaris* var. Black Valentine) and to a local lesion host (*P. vulgaris* var. Pinto). It would be interesting to see whether Black Valentine has slightly larger intervessel pit pores than Pinto, so that the virus is freely mobile in the xylem of the former, but not of the latter bean variety (see also Valverde and Fulton 1982).

Many of the pathogens that spread through the xylem, as fungal spores or bacteria, are very much larger than intervessel pit membrane pores. These organisms can only move from one vessel into the next by destructive action, e.g., by enzymatic dissolution of the vessel wall, probably the pit membrane, the weakest part of the wall. This kind of movement is not passive as in the case of virus particle movement described above, but it is an active penetration from vessel to vessel.

The interesting question now arises of whether some pathogens are able to enter vessels without causing cavitation, and can thus spread with the transpiration stream. If this is the case, it is a remarkable adaptation. We do know that certain insects ("sharpshooters") have accomplished the feat: some of them not only penetrate, with their mouth parts, vessels containing water under tension without cavitation, but seem to have no difficulties pumping water out (Mittler 1967). As the insects insert their mouth parts they salivate a polymer that rapidly sets; this allows the insect to suck sap out without causing a cavitation. Curiously, the insect salivates during the exit so we might presume that the vessel remains unembolized.

This brings us to the question whether fungal hyphae can penetrate vessel walls, produce conidia, and have the conidiospores float with the transpiration stream to the vessel ends, germinate and repeat the process, thus accomplishing distribution in the transpiration stream? This question is a very important one, and we do not know the answer. Any evidence I have seen is at best circumstantial. An important criterion has already been mentioned. Movement of spores with the transpiration stream *as well as against it* (i.e., downward in a tree) is evidence for a break of the water column at the point of entry. In Dutch elm disease, for example, upward movement of spores of *Ceratocystis ulmi* is faster than downward movement (Campana 1978). This might indicate not immediate, but somewhat delayed air seeding. It is known that the fungi penetrate cell walls often with microhyphae, which dissolve the cell wall locally (e.g., Chou and Levi 1971), a mechanism that might accomplish a tight seal. The port of entry must remain sealed long enough to enable the fungus to produce conidiospores in order to have them released into the transpiration stream. It is likely that penetrating hyphae continue to dissolve cell wall material (after all, they have to do this in order to exist and grow), and the moment of air seeding, if not immediate, could be somewhat delayed. It must be remembered that a crack of the order of 0. 1 µm wide is sufficient to admit air under stress conditions. This is well below the resolution of the light microscope! To look for the first inoculation injury with the electron microscope is like looking for a needle in a haystack.

Electron microscope inspection of wall penetration is interesting. However, when looking at such pictures, we never know whether we see the first or the fiftieth penetration. During the first penetration, the pit membrane may be under enormous physical stress, and tear if weakened. However, if the water column in the vessel has already been broken, the pressure difference is not very great, i.e., somewhere between 0 and 97.7 kPa (the difference between atmospheric and water-vapor pressure). Any hyphal penetration into a vessel whose water content has already cavitated could therefore look "intact". We must conclude from this brief discussion that fungal hyphae *might* penetrate functioning vessels, but evidence in support of this is no more than circumstantial. Furthermore, it would be important to know how long a tight seal can be maintained after entry. We should expect considerable differences between species. It is very difficult to test for embolism experimentally; we shall have to come back to this problem in the section on flow resistance (Sect. 9.4).

Bacteria are very much larger than intervessel pit pores, and also larger than the pore size required to admit air under ordinary xylem tensions. It is practically unavoidable that the first bacterium penetrating a vessel must produce a hole in the vessel wall (pit membrane) large enough to induce cavitation. The entering bacterium would then be swept either up or down to the end of the vessel where it would lodge against the next vessel-to-vessel pit membrane, multiply in the humid space of the vapor-filled vessel, and eventually enter the next vessel destructively. *Xanthomonas campestris*, a bacterium causing black rot lesions in cabbage (*Brassica oleracea*), is an interesting example. It enters the hydathodes and moves basipetally in the veins, i.e., against the normal xylem flow. It has been reported to plug xylem vessels (Sutton and Williams 1970), but plugging must be secondary, because the plugging material could not move against the transpiration stream in the xylem. In other words, the bacteria must have moved basipetally first by the gradient reversal caused by air seeding, and gum plugging would then be secondary. In the case of herbaceous plants, we must consider the possibility of repeated refilling of damaged vessels at night if xylem pressures are positive, and re-emptying during the day, when pressures drop below atmospheric.

When multiplying, some bacteria form tight clumps (R.N. Goodman, pers. comm.). Therefore, if a bacterium is flushed against a pit membrane, it may multiply and plug the entire pit cavity before dissolving the pit membrane. If parts of this clump grow into the next vessel and finally detach, they could be carried in the transpiration stream without causing cavitation. This is the only mechanism by which I would visualize spread of bacteria in the intact transpiration stream. The matter obviously needs to be investigated experimentally.

As far as distribution is concerned, the question of whether embolism takes place at the time of entry may appear to be merely an academic one. However, it becomes more important when we discuss the cause of increased flow resistance, i.e., the cause of the plant's ultimate failure and death. Embolism at the time of pathogen entry puts the vessel out of function immediately, and any visible later blockage must be considered secondary (Sect. 9.4).

Finally, in order to move in the xylem, a pathogen has to be able to exist in xylem sap, i.e., in an osmotic potential of a fraction of an atmosphere. A protoplast containing a vacuole could possibly not contain a turgor pressure in the absence of a cell wall.

9.3 Effect of Disease on Supply and Demand of Xylem Water

It should be quite clear by now that a drop in xylem pressure below a critical level causes cavitation and normally puts the xylem out of function permanently. There do, however, appear to be instances where xylem refills while under negative pressure (Chap. 4.7) and it will be interesting to see if this observation "stands the test of time". The cause of the negative pressure inducing cavitation can be either a failing supply of water to the xylem by the roots, or excessive demand by transpiration, i.e., either hydrostatic or hydrodynamic pressure, respectively.

Insufficient supply of water to the xylem, such as a prolonged drought, can cavitate xylem water, interrupt the water supply to leaves, and thereby kill the plant. The water supply to the stem can also be curtailed by failures of the root system, e.g., by damage caused by root parasites such as nematodes, etc.

Xylem pressure can also be dropped below the point of cavitation by excessive transpirational demand for water. The regulation of transpiration is rather complex, it is influenced directly by the water potential in the leaf (the hydropassive mechanism of regulation): stomata close when the water potential in the leaf drops below a certain level. However, if the plant is sensitized, stomatal closure is induced hormonally (by abscisic acid); the plant can thus be drought-adapted (Raschke 1975). Abscisic acid can induce permanent closure and leaf senescence. It has been reported that other chemicals have the same effect (Thimann and Satler 1979). Senescence-inducing substances have to be present in mere hormonal concentrations to be effective. They do not have to destroy the leaf directly, but, by closing stomata permanently, the leaf cannot carry on photosynthesis and is thus put out of function. Infection may cause a senescence-inducing substance to be produced at any point where it can enter the transpiration stream, be it along the stem, or in the roots. It may be a metabolite produced by a pathogen, or by the host plant under the influence of a pathogen. In this case, the plant does not die from a failure of the xylem, but from lack of photosynthesis, although yellow, drooping leaves may simulate wilting. Lethal yellowing of coconut palms (and other palm species) is an example, although the death of the plant is not caused only by leaf senescence, but also by phloem invasion by the pathogen. Lethal yellowing is a mycoplasma disease that appears to be transmitted to leaves by a leafhopper (Howard et al. 1979; Howard and Thomas 1980). The mycoplasma-like organisms multiply in the phloem and are transported to phloem sinks, such as roots, young leaves and inflorescences, and fruits. They can be found in great numbers in the sieve tubes of young inflorescences just proximal to necrotic tips (Parthasarathy 1974; McCoy 1978; Thomas 1979). Premature abscission of young fruits is the first visible symptom (phloem failure); at a later stage of the disease, mature leaves turn yellow and die. However, yellowing is only a very late expression of leaf senescence. The beginning of senescence can be detected at least 2 weeks before fruits drop (the first visible symptom) by the consistently high xylem pressure in mature leaves, indicating stomatal closure (McDonough and Zimmermann 1979). In other words, even though the yellow leaves droop, lethal yellowing is not a wilt disease.

There are many diseases that result in yellowing leaves. It would be interesting to check diurnal xylem pressures in these and compare them with healthy controls, in order to see whether stomata are closed and thus the yellowing leaves are senescent. If this is the case, then xylem pressure measurements would indicate stomatal closure long before yellowing becomes obvious.

Under the heading, *Shoot Diseases and Foliar Diseases,* Ayres (1978) reviewed the effect of certain pathogen metabolites, such as fusicoccin, that open stomata. This causes transpiration in excess of available water and may not only cause wilting, but possibly also cavitation, i.e., permanent blockage of xylem.

9.4 Xylem Blockage

This section concerns xylem failure due to excessive flow resistance. It should not be regarded as a review of wilt diseases because there are many aspects of wilt diseases, such as permeability changes in living leaf cells, that have nothing to do with xylem structure and function. Readers are referred to the appropriate literature for this (e.g., Dimond 1970; Ayres 1978; Talboys 1978). The main purpose of this section is to point out an important aspect of xylem blockage that has received very little attention in the pathology literature: vapor blockage. In any healthy plant most tracheids and vessels eventually go out of function by cavitation, probably air seeding (see Chaps. 3.4, 4.2, and 7.1). Water columns under tension are so vulnerable that cavitation is a likely result of any kind of disturbance, be it mechanical stress (e.g., by the force of wind), bubble formation by freezing, microbial wall degradation, etc. From a physiological point of view, cavitation is the first failure that has to be *assumed.* Penetration of a functional element that does not involve cavitation has to be regarded as a remarkable achievement by a microorganism that must be *proved.*

The probability of cavitation during inoculation has already been discussed in Section 9.2. What has not yet been discussed is how the problem can be investigated. A vapor-blocked section of a plant stem has an increased flow resistance, as beautifully explained by Scholander in a number of papers (e.g., Scholander 1958, Figs. 1-2 and 1-4; see also Fig. 6.3). In the case of Scholander's experiments, the pressure drop in the xylem distal to the blockage was produced by air intake at the cut end or by air seeding, caused by freezing. The question now arises how we can distinguish between cavitation and physical plugging caused by hyphae, bacteria, or some high-molecular solute.

In experimental tests, the problem is very easily missed. First, wilting of cuttings whose xylem has been supplied with a solution of a "toxin" (e.g., Ipsen and Abul-Haji 1982) certainly does not give us an answer; just about any suspension or solution of a high-molecular substance will plug the xylem simply because the pores in the intervessel pit membranes are so small (ca. 25 nm!) and the investigator cuts the xylem first! Second, let us assume that the xylem path of a plant has suffered extensive embolization. As stem xylem is normally quite overefficient, wilting may not be immediately evident. If we now cut the stem and put the plant into a dye solution, the vapor-blocked vessels will refill as soon as the cut end of the shoot is brought into contact with the dye solution, which is at atmospheric pressure. Even when oak branches are cut in the air and allowed to transpire in full sunlight for 20 min, the stems can be put into water and the leaves will rehydrate especially in the low-light conditions of a laboratory (Tyree, pers. observ.). Capillarity is presumed sufficient to draw water far enough into embolized vessel to make contact with uncut vessels. Neither the dye pattern nor the transpiration rate shows the blockage. In other words, all these procedures do not show conclusively damage to xylem function that has taken place by cavitation. Dye movements under ordinary atmospheric conditions in the laboratory are therefore of limited value. Measurement of percent loss of hydraulic conductance on excised stem segments would provide some information on properly designed experiments (Chap. 4.3). When a conductance decrease finally does become evident, it very likely is the result of much later, secondary plugging.

In the late nineteenth century, many botanists described transpiration experiments, whereby the cut end of a shoot had to take up water from a container in which air pressure was lowered by a vacuum pump; air pressure can, of course, only be lowered to slightly above the vapor pressure of water (2.3 kPa), otherwise the water would boil. However, this is low enough to remove most of the water from vessels in which cavitation has taken place. Ursprung (1913) found that while even a clay wick continued to evaporate water with undiminished intensity when vacuum was applied to the container from which it was supplied with water, branches of *Robinia pseudoacacia* wilted. This situation is actually almost exactly identical to that of picking flowers in the garden or field, and putting them later into a vase. When one severs a flower (or a branch) from the intact plant, transpiration continues, thus withdrawing all water from the cut vessels into which, after some time, air is drawn through the cut end. From that time on, water has to come out of storage (Chap. 4.8), stomata close, and if one waits long enough before putting the flowers into a vase, wilting will begin. As soon as the cut ends are put into water, submersed cell walls take up water by capillarity and pass it on to the nearest intact tracheids and vessels. If the capacity of this apoplast pathway to the nearest, distal intact xylem is sufficient, the flower (or branch) will not wilt. If it is not sufficient to support transpiration, the flower (or branch) wilts. The depth to which the cut end of the stem is immersed is very important (see Fig. 2.12). Most flowers survive the ordeal of picking and embolism at the cut end, but it is well known that some do not (*Anemone nemorosa* of the western European forests, branches of *Fraxinus* sp., etc.).

Transpiration from a vacuum container is an excellent tool to test for embolized vessels. Figure 9.1 shows the set-up that was used by the early botanists, but it has been elaborated in several ways to serve our purpose. First, in order to maintain a reasonable vacuum even in the presence of small leaks, it is desirable to run the vacuum pump periodically throughout the experiment. A timer turns the pump on at intervals. A valve between the system and the pump is opened after the pump is turned on and closed before the pump is turned off to avoid vacuum

Fig. 9.1. Experimental arrangement to allow a cutting to transpire while taking up water (or a dye solution) from a container under vacuum (reduced pressure). Certain plants, such as a ring-porous tree species, require that the entire root system of a seedling be immersed. If the experiment lasts for a long time, the vacuum pump is run intermittently; a timer closes the valve before the pump is shut off, and opens it after the pump has been turned on

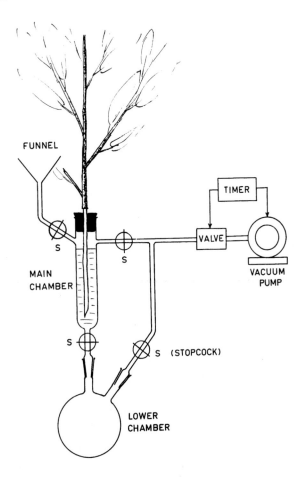

FUNNEL

TIMER

VALVE

S

S

MAIN
CHAMBER

VACUUM
PUMP

S

S   (STOPCOCK)

LOWER
CHAMBER

loss via the pump during its rest periods. Second, the vacuum system includes a lower chamber that permits drainage of the main chamber, which contains the plant. Liquid can thus be changed (i.e., successive liquids can be offered to the plant) without release of the vacuum.

In order to test the effect of a disease on cavitation, the following procedure was adopted (Newbanks et al. 1983). A branch or seedling of the test plant was infected by puncturing the stem with a hypodermic syringe whose needle had had its tip cut off. A spore suspension was injected into the xylem; control stems received water. After successive periods of time, branches or seedlings were harvested, fitted tightly into the vacuum container and the pump was turned on. We must assume that soon after the vacuum is established, transpiration draws water from all vessels, which are cut at the basal end of the branch, as well as from all vessels that are ruptured by the inoculation needle. The main chamber was then filled carefully from the funnel with solutions of a basic dye. The best results (i.e., the sharpest dye tracks) were obtained with Schiff's reagent. The plant was first given

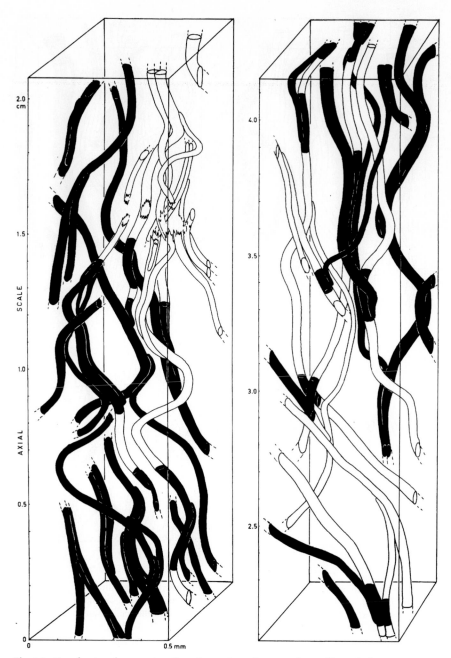

Fig. 9.2. Vessel network reconstruction of a section of a stem of a seedling of *Ulmus americana*. The block on the *right* belongs on the top of the one on the *left*, as the axial scale indicates. This is a control stem that has been punctured and injected with distilled water. The puncture can be recognized by the torn vessels at 1.5-cm height of the axial scale. All functional vessels are dye stained throughout their entire length (shown in *black*). Injured vessels were air-blocked and did not conduct dye solution. They appear unstained, except where their walls were stained by contact with conducting vessels. Note that the axial scale is foreshortened about ten times

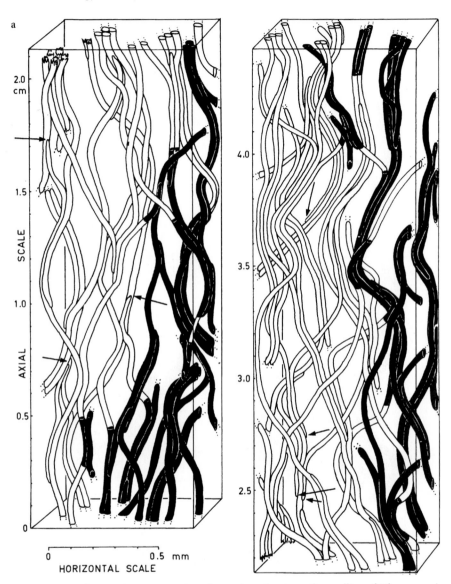

Fig. 9.3 a. Vessel network reconstruction of a section of a stem of a seedling of *Ulmus americana*, injected at 2.1 cm on the axial scale with a spore suspension of the Dutch elm disease fungus *Ceratocystis ulmi*, 4 days before harvest of the stem. Many of the vessels (marked by *arrows*) beyond the injection injury have already become non-conducting. As absolutely no microscopic evidence, such as fungal hyphae, etc., was visible, it must be assumed that the fungus damaged vessel walls enough to cause cavitation. The four pieces shown are all a single block of wood. From **a** and **b**, each piece on the right belongs on top of the piece on the left, as the axial scale indicates. Note that the axial scale is foreshortened about ten times

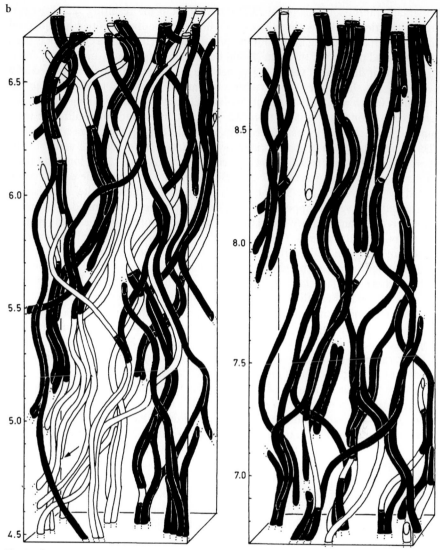

Fig. 9.3 b.

periodic acid for an hour or two; the main chamber was then drained via the lower chamber and Schiff's reagent introduced via a funnel into the main chamber, without ever losing the vacuum. After a period of uptake of Schiff's reagent, the main chamber was drained again without loss of vacuum and the plant left under vacuum until wilting. This precaution assures that embolized vessels are not refilled by water-vapor pressure when the vacuum is finally released. Serial sections were made of the stem xylem, and the vessel network reconstructed by cinematographic analysis (Chap. 2.1). As we look at the resulting vessel-network reconstruction we know that all vessels whose walls appear stained have been in contact with the moving xylem liquid during the experiment. If a cut shoot is used, the analyzed portion of the stem should be at least as far away from the cut end as the length of the longest vessel. Cut vessels are thus kept outside the analyzed area.

Figure 9.2 shows the reconstruction of part of the vessel network of an elm seedling, which was used as a control plant. Elm, by the way, is a ring-porous tree with very poor water uptake from the cut end under vacuum. Sufficient water uptake was achieved by using seedlings and submerging the entire root system. The root system was probably killed by the contact with periodic acid, but this does not matter. Figure 9.2 shows a piece of stem that was 4.2 cm long and had a transverse sectional area of 0.25 mm². The puncture of the needle can be recognized by torn vessels at 1.5 cm of the vertical scale. These "white" vessels were unstained. The vessels appearing black in the drawing were stained. It is important to realize that what we see in the microscope is only vessel walls; the method does not show the presence of liquid in the vessel lumen. Wall staining can take place not only by liquid flowing through the vessel lumen, but also by liquid seeping into the wall of an embolized vessel from a neighboring, conducting one. We refer to this as contact staining. Figure 9.2 shows that all embolized ("white") vessels are stained where they are in contact with dye-conducting vessels. In other words, any conducting vessel looks stained (black in the illustration) *throughout its entire length*; any vessel appearing partly white in the drawing must be regarded as an embolized vessel. If such an experiment is made without application of vacuum, results are much more ambiguous: dye is drawn into embolized vessels by the contraction of emboli and or by capillarity.

Figure 9.3 shows the reconstructed vessel network of an elm seedling that was infected with a spore suspension of the Dutch elm disease fungus *Ceratocystis ulmi*. The dye ascent was made under vacuum 4 days after inoculation. The reconstructed xylem block measured ca. 0.5×0.8×89 mm. The inoculation puncture appears on the left side at about 2.1 cm of the axial scale, the torn vessels are visible at the top of the first and the bottom of the second block. All punctured vessels are again "white" except for the usual points of contact staining. However, we can now see many more "white", i.e., embolized, vessels located entirely outside the area of physical injury. Upper ends of embolized vessels located below the injury and lower ends of embolized vessels above the injury are marked with arrows in Fig. 9.3. These vessels had been vapor-blocked, but not visibly injured. They show absolutely no microscopic evidence of fungal activity (hyphae, gums, etc.), but they must have been submicroscopically leaky when the dye ascent was made, probably by enzymatic action of the fungus before fungal hyphae and gums become visible microscopically and before visible symptoms such as wilting develop. Many visible symptoms have been known for years, but the real first damage, vapor blockage, has received very little attention (VanAlfen and MacHardy 1978).

The above description of xylem dysfunction caused by the Dutch elm disease fungus is certainly not complete. Readers interested in a more comprehensive summary are referred to published reviews (Sinclair and Campana 1978; Stipes and Campana 1981). The purpose of the present description is to stress the immediate and difficult-to-detect early damage caused by vapor blockage.

Another interesting case of damage by vapor blockage is the chestnut blight, the canker formation in the bark of chestnut trees (*Castanea* spp.) by the fungus *Endothia parasitica*. The fungus eventually girdles the tree stem, and thus kills the tree. Evidence of this is the sudden wilting of the entire part of the aerial stem distal to the girdle. There are also canker diseases of elm, butternut, and other ring-porous species that have similar effects. This situation may be confusing, because the immediate effect of girdling is xylem, not phloem interruption.

Perennial shoots, particularly trees of advanced age, do not die immediately from the interruption of the phloem. The interruption of phloem kills only very young, phloem-importing shoots quickly. Every forester knows that diffuse-porous trees and conifers can survive girdling for many years. This has been documented many times. One of the famous cases is the Y-shaped, 100-year-old *Pinus sylvestris* mentioned earlier in this volume: one-half of it was girdled in 1871 for experimental purposes by Theodor Hartig (who had discovered the sieve tubes in 1837). Before he died, he asked his son Robert (who is known for the Hartig net of mycorrhizae and is often called the Father of Forest Pathology) to observe how long the girdled part would survive. It never came to this, because the tree had to be felled for other reasons during the winter of 1888–1889. The girdled part had survived for 18 years and still looked healthy, although its foliage was somewhat thinner than that of the ungirdled crown. Robert Hartig published a very detailed analysis of the growth distribution in 1889. Another famous case is a horse chestnut (*Aesculus hippocastanum*) in Belgium that survived complete girdling for over 50 years (Martens 1971). Girdling has also been used experimentally in order to produce dwarf apple trees by the inversion of bark rings (Sax 1954). There are even examples of survival after pathological girdling: beeches (*Fagus grandifolia*) are often completely girdled by the fungus of the beech bark disease (a *Nectria* sp.) but they survive for a long time, often until the tree is blown over when, during a strong wind, the weakened trunk breaks. It can then be seen that the outer sapwood has been rotted by (secondary?) fungal infections, and thus weakened mechanically. The fact that the crown still contains healthy, green leaves when the tree falls down indicates that there was enough inner sapwood left to supply the crown with water.

Thus. any immediate damage of girdling is almost always the result of xylem, not phloem interruption. It is evident that the trees most vulnerable to girdling are the ring-porous species, which depend almost exclusively on the most recent, superficially located earlywood vessels for their water supply. When a tree is mechanically girdled, these superficially located vessels are invariably damaged and the tree dies from xylem interruption (perhaps with some exceptions, Chap. 6.1). In the case of the chestnut blight, the fungus appears to make the earlywood vessels leaky either by enzymatic action or simply by drying when the bark dies. Although the circumstantial evidence in support of this view is very strong, direct experimental evidence is still missing.

It is interesting to look at xylem failure, particularly of ring-porous trees, again in view of Braun's (1963, 1970) *Hydrosystem*. Braun assumes xylem evolution in dicotyledons from wood with tracheids only, to wood with vessels within a matrix of tracheids and fibers (such as chestnut), to finally the types with vessels surrounded by parenchyma (Fig. 4.8). Vessels within dead tissue such as tracheids can be regarded as most vulnerable; vessels surrounded (and thereby protected) by living parenchyma are less vulnerable to fungal attack. Superimposed upon this is the much greater vulnerability of ring-porous trees in which the water-conducting vessels are very superficially located in the tree stem and because vessels are very large, efficient and therefore less numerous (Chap. 1.4). It has been mentioned before that elm species with smaller vessels have been found to be less vulnerable to Dutch elm disease than those with larger vessels (Elgersma 1970).

Vapor blockage of vessels has received very little attention from investigators studying vascular wilt diseases. It is interesting to read the literature in view of this, and look for circumstantial evidence, but there are usually some critical pieces of information missing. Exceptions are rather rare. Thus, Mathre (1964) described the pathogenicity of *Ceratocystis* species to *Pinus ponderosa*. He found that water conduction failed in an area beyond the visible infection site. This is most easily explained by vapor-blocked tracheids. Coutts (1976) reported dry zones in sapwood that resulted from fungal infections.

Although I have made a very strong case for vapor blockage on these pages, this does not mean that I do not believe other types of blockage are possible under certain circumstances, especially when xylem pressures are high. Vapor blockage is undoubtedly the first and thereby real cause of failure of water transport in many cases, especially in ring-porous trees. However, it is, of course, possible that certain pathogens, especially fungi, manage to enter vessels that contain water under tension. In fact, some evidence of this was obtained by Newbanks (pers. comm.) when he found hyphae of *Verticillium* in vessels of *Acer saccharum* that had conducted dye, taken up from a vacuum container. But, as *Acer* has a very closely knit vessel network, contact-staining has not yet been entirely ruled out. How long tensile water can exist after hyphal penetration is unknown. We must assume that the hyphae keep consuming wall material and that the wall sooner or later becomes leaky.

An interesting case is the *Phialophora* (*Verticillium*) *cinerescens* disease of carnation (*Dianthus* sp.). The fungus enters the roots without visible injury and causes simultaneous yellowing of leaves and gummosis of vessels (Péresse and Moreau 1968). The water status of the leaves at the time of yellowing seems to be unknown. If, for example, yellowing were a senescence effect, caused by the closure of stomata, triggered by a xylem solute from the infection site (as described above for lethal yellowing of coconut palms), pressures would remain high in the xylem, and the fungus could enter vessels without air seeding. We would then have a case where vessels are damaged beyond repair, but the damage would remain undetectable, because of pathologically induced high xylem pressure. Teleologically speaking, if the fungus induces the stomata to close first, it can send its spores up (but not down) in the intact, residual transpiration stream! Likewise, plants grown under very favorable conditions in the greenhouse, or in rainy weather outdoors, may possibly not show xylem damage due to unusually high xylem pressures. Such situations can be quantitatively considered by monitoring xylem pressure and applying Eq. (4.1).

In summary, it is perfectly conceivable that in a plant whose xylem undergoes the normal diurnal pressure fluctuation, i.e., experiences low pressures during the day, the xylem is easily made "leaky" by pathogen action, hence vessels embolize easily and are sealed off by gums and tyloses. If, on the other hand, xylem pressure is maintained high by external factors such as persistent rain, identical action of the pathogen would not embolize the vessels; the protective tylosis formation and/or gummosis would not take place in water-filled vessels and might provide greater distribution of the pathogen. By the same token, embolism is somewhat less likely in root than in shoot infections, because pressures are always higher in the roots. While such considerations are purely theoretical, they show infection processes from new angles, and may suggest new approaches for the investigation of xylem dysfunction.

9.5 Effects of Pathogens on Xylem Differentiation

In Chapter 8.3 we have already seen examples of the plant's possible reaction to the presence of pathogens in the xylem: the secretion of gums and the growth of tyloses from neighboring living parenchyma cells. As discussed in that chapter, all evidence available so far indicates that these events *follow* embolism. At least I am not aware of any firm evidence to the contrary. This is evidently a mechanism by which the plant limits the spread of pathogens. It is interesting that some authors suggest that gums are the result of vessel wall lysis; other authors, on the other hand, consider them a product of secretion from neighboring parenchyma cells (Catesson et al. 1976; Moreau et al. 1978). Whether gummosis can precede embolism in this case is not known.

Infection of carnations (*Dianthus* sp.) by *Phialophora cinerescens* also influences the differentiation of young xylem elements. While in healthy control plants membranes of intervessel pits are rendered more permeable by the hydrolysis of certain polysaccharides during the course of differentiation, this hydrolysis is stopped under the influence of fungal attack. When these tracheary elements mature, vessel-to-vessel conductance is reduced (Catesson et al. 1972).

The biochemistry of cell-wall carbohydrates and its modification by pathogens has been summarized by Fincher and Stone (1981).

Damaged xylem is sealed off from the functioning parts in other ways, namely, by a layer of what early German botanists called Schutzholz (protective wood). The apoplast of this barrier zone is made impermeable by impregnation of the cell walls. Mullick (1977) found that the bordered pits of coniferous tracheids can become encrusted, and thus non-conducting, 19 days after the nearby bark has been wounded. This is interesting because it appears to be independent of embolism. Another reaction to xylem wounding, also non-specific, is the production of living parenchyma cells by the cambium which can thus effectively isolate new from old, injured or necrotic layers of xylem (Tippet and Shigo 1981).

9.6 The Problem of Injecting Liquids

Injecting chemicals such as fungicides into trees is a purely practical problem that actually is beyond the scope of this volume. The reason why a short section is devoted to injection is to summarize briefly the relevant anatomical and physiological properties that have been discussed before and are absolutely essential if one wants to make a successful injection.

First of all, it is worthwhile to study the dye ascent patterns in the tree species one is concerned with (Chap. 2.1). Each species has a characteristic tangential spread that usually amounts to an angle of 1–3°. Obviously, the injected solution will spread more around the trunk, the longer the axial distance available along the trunk. For this reason, fewer injection holes are needed around the trunk, the taller the trunk and the lower the injection holes are placed on it. On the one hand, we probably want to supply the entire crown with liquid, on the other, we want to drill as few injection holes as possible, because each one injures the stem.

The question where to inject does not depend on desired (maximum) distribution versus (minimum) injury alone, it also depends on where the tree is located. In an orchard tree we can do as we please, along a city street we have to consider the presence of pedestrians. For various reasons, small roots are the best injection points. Roots are relatively "disposable" organs of a tree; injuries to roots are in many cases less serious than injuries to the stem. Root diameters can be chosen so that rubber tubing can be fitted over them. The entire transverse section is thus available to injection, which is particularly important in ring-porous species. This brings us to the second important point.

We should know the extent of water conduction in the sapwood of the species we are dealing with. Diffuse-porous species such as most fruit trees are no problem, because many growth rings of the sapwood contain conducting vessels and each injection point can thus reach a portion of the crown. The real difficulty arises with ring-porous species such as elm (*Ulmus* spp.), where it is almost exclusively the earlywood vessels of the most recent growth ring that conduct. Root injection is then particularly useful. Trunk injection is difficult to accomplish successfully, because the injection nozzle must be shaped in such a way (e.g., with lateral slits) that the liquid reaches the most recent growth ring, which is located immediately beneath the cambium. It obviously makes no sense to drive an ordinary nozzle into the trunk of elm, thus blocking the conducting vessels, and then pumping gallons of expensive liquid into the central, non-conducting part of the wood. Radial xylem paths are minimally developed.

The third point to remember is the fact that xylem water is usually under tension during the day. When the hole is drilled, and the drill bit removed from the wood, air is being pulled into the ruptured vessels, thus blocking the xylem path. Subsequent injection may suffer from the air pockets between the applied liquid and the intact vessels beyond the injured ones. This problem has been discussed in Section 9.4. Species differ considerably in the efficiency of the extravascular water path from the applied liquid to the intact vessels. Vacuum infiltration immediately following drilling solves this problem partly, but I am not aware that anyone has used this method for therapeutic purposes, although we often use this method for experimental purposes. The advantage of vacuum infiltration is that there are small, hand-operated vacuum pumps available (no electricity needed). An alternative way of decreasing the bubble size is to apply the liquid under positive pressure. The air block diminishes in size when positive pressure is applied, according to the gas equation (the bubble size is inversely proportional to the absolute pressure). Some of these problems were investigated systematically by Stephen Day for a Master's Thesis at the University of Maine in Orono (Day 1980).

The fourth problem is the pH of the injected solution. This has also been discussed briefly before: alkaline solutes are absorbed by the vessel walls and, as a result, have very poor mobility. In other words, the solvent water moves much faster in the xylem than the solute. This is an advantage if one wants to mark the path of water, because subsequent dissection does not cause a loss of the dye; the dye track is permanently marked and can be observed after sectioning. An acidic solute, on the other hand, is not absorbed on the vessel walls, i.e., it moves much faster and is better distributed. A pH slightly below neutral would probably be ideal. Unfortunately, some of the fungicides used against Dutch elm disease have such a low pH that they damage the xylem (see Newbanks et al. 1982).

Finally, especially when dealing with trees with very long vessels, such as elms, we must realize that injected liquid may never reach the visibly diseased parts of the crown, because much of the xylem path leading to the diseased parts of the crown may already be non-functional. Similarly, we must realize that when, during sanitation work, dead branches are removed, it may be necessary to remove a good portion of the proximal healthy-looking part of the branch as well. Cavitation carries fungal spores downward for considerable distances in species with long vessels such as elm. It is therefore very likely that fungal spores are already at the basal end of embolized vessels long before any symptom can be seen at that point. For practical purposes it is best to follow the advice of cooperative extension agents or experienced arborists who have learned by trial and error how much of the living part should be removed.

The safest test for the efficiency of an injection method is the chemical analysis of actual crown parts. Such work has been done by D.N. Roy and collaborators in Canada (see Newbanks et al. 1982).

It is obviously not possible to give brief instructions to make therapeutic injections successful, or is it the purpose of this book to give advice for such practical matters. In medicine, it is assumed that the operating surgeon is fully trained in human anatomy, physiology, pathology, etc. It would be far too expensive to have fully trained tree physiologists treat trees in backyards and on roadsides. However, it is hoped that the few points made in this last section will help to point out the importance of understanding the tree's structure and function when dealing with practical problems.

References

Addicott FT 1978 Abscission strategies in the behavior of tropical trees. In: Tomlinson PB, Zimmermann MH (eds) Tropical trees as living systems. Cambridge University Press, Cambridge, 381–398

Aloni R, Pradel KS, Ullrich CI 1995 The three-dimensional structure of vascular tissues in *Agrobacterium tumefaciens*-induced crown galls and in the host stems of *Ricinus communis* L. Planta 196:597–605

André JP 1993 Micromoulage des espaces vides intercellulaires dans les tissus végétaux. CR Acad Sci Paris 316:1336–1341

André JP 1998 A study of the vascular organization of bamboos (Poaceae–Bambuseae) using a microcasting method. IAWA J 19:265–278

André JP 2000 Heterogeneous, branched, zigzag and circular vessels: unexpected but frequent forms of tracheary element files: description, localization, form. In: Savidge R, Barnett J, Napier R (eds) Cell and molecular biology of wood formation. Bios Scientific Publishers, Oxford, 387–395

André JP 2002 L´ organisation vasculaire des Angiospermes: une nouvelle vision. Editions INRA, Versailles, 144 p

André JP, Catesson AM, Liberman M 1999 Characters and origin of vessels with heterogeneous structure in leaf and flower abscission zones. Can J Bot 77:253–261

Andrews HN 1961 Studies in paleobotany. John Wiley, New York, 487 pp

Apfel RE 1972 The tensile strength of liquids. Sci Am 227(2):58–71

Arber A 1920 Water plants – a study of aquatic angiosperms. Cambridge University Press, Cambridge. Reprinted by Cramer, Weinheim, Germany, Wheldon and Wesley, Hafner Codicote Herts, New York, 436 pp

Ayres PG 1978 Water relations of diseased plants. In: Kozlowski TT (ed) Water deficits and plant growth, vol 5. Academic Press, New York, 1–60

Baas P 1976 Some functional and adaptive aspects of vessel member morphology. Leiden Bot Ser 3:157–181

Baas P (ed) 1982a New perspectives in wood anatomy. Nijhoff, Junk, The Hague, 252 pp

Baas P 1982b Systematic, phylogenetic, and ecological wood anatomy. In: Baas P (ed) New perspectives in wood anatomy. Nijhoff, Junk, The Hague, 23–58

Baas P, Zweypfenning RCVJ 1979 Wood anatomy of the Lythraceae. Acta Bot Neerl 28:117–155

Bailey IW 1913 The preservative treatment of wood II. For Q 11:12–20 (Reprinted as Chap 19 in: Bailey IW(ed) Contributions to plant anatomy. Chron Bot Waltham Massachusetts 1954)

Bailey IW 1933 The cambium and its derivative tissues. VIII. Structure, distribution, and diagnostic significance of vestured pits in dicotyledons. J Arnold Arbor 14:259–273

Bailey IW 1953 Evolution of the tracheary tissue of land plants. Am J Bot 40:4–8 (Reprinted as Chap 13 in: Bailey IW (ed) Contributions to plant anatomy. Chron Bot Waltham Massachusetts 1954)

Bailey IW 1958 The structure of tracheids in relation to the movement of liquids, suspensions, and undissolved gases. In: Thimann KV (ed) The physiology of forest trees. Ronald Press, New York, 71–82

Bailey IW, Tupper WW 1918 Size variation in tracheary cells. 1. A comparison between the secondary xylems of vascular cryptogams, gymnosperms, and angiosperms. Am Acad Arts Sci Proc 54:149–204

Balling A, Zimmermann U 1990 Comparative measurements of the xylem pressure of *Nicotiana* plants by means of the pressure bomb and pressure probe. Planta 182:325–338

Banks HP 1964 Evolution and plants of the past. Wadsworth, Belmont, 170 pp

Barghoorn ES 1964 Evolution of cambium in geologic time. In: Zimmermann MH (ed) The formation of wood in forest trees. Academic Press, New York, 3–17

Bauch J 1973 Biologische Eigenschaften des Tannennaskerns. Mitt Bundesforschungsanst Forst Holzwirtsch 93:213–224

Bauch J, Schweers W, Berndt H 1974 Lignification during heartwood formation: comparative study of rays and bordered pit membranes in coniferous woods. Holzforschung 28:86–91

Becker P, Tyree MT, Tsuda M 1999 Hydraulic conductances of angiosperms versus conifers: similar transport sufficiency at the whole-plant level. Tree Physiol 19:445–452

Begg JE, Turner NC 1970 Water potential gradients in field tobacco. Plant Physiol 46:343–346

Bell A 1980 The vascular pattern of a rhizomatous ginger (*Alpinia speciosa* L., Zingiberaceae). Ann Bot 46:203–220

Benayoun J, Catesson AM, Czaninski Y 1981 Cytochemical study of differentiation and breakdown of vessel end walls. Ann Bot 47:687–698

Benkert R, Zhu JJ, Zimmermann G, Turk R, Bentrup FW, Zimmermann U 1995 Long-term pressure measurements in the liana *Tetrastigma voinierianum* by means of the xylem pressure probe. Planta 196:804–813

Berger W 1931 Das Wasserleitungssystem von krautigen Pflanzen, Zwergsträuchern und Lianen in quantitativer Betrachtung. Beih Bot Cbl 48:363–390

Berthelot M 1850 Sur quelques phénoménes de dilatation forcée des liquides. Ann Chim Phys 3e Sér 30:232–237

Bierhorst DW 1958 Vessels in Equisetum. Am J Bot 45:534–537

Bierhorst DW, Zamora PM 1965 Primary xylem elements and element associations of angiosperms. Am J Bot 52:657–710

Biermans MBGM, Dukema KM, de Vries DA 1976 Water movement in porous media towards an ice front. Nature 264:166–167

Bode HR 1923 Beiträge zur Dynamik der Wasserbewegungen in den Gefässpflanzen. Jahrb Wiss Bot 62:91–127

Böhm J 1867 Über die Funktion und Genesis der Zellen in den Gefässen des Holzes. Sitzber Akad Wiss Vienna II Abt 55:851

Böhm J 1893 Capillarität und Saftsteigen. Ber Dtsch Bot Ges 11:203–212

Bolton AJ, Jardine P, Jones GL 1975 Interstitial spaces. A review and observations on some Araucariaceae. IAWA Bull 1975/1:3–12

Borghetti M, Edwards WRN, Grace J, Jarvis PG, Raschi A 1991 The refilling of embolised xylem in *Pinus sylvestris* L. Plant Cell Environ 14:357–369

Bormann FH, Berlyn G (eds) 1981 Age and growth rate of tropical trees: new directions for research. Yale University School for Environmental Studies, New Haven, Conn, Bulletin No 94, 137 pp

Bosshard HH 1955 Zur Physiologic des Eschenbraunkerns. Schweiz Z Forstwes 106:592–612

Bosshard HH 1974 Holzkunde, vols 1 and 2. Birkhäuser, Basel, 224 pp, 312 pp

Bosshard HH 1976 Jahrringe und Jahrringbrücken. Schweiz Z Forstwes 127:675–693

Bosshard HH, Kučera L 1973 Die dreidimensionale Strukturanalyse des Holzes. I. Die Vernetzung des Gefässsystems in *Fagus sylvatica* L. Holz Roh Werkst 31:437–445

Bosshard HH, Kučera L, Stocker U 1978 Gewebe-Verknüpfungen in *Quercus robur* L. Schweiz Z Forstwes 129:219–242

Botosso PC, Gomes AV 1982 Radial vessels and series of perforated ray cells in Annonaceae. IAWA Bull 3:39–44

Boyer JS 1967 Leaf water potentials measured with a pressure chamber. Plant Physiol 42:133–137

Boyer JS 1985 Water transport. Ann Rev Plant Physiol 20:351–364

Braun HJ 1959 Die Vernetzung der Gefässe bei Populus. Z Bot 47:421–434

Braun HJ 1963 Die Organisation des Stammes von Bäumen und Sträuchern. Wissenschaft Verl Ges, Stuttgart, 162 pp

Braun HJ 1970 Funktionelle Histologic der sekundären Sprossachse. I. Das Holz. In: Zimmermann W, Ozenda P, Wulff HD (eds) Encyclopedia of plant anatomy, 2nd edn. Bornträger, Berlin, 190 pp

Breitsprecher A, Hughes W 1975 A recording dendrometer for humid environments. Biotropica 7:90–99

Briggs LJ 1950 Limiting negative pressure of water. J Appl Phys 21:721–722

Bristow JM 1975 The structure and function of roots in aquatic vascular plants. In: Torrey JG, Clarkson DT (eds) The development and function of roots. Academic Press, New York, 221–236

Brodribb TJ, Field TS 2000 Stem hydraulic supply is linked to leaf photosynthetic capacity: evidence from New Caledonian and Tasmanian rainforests. Plant Cell Environ 23:1381–1388

Buchholz M 1921 Über die Wasserleitungsbahnen in den interkalaren Wachstumszonen monokotyler Sprosse. Flora 114:119–186

Budgett HM 1912 The adherence of flat surfaces. Proc R Soc Lond A 86:25–35

Burger H 1953 Holz, Blattmenge und Zuwachs. XIII. Fichten im gleichaltrigen Hochwald. Mitt Schweiz Anst Forst Versuchswes 9:38–130

Burggraaf PD 1972 Some observations on the course of the vessels in the wood of *Fraxinus excelsior* L. Acta Bot Neerl 21:32–47

Burke MJ, Gusta LV, Quamme HA, Weiser CJ, Li PH 1976 Freezing and injury in plants. Annu Rev Plant Physiol 27:507–528

Butterfield BG, Meylan BA 1982 Cell wall hydrolysis in the tracheary elements of the secondary xylem. In: Baas P (ed) New perspectives in wood anatomy. Nijhoff, Junk, The Hague, 71–84

Calkin HW, Gibson AC, Nobel PS 1986 Biophysical model of xylem conductance in tracheids of the fern *Pteris vittata*. J Exp Bot 37:1054–1064

Campana RJ 1978 Inoculation and fungal invasion of the tree. In: Sinclair WA, Campana RJ (eds) Dutch elm disease: perspectives after 60 years. Cornell Univ Agric Expt Stn Search Agric 8(5):17–20

Canny MJ 1995a Potassium cycling in *Helianthus*: ions of the xylem sap and secondary vessel formation. Philos Trans R Soc Lond B 348:457–469

Canny MJ 1995b A new theory for the ascent of sap: cohesion supported by tissue pressure. Ann Bot 75:343–357

Canny MJ 1997 Vessel contents during transpiration, embolism and refilling. Am J Bot 84:1223–1230

Canny MJ, Huang CX, McCully ME 2001a The cohesion theory debate continues. Trends Plant Sci 6:454–456

Canny MJ, McCully ME, Huang CX 2001b Cryo-scanning electron microscopy observations of vessel content during transpiration in walnut petioles. Fact or artifacts? Plant Physiol Biochem 39:551–563

Carlquist S 1975 Ecological strategies of xylem evolution. University of California Press, Berkeley, 259 pp

Carlquist S 1977a Ecological factors in wood evolution: a floristic approach. Am J Bot 64:887–896

Carlquist S 1977b Wood anatomy of Onagraceae: additional species and concepts. Ann Mo Bot Gard 64:627–637

Carlquist S 1988 Comparative wood anatomy: systematic, ecological, and evolutionary aspects of dicotyledon wood. Springer, Berlin Heidelberg New York

Carlquist S 1989 Adaptive wood anatomy of chaparral shrubs. In: Keeley SC (ed) The California chaparral: paradigms reexamined. Natural History Museum of Los Angeles County, Los Angeles, CA, Science Series 34, 25–36

Carlquist S, Hoekman DA 1986 Ecological wood anatomy of the woody southern California flora. IAWA Bull NS 6:319–347

Carte AE 1961 Air bubbles in ice. Proc Phys Soc 77:757–768

Carter JC 1945 Wetwood of elms. Nat Hist Survey 23:407–448

Catesson AM, Czaninski Y, Péresse M, Moreau M 1972 Modifications des parois vasculaires de l'oeillet infecté par le *Phialophora cinerescens* (Wr) van Beyma. CR Acad Sci Paris Ser D 275:827–829

Catesson AM, Czaninski Y, Moreau M, Péresse M 1979 Conséquences d'une infection vasculaire sur la maturation des vaisseaux. Rev Mycol 43:239–243

Catesson AM, Czaninski Y, Péresse M, Moreau M 1976 Sécrétions intravasculaires de substances "gommeuses" par les cellules associées aux vaisseaux en réaction à une attaque parasitaire. Soc Bot Fr Coll Sécrét Végét 123:93–107

Chattaway MM 1948 Note on the vascular tissue in the rays of *Banksia*. J Council Sci Ind Res 21:275–278

Chattaway MM 1949 The development of tyloses and secretion of gum in heartwood formation. Aust J Sci Res Ser B Biol Sci 2:227–240

Chattaway MM 1952 The sapwood-heartwood transition. Aust For 16:25–34

Cheadle VI 1953 Independent origin of vessels in the monocotyledons and dicotyledons. Phytomorphology 3:23–44

Chen PYS, Sucoff EI, Hossfeld R 1970 The effect of cations on the permeability of wood to aqueous solutions. Holzforschung 24:65–67

Chou CK, Levi MP 1971 An electron microscopical study of the penetration and decomposition of tracheid walls of *Pinus sylvestris* by *Poria vaillantii*. Holzforschung 25:107–112

Cichan MA, Taylor TN 1982 Vascular cambium development in Sphenophyllum: a Carboniferous arthrophyte. LWA Bull 3:155–160

Clarkson DT, Robards AW 1975 The endodermis, its structural development and physiological role. In: Torrey JG, Clarkson DT (eds) The development and function of roots. Academic Press, New York, 415–436

Cochard H, Tyree MT 1990 Xylem dysfunction in *Quercus*: vessel size, tyloses, cavitation and seasonal changes in embolism. Tree Physiol 6:393–407

Cochard H, Cruiziat P, Tyree MT 1992 Use of positive pressures to establish vulnerability curves: further support for the air-seeding hypothesis and implications for pressure-volume analysis. Plant Physiol 100:205–209

Cochard H, Ewers FW, Tyree MT 1994 Water relations of a tropical vinelike bamboo (*Rhipdocladum racemiflorum*): root pressures, vulnerability to cavitation and seasonal changes in embolism. J Exp Bot 45:1085–1089

Cochard H, Bréda N, Granier A 1996a Whole tree hydraulic conductance and water loss regulation in *Quercus* during drought: evidence for stomatal control of embolism? Ann Sci For 53:197–206

Cochard H, Ridolfi M, Dreyer E 1996b Response to water stress in an ABA-unresponsive hybrid poplar (*Populus koreana × trichocarpai* cv. 'Peace') II: hydraulic properties and xylem embolism. New Phytol 134:455–461

Cochard H, Peiffer M, Le Gall K, Granier A 1997 Developmental control of xylem hydraulic resistances and vulnerability to embolism in *Fraxinus excelsior* L. Impacts on water relations. J Exp Bot 48:655–663

Cochard H, Lemoine D, Dreyer E 1999 The effects of acclimation to sunlight on the xylem vulnerability to embolism in *Fagus sylvatica* L. Plant Cell Environ 22:101–108

Cochard H, Bodet C, Améglio T, Cruiziat P 2000 Cryo-scanning electron microscopy observations of vessel contents during transpiration in walnut petioles. Fact or artifacts. Plant Physiol 124:1191–1202

Cochard H, Thierry A, Cruiziat P 2001 The cohesion theory debate continues. Trends Plant Sci 6:456

Cohen Y, Fuchs M, Green GC 1981 Improvement of the heat pulse method for determining sap flow in trees. Plant Cell Environ 4:391–397

Comstock GL, Côté WA 1968 Factors affecting permeability and pit aspiration in coniferous sapwood. Wood Sci Technol 2:279–291

Comstock JP, Sperry JS 2000 Theoretical considerations of optimal conduit length for water transport in vascular plants. New Phytol 148:195–218

Connor DJ, Legge NJ, Turner NC 1977 Water relations of mountain ash (*Eucalyptus regnans* F Muell) forests. Aust J Plant Physiol 4:735–762

Conway VM 1940 Growth rates and water loss in *Cladium mariscus* R Br. Ann Bot NS 4:151–164

Core HA, Côté WA Jr, Day AC 1979 Wood structure and identification, 2nd edn. Syracuse University Press, Syracuse, 172 pp

Coster C 1927 Zur Anatomie und Physiologie der Zuwachszonen und Jahresringbildung in den Tropen. Ann Jard Bot Buitenzorg 38:1–114

Côté WA Jr, Day AC 1962 Vestured pits – fine structure and apparent relationship with warts. TAPPI 45:906–910

Coutts MP 1976 The formation of dry zones in the sapwood of conifers. I. Eur J For Pathol 6:372–381

Coutts MP, Rishbeth J 1977 The formation of wetwood in grand fir. Eur J For Pathol 7:13–22

Crombie DS, Hipkins MF, Milburn JA 1985 Gas penetration of pit membranes in the xylem of *Rhododendron* as the cause of acoustically detectable sap cavitation. Aust J Plant Physiol 12:445–453

Davis D, Sperry JS, Hacke UG 1999 The relationship between xylem conduit diameter and cavitation caused by freezing. Am J Bot 86:1367–1372

Day SJ 1980 The influence of sapstream continuity and pressure on distribution of systemic chemicals in American elm (*Ulmus americana* L.). Master's Thesis, University of Maine, Orono

de Vries H 1886 Studien von Zuigwortels. Maandbl Natuurwet 13:53–68 (Summary in Bot Z 44:788–790)

Dimond AE 1966 Pressure and flow relations in vascular bundles of the tomato plant. Plant Physiol 41:119–131

Dimond AE 1970 Biophysics and biochemistry of the vascular wilt syndrome. Annu Rev Phytopathol 8:301–322

Dixon HH 1914 Transpiration and the ascent of sap in plants. MacMillan, London, 216 pp

Dixon HH, Joly J 1894 On the ascent of sap. Philos Trans R Soc Lond Ser B 186:563–576

Dixon MA, Tyree MT 1984 A new temperature corrected stem hygrometer and its calibration against the pressure bomb. Plant Cell Environ 7:693–697

Dobbs RC, Scott DRM 1971 Distribution of diurnal fluctuations in stem circumference of Douglas fir. Can J For Res 1:80–83

Donny J 1846 Sur la cohésion des liquides et sur leur adhésion aux corps solides. Ann Chim Phys 3e Sér 16:167

Doyle AC 1986 Sherlock Holmes: the complete novels and stories, vol 1. Bantam Books, New York

Duchartre P 1858 Recherches expérimentales sur la transpiration des plantes dans les milieux humides. Bull Soc Bot France 5:105–111

Edwards WRN, Jarvis PG, Grace J, Moncrieff JB 1994 Reversing cavitation in tracheids of *Pinus sylvestris* L. under negative water potentials. Plant Cell Environ 17:389–397

Elgersma DW 1970 Length and diameter of xylem vessels as factors in resistance of elms to *Ceratocystis ulmi*. Neth J Plant Path 76:179–182

Enns LC, Canny MJ, McCully ME 2000 An investigation of the role of solutes in the xylem sap and in the xylem parenchyma as the source of root pressure. Protoplasma 211:183–197

Esau K 1965 Plant anatomy, 2nd edn. John Wiley, New York, 767 pp

Eschrich W 1975 Sealing systems in phloem. In: Zimmermann MH, Milburn JA (eds) Transport in plants I. Encyclopedia of plant physiology, new series, vol 1. Springer, Berlin Heidelberg New York, 39–56

Everitt DH 1961 The thermodynamics of frost damage to porous liquids. Trans Faraday Soc 57:1541–1550

Ewart AJ 1905–1906/1907–1908 The ascent of water in trees. Philos Trans Soc Lond B 198:41–45, 199:341–392

Ewers FW 1985 Xylem structure and water conduction in conifer tress, dicot trees, and lianas. IAWA Bull NS 6:309–317

Ewers FW, Fisher JB 1989 Techniques for measuring vessel lengths and diameters in stems of woody plants. Am J Bot 76:645–656

Ewers FW, Zimmermann MH 1984a The hydraulic architecture of eastern hemlock (*Tsuga canadensis*). Can J Bot 62:940–946

Ewers FW, Zimmermann MH 1984b The hydraulic architecture of balsam fir (*Abies balsamea*). Physiol Plant 60:453–458

Ewers FW, Fisher JB, Chiu ST 1989 Water transport in the liana *Bauhinia fassoglensis* (Fabaceae). Plant Physiol 91:1625–1631

Ewers FW, Fisher JB, Fichtner K 1991 Water flux and xylem structure in vines. In: Putz FE, Mooney HA (eds) Biology of vines. Cambridge University Press, Cambridge, 119–152

Ewers FW, Cochard H, Tyree MT 1997 A survey of root pressures in vines of a tropical lowland forest. Oecologia 110:190–196

Fahn A 1964 Some anatomical adaptations of desert plants. Phytomorphology 14:93–102

Fahn A 1974 Plant anatomy, 2nd edn. Pergamon, Oxford, 611 pp

Fegel AC 1941 Comparative anatomy and varying physical properties of trunk, branch, and root wood in certain northeastern trees. Bull NY State Coll for Syracuse Univ, vol 14, No 2b. Tech Publ No 55:1 20

Filzner P 1948 Ein Beitrag zur ökologischen Anatomic von Rhynia. Biol Zbl 67:13–17

Fincher GB, Stone BA 1981 Metabolism of non-cellulosic polysaccharides. In: Tanner W, Loewus FA (eds) Plant carbohydrates 11. Encyclopedia of plant physiology, new series, vol 13B. Springer, Berlin Heidelberg New York, pp 68–132

Firbas F 1931a Untersuchungen über den Wasserhaushalt der Hochmoorpflanzen. Jahrb Wiss Bot 74:459–696

Firbas F 1931b Über die Ausbildung des Leitungssystems und das Verhalten der Spaltöffnungen im Frühjahr bei Pflanzen des Mediterrangebietes und der tunesischen Steppen und Wüsten. Beih Bot Cbl 48:451–465

Fisher JB, Ewers FW 1995 Vessel dimension in Liana and tree species of Gnetum (Gnetales). Am J Bot 82:1350–1357

Fisher JB, Guillermo AA, Ewers FW, Lópex-Portillo J 1997 Survey of root pressure in tropical vines and woody species. Int J Plant Sci 158:44–50

Foster AS 1956 Plant idioblasts: remarkable examples of cell specialization. Protoplasma 46:183–193

French JC, Tomlinson PB 1981a Vascular patterns in stems of Araceae: subfamilies Calloideae and Lasioideae. Bot Gaz 142:366–381

French JC, Tomlinson PB 1981b Vascular patterns in stems of Araceae: subfamily Pothoideae. Am J Bot 68:713–729

French JC, Tomlinson PB 1981c Vascular patterns in stems of Araceae: subfamily Monsteroideae. Am J Bot 68:1115–1129

French JC, Tomlinson PB 1981d Vascular patterns in stems of Araceae: subfamily Philodendroideae. Bot Gaz 142:550–563

Frensch J, Steudle E 1989 Axial and radial hydraulic resistance to roots of maize (*Zea mays* L.) Plant Physiol 91:719–726

Frensch J, Hsiao TC, Steudle E 1996 Water and solute transport along developing maize roots. Planta 198:348–355

Frenzel P 1929 Über die Porengrössen einiger pflanzlicher Zellmembranen. Planta 8:642–665

Frey-Wyssling A 1959 Die pflanzliche Zellwand. Springer, Berlin Heidelberg New York, 367 pp

Frey-Wyssling A 1982 Introduction to the symposium: cell wall structure and biogenesis. IAWA Bull 3:25–30

Frey-Wyssling A, Bosshard HH 1953 Über den Feinbau der Schliesshäute in Hoftüpfeln. Holz Roh Werkst 11:417–420

Frey-Wyssling A, Mühlethaler K, Bosshard HH 1955/1956 Das Elektronenmikroskop im Dienste der Bestimmung von Pinusarten. Holz Roh Werkst 13:245–249, 14:161–162

Frey-Wyssling A, Mühlethaler K, Bosshard HH 1959 Über die mikroskopische Auflösung der Haltefäden des Torus in Hoftüpfeln. Holzforsch Holzverwert 11:107–108

Friedrich J 1897 Über den Einfluss der Witterung auf den Baumzuwachs. Zbl Ges Forstwes 23:471–495

Fritts HC, Fritts EC 1955 A new dendrograph for recording radial changes of a tree. For Sci 1:271–276

Fuchs EE, Livingston NJ 1996 Hydraulic control of stomatal conductance in Douglas-fir (*Pseudotsuga menziesii* (Mirb.) Franco) and alder (*Alnus rubra* (Bong)) seedlings. Plant Cell Environ 19:1091–1098

Fujii T 1993 Application of a resin casting method to wood anatomy of some Japanese Fagaceae species. IAWA J 14:273–288

Fujita M, Nakagawa K, Mori N, Harada H 1978 The season of tylosis development and changes in parenchyma cell structure in *Robinia pseudoacacia* L. Bull Kyoto Univ For 50:183–190

Gardiner W 1883 On the physiological significance of water glands and nectaries. Proc Cambridge Philos Soc 5:35–50

Gessner F 1951 Untersuchungen über den Wasserhaushalt der Nymphaeaceen. Biol Generalis 19:247–280

Gibbs RD 1958 Patterns of the seasonal water content of trees. In: Thimann KV (ed) The physiology of forest trees. Ronald Press, New York, 43–69

Grace J 1993 Refilling of embolized xylem. In: M Borghetti M, Grace J, Raschi A (eds) Water transport in plants under climate stress. Cambridge University Press, Cambridge, 51–62

Granier A, Anfodillo T, Sabatti M, Cochard H, Dreyer E, Tomasi M, Valentini R, Bréda N 1994 Axial and radial water flow in the trunks of oak trees: a quantitative and qualitative analysis. Tree Physiol 14:1383–1396

Greenidge KNH 1952 An approach to the study of vessel length in hardwood species. Am J Bot 39:570–574

Greenidge KNH 1957 Ascent of sap. Annu Rev Plant Physiol 8:237 256

Greenidge KNH 1958 Rates and patterns of moisture movement in trees. In: Thimann KV (ed) The physiology of forest trees. Ronald Press, New York, 19–41

Grier CC, Waring RH 1974 Conifer foliage mass related to sapwood area. For Sci 20:205–206

Haberlandt G 1914 Physiological plant anatomy. Translated from the 4th German edn by Drummond M, MacMillan, London, 777 pp

Hacke UG, Sperry JS, Ewers BE, Ellsworth DS, Schäfer KVR, Oren R 2000 Influence of soil porosity on water use in *Pinus taeda*. Oecologia 124:495–505

Hacke UG, Stiller V, Sperry JS, Pittermann J, McCulloh KA 2001 Cavitation fatigue: embolism and refilling cycles can weaken cavitation resistance of xylem. Plant Physiol 125:770–786

Häusermann E 1944 Über die Benetzungsgrösse der Mesophyllinterzellularen. Ber Schweiz Bot Ges 54:541–578

Haider K 1954 Zur Morphologie und Physiologie der Sporangien leptosporangiater Farne. Planta 44:370–411

Hallé F, Oldeman RAA, Tomlinson PB 1978 Tropical trees and forests, an architectural analysis. Springer, Berlin Heidelberg New York, 420 pp

Hammel HT 1967 Freezing of xylem sap without cavitation. Plant Physiol 42:55–66

Hammel HT, Scholander PF 1976 Osmosis and tensile solvent. Springer, Berlin Heidelberg New York, 133 pp

Handley WRC 1936 Some observations on the problem of vessel length determination in woody dicotyledons. New Phytol 35:456–471

Hargrave KE, Kolb KJ, Ewers FW, Davis SD 1994 Conduit diameter and drought-induced embolism in *Salvia mellifera* Greene (Labiatae). New Phytol 126:695–805

Harlow WM 1970 Inside wood, masterpiece of nature. American Forestry Association, Washington, DC, 120 pp

Hartig R 1889 Ein Ringelungsversuch. Allg Forst Jagdztg 365–373, 401–410

Hartig T 1878 Anatomic und Physiologie der Holzpflanzen. Springer, Berlin Heidelberg New York, 412 pp

Hartley C, Davidson RW, Crandall BS 1961 Wetwood, bacteria, and increased pH in trees. US For Prod Lab Report 2215:1–34

Harvey HP, van den Driessche 1999 Nitrogen and potassium effects on xylem cavitation and water-use efficiency in poplars. Tree Physiol 943–950

Hatheway WH, Winter DE 1981 Water transport and storage in Douglas fir: a mathematical model. Mitt Forst Bundes Versuchsanst Wien 142:193–222

Heath J, Kerstiens G, Tyree MT 1997 Stem hydraulic conductance of European beech (*Fagus sylvatica* L.) and pedunculate oak (*Quercus robur* L.) grown in elevated CO_2. J Exp Bot 48:1487–1489

Hejnowicz Z, Romberger JA 1973 Migrating cambial domains and the origin of wavy grain in xylem of broadleaved trees. Am J Bot 209–222

Hellqvist J, Richards GP, Jarvis PG 1974 Vertical gradients of water potential and tissue water relations in Sitka spruce trees measured with the pressure chamber. J Appl Ecol 11:637–668

Henzler T, Waterhouse RN, Smyth AJ, Carvajal M, Cooke DT, Schäffenr AR, Steudle E, Clarkson DT 1999 Diurnal variations in hydraulic conductivity and root pressure can be correlated with the expression of putative aquaporins in the roots of Lotus japonicus. Planta 210:50–60

Hillis WE 1977 Secondary changes in wood. In: Loewus FA, Runeckles VC (eds) The structure, biosynthesis, and degradation of wood. Plenum, New York, 247–309

Höll W 1975 Radial transport in rays. In: Zimmermann MH, Milburn JA (eds) Transport in plants 1. Encyclopedia of plant physiology, new series, vol 1. Springer, Berlin Heidelberg New York, 432–450

Holbrook NM, Sinclair TR 1992a Water balance in the arborescent palm, Sabal palmetto. I. Stem structure, tissue water release properties and leaf epidermal conductances. Plant Cell Environ 15:393–399

Holbrook NM, Sinclair TR 1992b Water balance in the arborescent palm, Sabal palmetto. II. Transpiration and stem water storage. Plant Cell Environ 15:401–409

Holbrook NM, Zwieniecki MA 1999 Embolism repair and xylem tension. Do we need a miracle? Plant Physiol 120:7–10

Holbrook NM, Burns MJ, Sinclair TR 1992 Frequency and time-domain dielectric measurements of stem water content in the arborescent palm, Sabal palmetto. J Exp Bot 43:111–119

Holbrook NM, Burns MJ, Field CB 1995 Negative xylem pressures in plants: a test of the balancing pressure technique. Science 270:1193–1194

Holbrook NM, Ahrens ET, Burns MJ, Zwieniecki MA 2001 In vivo observation of cavitation and embolism repair using magnetic resonance imaging. Plant Physiol 126:27–31

Holle H 1915 Untersuchungen über Welken, Vertrocknen und Wiederstraffwerden. Flora 108:73–126

Hong SG, Sucoff E 1982 Rapid increase in deep supercooling of xylem parenchyma. Plant Physiol 69:697–700

Hook DD, Brown CL, Wetmore RH 1972 Aeration in trees. Bot Gaz 133:443–454

Howard FW, Thomas DL 1980 Transmission of palm lethal decline to Veitchia merrillii by a planthopper Myndus crudus. J Encon Entomol 73:715–717

Howard FW, Thomas DL, Donselman HM, Collins ME 1979 Susceptibilities of palm species to mycoplasma-like organism-associated diseases in Florida. Plant Prot Bull 27:109–117

Howard RA 1974 The stem-node-leaf continuum of the Dicotyledonae. J Arnold Arbor 55:125–181

Hubbard RM, Ryan MG, Stiller V, Sperry JS 2001 Stomatal conductance and photosynthesis vary linearly with plant hydraulic conductance in ponderosa pine. Plant Cell Environ 24:113–121

Huber B 1928 Weitere quantitative Untersuchungen über das Wasserleitungssystem der Pflanzen. Jahrb Wiss Bot 67:877–959

Huber B 1932 Beobachtung und Messung pflanzlicher Saftströme. Ber Dtsch Bot Ges 50:89–109

Huber B 1935 Die physiologische Bedeutung der Ring- und Zerstreutporigkeit. (Physiological significance of ring- and diffuse-porousness). Ber Dtsch Bot Ges 53:711–719 (Xerox copies of English translation available from National Translation Center, 35 West 33rd St, Chicago, IL 60616, USA)

Huber B 1956 Die Gefässleitung. In: Ruhland W (ed) Encyclopedia of plant physiology, vol 3. Springer, Berlin Heidelberg New York, 541–582

Huber B, Metz W 1958 Über die Bedeutung des Hoftüpfelverschlusses for die axiale Wasserleitfähigkeit von Nadelhölzern. Planta 51:645–672

Huber B, Schmidt E 1936 Weitere thermo-elektrische Untersuchungen über den Transpirationsstrom der Bäume. (Further thermoelectric investigations on the transpiration stream in trees) Tharandt Forst Jahrb 87:369–412 (Xerox copies of English translation available from National Translation Center, 35 West 33rd St, Chicago, IL 60616, USA)

Huber B, Schmidt E 1937 Eine Kompensationsmethode zur thermo-elektrischen Messung lang-samer Saftströme. Ber Deutsch Bot Ges 55:514–529 (Xerox copies of English translation available from National Translation Center, 35 West 33rd St, Chicago, IL 60616 USA)

Hudson MS, Shelton SV 1969 Longitudinal flow of liquids in southern pine poles. For Prod J 19:25–32

Ipsen JD, Abul-Haji YJ 1982 Fluorescent antibody technique as a means of localizing *Ceratocystis ulmi* toxins in elm. Can J Bot 60:724–729

Irvine J, Grace J 1997 Continuous measurements of water tensions in the xylem of trees based on the elastic properties of wood. Planta 202:455–461

Isebrands JG, Larson PR 1977 Vascular anatomy of the nodal region in eastern cottonwood. Am J Bot 64:1066–1077

Jaccard P 1913 Eine neue Auffassung über die Ursachen des Dickenwachstums. Naturwiss Z Forst Landwirtsch 11:241–279

Jaccard P 1919 Nouvelles recherches sur l'accroissement en épaisseur des arbres. Publ No 23 Fondation Schnyder von Wartensee, Zürich

Jeje AA, Zimmermann MH 1979 Resistance to water flow in xylem vessels. J Exp Bot 30:817–827

Jenik J 1978 Discussion. In: Tomlinson PB, Zimmermann MH (eds) Tropical trees in living systems. Cambridge University Press, Cambridge, 529 pp

Johnson LPV 1945 Physiological studies on sap flow in the sugar maple, *Acer saccharum* Marsh. Can J Res 23:192–197

Johnson RW, Tyree MT 1992 Effect of stem water content on sap flow from dormant maple and butternut stems: induction of sap flow in butternut. Plant Physiol 100:853–858

Johnson RW, Tyree MT, Dixon MA 1987 A requirement for sucrose in xylem sap flow from dormant maple trees. Plant Physiol 84:495–500

Jones CH, Edson AW, Morse WJ 1903 The maple sap flow. Vermont Agric Exp Stn Bull 103

Jones CS, Lord EM 1982 The development of split axes in *Ambrosia dumonsa* (Gray) Payne (Asteraceae). Bot Gaz 143:446–453

Jost L 1916 Versuche über die Wasserleitung in der Pflanze. Z Bot 8:1–55

Kaiser P 1879 Über die tägliche Periodizität der Dickendimensionen der Baumstämme. Inaug Diss, Halle, 38 pp

Kaufman MR 1968 Evaluation of the pressure chamber technique for estimating plant water potential of forest tree species. For Sci 14:369–374

Kelso WC, Gertjejansen CO, Hossfeld RL 1963 The effect of air blockage upon the permeability of wood to liquids. Univ Minnesota Agric Res Stn Tech Bull 242:1–40

Kitajima K 1994 Relative importance of photosynthetic traits and allocation patterns as correlates of seedling shade tolerance of 13 tropical trees. Oecologia 98:419–428

Kitin P, Sano PY, Funada R 2001 Analysis of cambium and differentiating vessel elements in *Kalopanax pictus* using resin cast replicas. IAWA J 22:15–28

Klein G 1923 Zur Aetiologie der Thyllen. Z Bot 15:418–439

Klepper B, Browning VD, Taylor HM 1971 Stem diameter in relation to plant water status. Plant Physiol 48:683–685

Klotz LH 1978 Form of the perforation plates in the wide vessels of metaxylem in palms. J Arnold Arbor 59:105–128

Kolb KJ, Sperry JS 1999 Transport constraints on water use in the Great Basin shrub, *Artemisia tridentate*. Plant Cell Environ 22:925–935

Kórán Z, Côté WA 1965 The ultrastructure of tyloses. In: Côté WA (ed) Cellular ultrastructure of woody plants. Syracuse University Press, Syracuse, NY, 319–333

Krahmer RL, Côté WA 1963 Changes in coniferous wood cells associated with heartwood formation. TAPPI 46:42–49

Kramer K 1974 Die tertiären Hölzer Südost-Asiens, I und II. Palaeontogr Abt B 144:45–181, 145:1–150

Kramer PJ 1937 The relation between the rate of transpiration and the rate of absorption of water in plants. Am J Bot 24:10–15

Kraus G 1877 Die Verteilung und Bedeutung des Wassers bei Wachstums- und Spannungsvorgängen in der Pflanze. Bot Ztg 35:595–597

Kraus G 1881 Die tägliche Schwellungsperiode der Pflanzen. Abhandl Naturf Gesellsch Halle 15

Kučera L 1975 Die dreidimensionale Strukturanalyse des Holzes. 11. Das Gefässstrahlnetz bei der Buche (*Fagus sylvatica* L.). Holz Roh Werkst 33:276–282

Kursanov AL 1957 The root system as an organ of metabolism. UNESCO Intern Conf Radioisotopes Sci Res, UNESCO/NS/RIC/128, Pergamon, London, 12 pp

Läuchli A, Bieleski RL 1983 Inorganic plant nutrition. In: Encyclopedia of plant physiology, new series, vol 15. Springer, Berlin Heidelberg New York

Laming PB, ter Welle BJH 1971 Anomalous tangential pitting in *Picea abies* Karst (European spruce). IAWA Bull 1971/4:3–10

Langan SJ, Ewers FW, Davis SD 1997 Xylem dysfunction caused by freezing and water stress in two species of co-occurring chaparral shrubs. Plant Cell Environ 20:425–437

Larson DW 1994 Radially sectored hydraulic pathways in the xylem of *Thuja occidentalis* as revealed by the use of dyes. Int J Plant Sci 155:569–582

Larson PR, Isebrands JG 1978 Functional significance of the nodal constricted zone in *Populus deltoides*. Can J Bot 56:801–804

Lewis AM, Harnden VD, Tyree MT 1994 Collapse of water-stress emboli in the tracheids of *Thuja occidentalis* L. Plant Physiol 106:1639–1646

Liese W, Bauch J 1964 Über die Wegsamkeit der Hoftüpfel von Coniferen. Naturwissenschaften 21:516

Liese W, Ledbetter MC 1963 Occurrence of a warty layer in vascular cells of plants. Nature 197:201–202

Loch JPG, Kay BD 1978 Water redistribution in partially frozen, saturated silt under several temperature gradients and overburden loads. Soil Sci Soc Am J 42:400–406

Lybeck BR 1959 Winter freezing in relation to the rise of sap in tall trees. Plant Physiol 34:482–486

MacDougal DT 1924 Dendrographic measurements. Carnegie Inst Washington Publ 350:1–88

MacDougal DT, Overton JB, Smith GM 1929 The hydrostatic-pneumatic system of certain trees: movements of liquids and gases. Carnegie Inst Washington Publ 397:1–99

Machado J-L, Tyree MT 1994 Patterns of hydraulic architecture and water relations of two tropical canopy trees with contrasting leaf phenologies: *Ochroma pyramidale* and *Pseudobombax septenatum*. Tree Physiol 14:219–240

Mackay JFG, Weatherley PE 1973 The effects of transverse cuts through the stems of transpiring woody plants on water transport and stress in leaves. J Exp Bot 24:15–28

Mantai KE, Newton ME 1982 Root growth in Myriophyllum: a specific plant response to nutrient availability? Aquatic Bot 13:45–55

Marshall DC 1958 Measurement of sap flow in conifers by heat transport. Plant Physiol 33:385–396

Martens P 1971 Un marronnier centenaire privé d'écorce à sa base depuis plus d'un demi-siècle. Bull Acad Roy Belgique Classe Sci 57:65–84

Martre P, Durand JL, Cochard H 2000 Changes in axial hydraulic conductivity along elongating leaf blades in relation to xylem maturation in tall fescue. New Phytol 146:235 – 247

Marvin JW 1958 The physiology of maple sap flow. In: Thimann KV (ed) The physiology of forest trees. Ronald Press, New York, pp 95–124

Marvin JW, Ericson RO 1956 A statistical evaluation of some of the factors responsible for the flow of sap from the sugar maple. Plant Physiol 31:57–61

Marvin JW, Greene MT 1951 Temperature induced sap flow in excised stems of *Acer*. Plant Physiol 26:565–580

Marvin JW, Morselli MF, Laing FM 1967 A correlation between sugar concentration and volume yields in sugar maple: an 18-year study. For Sci 13:346–351

Mathre DE 1964 Pathogenicity of *Ceratocystis ips* and *Ceratocystis minor* to *Pinus ponderosa*. Contrib Boyce Thompson Inst 22:363–388

McCoy RE 1978 Mycoplasmas and yellows diseases. In: Barile MF, Razin S, Tully JG, Whitcomb RF (eds) The mycoplasmas, vol 2. Academic Press, New York

McCully ME, Huang CX, Ling LEC 1998 Daily embolism and refilling of xylem vessels in the roots of field-grown maize. New Phytol 138:327–342

McDonough J, Zimmermann MH 1979 Effect of lethal yellowing on xylem pressure in coconut palms. Principes 23:132–137

Meinzer FC, Goldstein G, Jackson P, Holbrook NM, Gutierrez MV, Cavelier J 1995 Environmental and physiological regulation of transpiration in tropical forest gap species: the influence of boundary layer and hydraulic conductance properties. Oecologia 101:514–522

Melcher PJ, Meinzer FC, Yount DE, Goldstein G, Zimmermann U 1998 Comparative measurements of xylem pressure in transpiring and non-transpiring leaves by means of the pressure chamber and the xylem pressure probe. J Exp Bot 49:1757–1760

Metzger K 1894, 1895 Studien über den Aufbau der Waldbäume und Bestände nach statischen Gesetzen. Mündener Forstl Hefte 5:61–74 (1894), 6:94–119 (1894), 7:45–97 (1895)

Meyer FJ 1928 Die Begriffe "stammeigene Bündel" und "Blattspurbündel" im Lichte unserer heutigen Kenntnisse vom Aufbau und der physiologischen Wirkungsweise der Leitbündel. Jahrb Wiss Bot 69:237–263

Meyer RW, Côté WA Jr 1968 Formation of the protective layer and its role in tylosis development. Wood Sci Technol 2:84–94

Meylan BA, Butterfield BG 1972 Three-dimensional structure of wood. Chapman and Hall, London, 80 pp

Meylan BA, Butterfield BG 1978a The structure of New Zealand woods. NZ Dep Sci Ind Res B 222:1–250

Meylan BA, Butterfield BG 1978b Occurrence of helical thickenings in the vessels of New Zealand woods. New Phytol 81:139–146

Milburn JA 1973a Cavitation in *Ricinus* by acoustic detection: induction in excised leaves by various factors. Planta 110:253–265

Milburn JA 1973b Cavitation studies on whole Ricinus plants by acoustic detection. Planta 112:333–342

Milburn JA, Johnson RPC 1966 The conduction of sap. II. Detection of vibrations produced by sap cavitation in *Ricinus* xylem. Planta 69:43–52

Milburn JA, McLaughlin ME 1974 Studies of cavitation in isolated vascular bundles and whole leaves of *Plantago major* L. New Phytol 73:861–871

Milburn JA, O'Malley PER 1984 Freeze-induced sap absorption in *Acer pseudoplantanus*: a possible explanation. Can J Bot 61:2101–2106

Miller DM 1985 Studies of root function in *Zea mays*. III. Xylem sap composition at maximum root pressure provides evidence of active transport into the xylem and a measurement of the reflection coefficient of the root. Plant Physiol 77:162–167

Mittler TE 1967 Water tensions in plants – an entomological approach. Ann Entomol Soc Am 60:1074–1076

Moreau M, Catesson AM, Péresse M, Czaninski Y 1978 Dynamique comparée des réactions cytologiques du xylème de l'Oeillet en présence de parasites vasculaires. Phytopath Z 91:289–306

Morselli MF, Marvin JW, Laing FM 1978 Image-analyzing computer in plant science: more and larger vascular rays in sugar males of high sap and sugar yield. Can J Bot 56:983–986

Münch E 1930 Die Stoffbewegungen in der Pflanze. Fischer, Jena, 234 pp

Münch E 1943 Durchlässigkeit der Siebröhren für Druckströmungen. Flora 136:223–262

Mullick DB 1977 The non-specific nature of defense in bark and wood during wounding, insect and pathogen attack. In: Loewus FA, Runeckles VC (eds) The structure, biogenesis, and degradation of wood. Plenum, New York, 395–442

Murdoch CW 1979 A selected bibliography on bacterial wetwood of trees. Univ Maine School For Res Tech Note 72:1–7

Murdoch CW 1981 Bacterial wetwood in elm. PhD thesis, University of Maine, Orono (Plant Sci), 145 pp

Nakashima H 1924 Über den Einfluss meteorologischer Faktoren auf den Baumzuwachs. J Coll Agric (Tokyo) 12:71–262

Nardini A, Tyree MT 1999 Root and shoot hydraulic conductance of seven *Quercus* species. Ann For Sci 56:371–377

Nardini A, Tyree MT, Salleo S 2001 Xylem cavitation in the leaf of *Prunus laurocerasus* L. and its impact on leaf hydraulics. Plant Physiol 125:1700–1709

Newbanks D, Roy DN, Zimmermann MH 1982 Dutch elm disease: what an arborist should know. Arnoldia 42:60–69

Newbanks D, Bosch A, Zimmermann MH 1983 Evidence for xylem dysfunction by embolization in Dutch elm disease *Ceratocystis ulmi* on *Ulmus americana*. Phytopathology 73:1060–1063

Newman EI 1973 Permeability to water of five herbaceous species. New Phytol 72:547–555

Niklas KJ 1993a The scaling of plant height: a comparison among major plant clades and anatomical grades. Ann Bot 72:165–172

Niklas KJ 1993b Influence of tissue density-specific mechanical properties on scaling of plant height. Ann Bot 72:173–179

Nobel PS 1983 Biophysical plant physiology and ecology. Freeman and Co, New York

Nobel PS, Sanderson J 1984 Rectifier-like activities of roots of two desert succulents. J Exp Bot 35:727–737

North GB, Ewers FW, Nobel PS 1992 Main root–lateral root junctions of two desert succulents: changes in axial and radial components of hydraulic conductivity during drying. Am J Bot 79:1039–1050

O'Brien TP, Carr DJ 1970 A suberized layer in the cell walls of the bundle sheath of grasses. Aust J Biol Sci 23:275–287

O'Brien TP, Thimann KV 1967 Observations on the fine structure of the oat coleoptile. III. Protoplasma 63:443–478

Oertli JJ 1971 The stability of water under tension in the xylem. Z Pflanzenphysiol 65:195–209

Ohtani J, Ishida S 1976 Study on the pit of wood cells using scanning electron microscopy. Report 5. Vestured pits in Japanese dicotyledonous woods. Res Bull Coll Expt For Coll Agric Hokkaido Univ 33:407–436

Olien WC, Bukovac MJ 1982 Ethephon-induced gummosis in sour cherry (*Prunus cerasus* L). Plant Physiol 70:547–555, 556–559

O'Malley PER 1979 Mechanism of sap flow in maple. PhD dissertation, University of Glasgow, Glasgow, Scotland

O'Malley PER, Milburn JA 1983 Freeze-induced fluctuations in xylem sap pressure in *Acer pseudoplatanus*. Can J Bot 61:3100–3106

Pammenter NW, van der Willigen C 1998 A mathematical and statistical analysis of the curves illustrating vulnerability of xylem to cavitation. Tree Physiol 18:589–593

Panshin AJ, de Zeeuw C 1980 Textbook of wood technology, 4th edn. McGraw-Hill, New York, 722 pp

Parthasarathy MV 1974 Mycoplasma-like organisms associated with lethal-yellowing disease of palms. Phytopathology 64:667–674

Passioura JB 1972 The effect of root geometry on the yield of wheat growing on stored water. Aust J Agric Res 23:745–752

Pate JS, Canny MJ 1999 Quantification of vessel embolisms by direct observation: a comparison of two methods. New Phytol 141:33–43

Patiño S, Tyree MT, Herre EA 1995 A comparison of the hydraulic architecture of woody plants of differing phylogeny and growth form with special reference to free-standing and hemiepiphytic *Ficus* species from Panama. New Phytologist 129:125–134

Pearcy RW, Yang W 1996 A three-dimensional crown architecture model for assessment of light capture and carbon gain in understory plants. Oecologia 108:1–12

Péresse M, Moreau M 1968 Essai d'expression numérique des symptômes de la verticilliose sur oeillets. CR Soc Biot 162:234

Peterson CA, Cholewa E 1998 Structural modifications of the apoplast and their potential impact on ion uptake. Pflanzenerhähr Bodenk 161:521–531

Peterson CA, Murrmann M, Steudle E 1993 Location of major barriers to water and ion movement in young roots of *Zea mays* L. Planta 190:127–136

Petty JA 1978 Fluid flow through the vessels of birch wood. J Exp Bot 29:1463–1469

Petty JA 1981 Fluid flow through the vessels and intervascular pits of sycamore wood. Holzforschung 35:213–216

Pickard WF 1981 The ascent of sap in plants. Prog Biophys Mol Biol 37:181–229

Pickard WF 1989 How might a tracheary element which is embolized by day be healed by night? J Theor Biol 141:259–279

Pomerleau R, Mehran AR 1966 Distribution of spores of *Ceratocystis ulmi* labelled with phosphorus-32 in green shoots and leaves of *Ulmus americana*. Nat Gen Que 93:577–582

Praël E 1888 Vergleichende Untersuchungen über Schutz- und Kernholz der Laubbäume. Jahrb Wiss Bot 19:1–81

Preston RD 1952 Movement of water in higher plants. In: Frey-Wyssling (ed) Deformation and flow in biological systems. Elsevier, Amsterdam, 257–321

Pridgeon AM 1982 Diagnostic anatomical characters in the Pleurothallidinae (Orchidaceae). Am J Bot 69:921–938

Priestley JH, Scott LI, Malins ME 1933 A new method of studying cambial activity. Proc Leeds Philos Soc 2:365–374

Priestley JH, Scott LI, Malins ME 1935 Vessel development in the angiosperms. Proc Leeds Philos Soc 3:42–54

Putz FE, Mooney HA 1991 The biology of vines. Cambridge University Press, Cambridge

Raschke K 1975 Stomatal action. Annu Rev Plant Physiol 226:309–340

Reichenbach H von 1845 Untersuchungen über die zellenartigen Ausfüllungen der Gefässe. (Published anonymously) Z Bot 3:225–231, 241–253

Reicher K 1907 Die Kinematographie in der Neurologic. Neurol Zentralbl 26:496

Rein H 1928 Die Thermo-Stromuhr. Ein Verfahren welches mit etwa ±10 Prozent Genauigkeit die unblutige langdauernde Messung der mittleren Durchflussmengen an gleichzeitigen Gefässen gestattet. Z Biol 87:394–418

Reiner M 1960 Deformation, strain, and flow. An elementary introduction to theology. Lewis, London, 347 pp

Ren Z, Sucoff E 1995 Water movement through *Quercus rubra* L. Leaf water potential and conductance during polyclic growth. Plant Cell Environ 18:447–453

Renner O 1911 Experimentelle Beiträge zur Kenntnis der Wasserbewegung. Flora 103:173–247

Renner O 1915 Theoretisches und Experimentelles zur Kohäsionstheorie der Wasserbewegung. Jahrb Wiss Bot 56:617–667

Renner O 1925 Die Porenweite der Zellhäute und ihre Beziehung zum Saftsteigen. Ber Dtsch Bot Ges 43:207–211

Richter H 1974 Erhöhte Saugspannungswerte und morphologische Veränderungen durch transversale Einschnitte in einen Taxus-Stamm. Flora 163:291–309

Richter H 2001 The cohesion theory debate continues: the pitfalls of cryobiology. Trends Plant Sci 6:456–457

Richter JP 1970 The notebooks of Leonardo da Vinci (1452–1519), compiled and edited from the original manuscripts. Dover, New York. (Reprint of a work originally published by Sampson Low Marston Searle and Rivington London, 1883)

Riedl H 1937 Bau und Leistungen des Wurzelholzes (Structure and function of root wood). Jahrb Wiss Bot 85:1–75 (Xerox copies of English translation available from National Translation Center, 35 West 33rd St, Chicago, IL 60616, USA)

Roberts J 1977 The use of tree-cutting techniques in the study of the water relations of mature *Pinus sylvestris* L. J Exp Bot 28:751–767

Rodger EA 1933 Wound healing in submerged plants. Am Midl Nat 14:704–713

Rood SB, Patiño S, Coombs K, Tyree MT 2000 Branch sacrifice: cavitation-associated drought adaptation of riparian cottonwoods. Trees 14:248–257

Rouschal E 1937 Die Geschwindigkeit des Transpirationsstromes in Macchiengehölzen (thermoelektrische Messungen). Sitzgsber Akad Wiss Wien 146:119–133

Rouschal E 1940 Fluoreszenzoptische Messungen der Geschwindigkeit des Transpirationsstromes an krautigen Pflanzen mit Berücksichtigung der Blattspurleitflächen. Flora 134:229–256

Rouschal E 1941 Beiträge zum Wasserhaushalt von Gramineen und Cyperaceen. 1. Die fasikuläre Wasserleitung in den Blättern und ihre Beziehung zur Transpiration. Planta 32:66–87

Rübel E 1919 Experimentelle Untersuchungen über die Beziehungen zwischen Wasserleitungsbahn und Transpirationsverhältnissen bei *Helianthus annuus* L. Beih Bot Cbl 37:1–62

Rundel PW, Stecker RE 1977 Morphological adaptations of tracheid structure to water stress gradients in the crown of *Sequoiadendron giganteum*. Oecologia 27:135–139

Saliendra NZ, Sperry JS, Comstock JP 1995 Influence of leaf water status on stomatal response to hydraulic conductance, atmospheric drought, and soil drought in *Betula occidentalis*. Planta 196:357–366

Salisbury EJ 1913 The determining factors in petiolar structure. New Phytol 12:281–289

Salleo S, LoGullo MA 1989 Xylem cavitation in nodes and internodes of *Vitis vinifera* L. plants subjected to water stress: limits of restoration of water conduction in cavitated xylem conduits. In: Kreeb KH, Richter H, Hinckley TM (eds) Structural and functional responses to environmental stresses: water shortage. SPB Academic Publishing, The Hague, 33–42

Salleo S, LoGullo MA, De Paoli D, Zippo M 1996 Xylem recovery from cavitation-induced embolism in young plants of *Laurus nobilis*: a possible mechanism. New Phytol 132:47–56

Salleo S, Nardini A, Pitt F, LoGullo MA 2000 Xylem cavitation and hydraulic control of stomatal conductance in laurel (*Laurus nobilis* L.) Plant Cell Environ 23:71–79

Salleo S, LoGullo MA, Raimondo F, Nardini A 2001 Vulnerability to cavitation of leaf minor veins: any impact on gas exchange? Plant Cell Environ 24:851–859

Sanio K 1872 Über die Grösse der Holzzellen bei der gemeinen Kiefer (*Pinus silvestris*). Jahrb Wiss Bot 8:401–420

Sauter JJ 1966 Untersuchungen zur Physiologie der Pappelholzstrahlen. II. Z Pflanzenphysiol 55:349–362

Sauter JJ 1967 Der Einfluss verschiedener Temperaturen auf die Reservestärke in parenchym atischen Geweben von Baumsprossachsen. Z Pflanzenphysiol 56:340–352

Sauter JJ 1974 Maple. In: McGraw-Hill yearbook of science and technology. McGraw-Hill, New York, 270–281

Sauter JJ 1976 Analysis of the amino acids and amides in the xylem sap of *Salix caprea* L. in early spring. Z Pflanzenphysiol 79:276–280

Sauter JJ 1980 Seasonal variation of sucrose content in the xylem sap of *Salix*. Z Pflanzenphysiol 98:377–391

Sauter JJ, Iten W, Zimmermann MH 1973 Studies on the release of sugar into the vessels of sugar maple (*Acer saccharum*). Can J Bot 51:1–8

Sax K 1954 The control of tree growth by phloem blocks. J Arnold Arbor 35:251–258

Schmid R, Machado RD 1968 Pit membranes in hardwoods – fine structure and development. Protoplasma 66:185 204

Schneider H, Wistuba N, Wagner H-J, Thürmer F, Zimmermann U 2000 Water rise kinetics in refilling xylem after desiccation in a resurrection plant. New Phytol 148:221–238

Schneider IR, Worley JF 1959 Upward and downward transport of infectious particles of southern bean mosaic virus through steamed portions of bean stems. Virology 8:230–242

Scholander PF 1958 The rise of sap in lianas. In: Thimann KV (ed) The physiology of forest trees. Ronald Press, New York, 3–17

Scholander PF 1972 Tensile water. Am Sci 60:584–590

Scholander PF, Flagg W, Hock RJ, Irving L 1953 Studies on the physiology of frozen plants and animals in the arctic. J Cell Comp Physiol 42 (Suppl 1):1–56

Scholander PF, Love WE, Kanwisher JW 1955 The rise of sap in tall grapevines. Plant Physiol 30:93–104

Scholander PF, Hemmingsen E, Garey W 1961 Cohesive lift of sap in the rattan vine. Science 134:1835–1838

Scholander PF, Hammel HT, Bradstreet ED, Hemmingsen EA 1965 Sap pressures in vascular plants. Science 148:339–346

Schoute JC 1912 Über das Dickenwachstum der Palmen. Ann Jard Bot Buitenzorg 2e Ser 11:1–209

Schwendener S 1890 Die Mestomscheiden der Gramineenblätter. Sitzungsber Preuss Akad Wiss Phys Math KI 22:405–426

Scott FM 1950 Internal suberization of tissues. Bot Gaz 111:378

Sculthorpe CD 1967 The biology of aquatic vascular plants. Arnold, London, 610 pp

Shain L, Mackay JFG 1973 Seasonal fluctuation in respiration of aging xylem in relation to heartwood formation in *Pinus radiata*. Can J Bot 51:737–741

Shibata N, Harada H, Saiki H 1981 Difference in the development of incubated tyloses within the sapwood of *Castanea crenata* Sieb. et Zucc. Bull Kyoto Univ For 53:231–240

Shinozaki K, Yoda K, Hozumi K, Kira T 1964 A quantitative analysis of plant form – the pipe model theory. 1. Basic analyses 11. Further evidence of the theory and its application in forest ecology. Jpn J Ecol 14:97–105, 133–139

Sinclair WA, Campana RJ (eds) 1978 Dutch elm disease: perspectives after 60 years. Cornell Univ Agric Exp Stn Search Agric 8(5):1–52

Skene DS, Balodis V 1968 A study of vessel length in Eucalyptus obliqua L'Hérit. J Exp Bot 19:825–830

Slatyer RO 1968 Plant water relationships. Academic Press, London

Smith JAC, Nobel PS 1986 Water movement and storage in a desert succulent: anatomy and rehydration kinetics of leaves of *Agave deserti*. J Exp Bot 37:1044–1053

Snell K 1908 Untersuchungen über die Nahrungsaufnahme bei Wasserpflanzen. Flora 98:213–249

Spanner DC 1951 The Peltier effect and its use in the measurement of suction pressure. J Exp Bot 2:145–168

Sperry JS 1983 Observations on the structure and function of hydathodes in *Blechnum lehmannii* ieron. Am Fern J 73:65–72

Sperry JS 1985 Xylem embolism in the palm *Rhapis excelsa*. IAWA Bull NS 6:283–292

Sperry JS 1986 Relationship of xylem embolism to xylem pressure potential, stomatal closure, and shoot morphology in the palm *Rhapis excelsa*. Plant Physiol 80:110–116

Sperry JS 2000 Hydraulic constraints on plant gas exchange. Agric For Meteor 104:13–23

Sperry JS, Pockman WT 1993 Limitation of transpiration by hydraulic conductance and xylem cavitation in *Betula occidentalis*. Plan Cell Environ 16:279–287

Sperry JS, Saliendra NZ 1994 Intra- and inter-plant variation in xylem cavitation in *Betula occidentalis*. Plant Cell Environ 17:1233–1241

Sperry JS, Sullivan JEM 1992 Xylem embolism in response to freeze-thaw cycles and water stress in ring-porous, diffuse-porous, and coniferous species. Plant Physiol 100:605–613

Sperry JS, Tyree MT 1988 Mechanism of water-stress-induced xylem embolism. Plant Physiol 88:581–87

Sperry JS, Tyree MT 1990 Water-stress-induced xylem embolism in three species of conifers. Plant Cell Environ 13:427–436

Sperry JS, Holbrook NM, Zimmermann MH, Tyree MT 1987 Spring filling of xylem vessels in wild grapevine. Plant Physiol 83:414–417

Sperry JS, Donnelly JR, Tyree MT 1988a A method of measuring hydraulic conductivity and embolism in xylem. Plant Cell Environ 11:35–40

Sperry JS, Donnelly JR, Tyree MT 1988b Seasonal occurrence of xylem embolism in sugar maple (*Acer saccharum*). Am J Bot 75:1212–1218

Sperry JS, Adler FR, Eastlack SE 1993 The effect of reduced hydraulic conductance on stomatal conductance and xylem cavitation. J Exp Bot 44:1075–1082

Sperry JS, Nichols LK, Sullivan JEM, Eastlack SE 1994 Xylem embolism in ring-porous, diffuse-porous, and coniferous trees of northern Utah and interior Alaska. Ecology 75:1736–1752

Sperry JS, Adler FR, Campbell GS, Comstock JP 1998 Limitation of plant water use by rhizosphere and xylem conductance: results of a model. Plant Cell Environ 21:347–359

Stahl E 1897 Über den Pflanzenschlaf und verwandte Erscheinungen. Bot Ztg 55:71–108

Steudle E 1993 Pressure probe techniques: basic principles and application to studies of water and solute relations at the cell, tissue, and organ level. In: Smith JAC, Griffith H (eds) Water deficits: plant responses from cell to community. Bios Scientific Publishers, Oxford, 5–36

Steudle E, Peterson CA 1998 How does water get through roots? J Exp Bot 49:775–788

Steudle E, Oren R, Schulze E-D 1987 Water transport in maize roots. Plant Physiol 84:1220–1232

Stevens CL, Eggert RL 1945 Observations on the causes of flow of sap in red maple. Plant Physiol 20:636–648

Stipes RJ, Campana RJ (eds) 1981 Compendium of elm diseases. American Phytopathology Society, 96 pp

Stocker O 1928 In: Goebel (ed) Der Wasserhaushalt ägyptischer Wüsten- und Salzpflanzen. Bot Abhandl vol 2 13:1–200

Stocker O 1952 Grundriss der Botanik. Springer, Berlin Heidelberg New York

Strugger S 1940 Studien über den Transpirationsstrom im Blatt von Secale cereale und Triticum vulgare. Z Bot 35:97–113

Sucoff E 1969 Freezing of conifer xylem and the cohesion-tension theory. Physiol Plant 22:424–431

Suhayda CG, Goodman RN 1981 Infection courts and systemic movement of 32P labeled *Erwinia amylovora* in apple petioles and stems. Phytopathology 71:656–660

Sutton JC, Williams PH 1970 Relation of xylem plugging to black rot lesion development in cabbage. Can J Bot 48:391–401

Swanson RH, Whitfield WA 1981 A numerical analysis of heat pulse velocity theory and practice. J Exp Bot 32:221–239

Talbot AJB, Tyree MT, Dainty J 1975 Some notes concerning the measurement of water potentials of leaf tissue with specific reference to *Tsuga canadensis* and *Picea abies*. Can J Bot 53:784–788

Talboys PW 1978 Dysfunction of the water system. In: Horsfall JG, Cowling EB (eds) Plant disease: an advanced treatise, vol 3. Academic Press, New York, 141–162

Taylor G 1953 Dispersion of soluble matter in solvent flowing slowly through a Tube. Proc R Soc Lond A 219:186–203

Thimann KV, Satler SO 1979 Relation between leaf senescence and stomatal closure: senescence in light. Proc Natl Acad Sci USA 76:2295–2298

Thoday D, Sykes MG 1909 Preliminary observations on the transpiration current in submerged water-plants. Ann Bot 23:635–637

Thomas DL 1979 Mycoplasma-like bodies associated with lethal declines of palms in Florida. Phytopathology 69:928–934

Thürmer F, Zhu JJ, Gierlinger N, Schneider H, Benkert R, Gessner P, Herrmann B, Bentrup FW, Zimmermann U 1999 Diurnal changes in xylem pressure and mesophyll cell turgor pressure of the liana *Testrastigma voinierianum*: the role of cell turgor in long-distance water transport. Protoplasma 206:152–162

Thut HF 1932 The movement of water through some submerged plants. Am J Bot 19:693–709

Tippett JT, Shigo AL 1981 Barrier zone formation: a mechanism of tree defense against vascular pathogens. IAWA Bull 2:163–168

Tobiessen P, Rundel PW, Stecker RE 1971 Water potential gradient in a tall Sequoiadendron. Plant Physiol 48:303–304

Tomlinson PB 1978 Some qualitative and quantitative aspects of New Zealand divaricating shrubs. N Z J Bot 16:299–309

Tomlinson PB 1982 Anatomy of the monocotyledons. VII. Helobiae (Alismatidae). Clarendon, Oxford, 559 pp

Tomlinson PB, Fisher JB 2000 Stem vasculature in climbing monocotyledons: a comparative approach. In Wilson KL, Morrison DA (eds) Monocots: systematics and evolution. CSIRO, Melbourne, 89–97

Tomlinson PB, Zimmermann MH 1969 Vascular anatomy of monocotyledons with secondary growth – an introduction. J Arnold Arbor 50:159–179

Tomlinson PB, Fisher JB, Spangler RE, Richer RA 2001 Stem vascular architecture in the rattan palm *Calamus* (Arecaceae-Calamoideae-Calaminae). Am J Bot 88:797–809

Tomos AD 1988 Cellular water relations of plants. In: Franks F (ed) Water science reviews 3. Cambridge University Press, New York, 186–277

Traube M 1867 Experimente zur Theorie der Zellenbildung und Endosmose. Arch Anat Physiol Wiss Med 1867:87–165

Trendelenburg R 1939 Das Holz als Rohstoff. Lehmann, München

Tsuda M, Tyree MT 2000 Plant hydraulic conductance measured by the high pressure flow meter in crop plants. J Exp Bot 51:823–828

Tyerman SD, Bohnert H, Maurel C, Steudle E, Smith JAC 1999 Plant aquaporins: their molecular biology, biophysics and significance for plant water relations. J Exp Bot 50:1055–1071

Tyree MT 1969 The thermodynamics of short-distance translocation in plants. J Exp Bot 20:341–349

Tyree MT 1983 Maple sap uptake, exudation and pressure changes correlated with freezing exotherms and thawing endotherms. Plant Physiol 73:277–285

Tyree MT 1988 A dynamic model for water flow in a single tree: evidence that models must account for hydraulic architecture. Tree Physiol 4:195–217

Tyree MT 1993 Theory of vessel-length determination: the problem of nonrandom vessel ends. Can J Bot 71:297–302

Tyree MT 1999a The forgotten component of plant water potential. A reply – tissue pressures are not additive in the way M.J. Canny suggests. Plant Biol 1:598–601

Tyree MT 1999b Water relations and hydraulic architecture. In: Pugnaire FI, Valladares F (eds) Handbook of functional plant ecology. Marcel Dekker, New York, 221–268

Tyree MT, Alexander JD 1993 Hydraulic conductivity of branch junctions in three temperate tree species. Trees 7:156–159

Tyree MT, Cameron SI 1977 A new technique for measuring oscillatory and diurnal changes in leaf thickness. Can J For Res 7:540–544

Tyree MT, Dixon MA 1986 Water stress induced cavitation and embolism in some woody plants. Physiol Plant 66:397–405

Tyree MT, Ewers FW 1991 The hydraulic architecture of trees and other woody plants. New Phytol 119:345–360

Tyree MT, Ewers FW 1996 Hydraulic architecture of woody tropical plants. In: Smith A, Winter K, Mulkey S (eds) Tropical plant ecophysiology. Chapman and Hall, New York, 217–243

Tyree MT, Jarvis PG 1982 Water in tissues and cells. In: Lange OL, Nobel PS, Osmond CB, Ziegler H (eds) Physiological plant ecology II. Water relations and carbon assimilation. Encyclopedia of plant physiology, new series, vol 12B. Springer, Berlin Heidelberg New York, 35–77

Tyree MT, Sperry JS 1988 Do woody plants operate near the point of catastrophic xylem dysfunction caused by dynamic water stress? Answers from a model. Plant Physiol 88:571–580

Tyree MT, Sperry JS 1989a Vulnerability of xylem to cavitation and embolism. Annu Rev Pl Physiol Mol Biol 40:19–38

Tyree MT, Sperry JS 1989b Characterization and propagation of acoustic emission signals in woody plants: towards an improved acoustic emission counter. Plant Cell Environ 12:371–382

Tyree MT, Tammes PML 1975 Translocation of uranin in the symplasm of staminal hairs of Tradescantia. Can J Bot 53:2038–2046

Tyree MT, Wilmot TR 1990 Errors in the calculation of evaporation and leaf conductance in steady-state porometry: the importance of accurate measurement of leaf temperature. Can J For Res 20:1031–1035

Tyree MT, Yang S 1990 Water-storage capacity of Thuja, Tsuga and Acer stems measured by dehydration isotherms. The contribution of capillary water and cavitation. Planta 182:420–426

Tyree MT, Yang Y 1992 Hydraulic conductivity recovery versus water pressure in xylem of Acer saccharum. Plant Physiol 100:669–676

Tyree MT, Yianoulis P 1980 The site of water evaporation from sub-stomatal cavities, liquid path resistances and hydroactive stomatal closure. Ann Bot 46:175–193

Tyree MT, Yianoulis P 1984 A model to investigate the effects of evaporative cooling on the pattern of evaporation in sub-stomatal cavities. Ann Bot 53:189–201

Tyree MT, Zimmermann MH 1971 The theory and practice of measuring transport coefficients and sap flow in the xylem of red maple (Acer rubrum). J Exp Bot 22:1–18

Tyree MT, Benis M, Dainty J 1973 The water relations of hemlock (Tsuga canadensis). III. The temperature dependence of water exchange in a pressure-bomb. Can J Bot 51:1537–1543

Tyree MT, Caldwell C, Dainty J 1975 The water relations of hemlock (Tsuga canadensis). V. The localization of resistances to bulk water flow. Can J Bot 53:1078–1084

Tyree MT, Graham MED, Cooper KE, Bazos LJ 1983 The hydraulic architecture of Thuja occidentalis. Can J Bot 61:2101–2111

Tyree MT, Dixon MA, Tyree EL, Johnson R 1984 Ultrasonic acoustic emissions from the sapwood of cedar and hemlock: an examination of three hypotheses regarding cavitations. Plant Physiol 75:988–992

Tyree MT, Snyderman DA, Wilmot TR, Machado JL 1991 Water relations and hydraulic architecture of a tropical tree (*Schefflera morototoni*): data, models and a comparison to two temperate species (*Acer saccharum* and *Thuja occidentalis*). Plant Physiol 96:1105–1113

Tyree MT, Alexander JD, Machado JL 1992 Loss of hydraulic conductivity due to water stress in intact juveniles of *Quercus rubra* and *Populus deltoids*. Tree Physiol 10:411–415

Tyree MT, Cochard H, Cruiziat P, Sinclair B, Ameglio T 1993 Drought-induced leaf shedding in walnut: evidence for vulnerability segmentation. Plant Cell Environ 16:879–882

Tyree MT, Davis SD, Cochard H 1994 Biophysical perspectives of xylem evolution: is there a tradeoff of hydraulic efficiency for vulnerability to dysfunction? IAWA J NS 15:335–360

Tyree MT, Patiño S, Bennink J, Alexander J 1995 Dynamic measurements of root hydraulic conductance using a high-pressure flowmeter for use in the laboratory or field. J Exp Bot 46:83–94

Tyree MT, Velez V, Dalling JW 1998 Growth dynamics of root and shoot hydraulic conductance in seedlings of five neotropical tree species: scaling to show possible adaptation to differing light regimes. Oecologia 114:293–298

Tyree MT, Salleo S, Nardini A, LoGullo MA, Mosca R 1999a Refilling of embolised vessels in young stems of laurel: do we need a new paradigm? Plant Physiol 120:11–21

Tyree MT, Sobrado MA, Stratton LJ, Becker P 1999b Diversity of hydraulic conductance in leaves of temperate and tropical species: possible causes and consequences. J Trop For Sci 11:47–60

Tyree MT, Nardini A, Salleo S 2001 Hydraulic architecture of whole plants and single leaves. In: Quentin I (ed) L'Arbre 2000. The tree. IQ Press, Montreal, 215–221

Unger F 1861 Beiträge zur Anatomic und Physiologie der Pflanzen. Sitzungsber Math Naturwiss KI Akad Wiss Wien 44 Bd 2:327–368

Ursprung A 1913 Über die Bedeutung der Kohäsion für das Saftsteigen. Ber Dtsch Bot Ges 31:401–412

Ursprung A 1915 Über die Kohäsion des Wassers im Farnanulus. Ber Dtsch Bot Ges 33:153–162

Valverde RA, Fulton JP 1982 Characterization and variability of strains of southern bean mosaic virus. Phytopathology 72:1265–1268

VanAlfen NK, MacHardy WE 1978 Symptoms and host-pathogen interactions. In: Sinclair WA, Campana RJ (eds) Dutch elm disease: perspectives after 60 years. Cornell Univ Exp Stn Search Agric 8(5):20–25

Van Bel A 1993 Strategies of phloem loading. Ann Rev Pl Physiol Mol Biol 44:253–281

van den Honert TH 1948 Water transport in plants as a catenary process. Disc Farad Soc 3:146–153

van der Graaff NA, Baas P 1974 Wood anatomical variation in relation to latitude and attitude. Blumea 22:101–121

van Leperen W, van Meeteren U, van Gelder H 2000 Fluid ionic composition influences hydraulic conductance of xylem conduits J Exp Bot 51:769–776

van Steenis CGGJ 1969 Plant speciation in Malesia, with special reference to the theory of non-adaptive saltatory evolution. Biol J Linn See 1:97–133

van Vliet GJCM 1976 Radial vessels in rays. IAWA Bull 3:35–37

van Vliet GJCM 1978 Vestured pits of Combretaceae and allied families. Acta Bot Neerl 27:273–285

Vignes M, Dukema KM 1974 A model for the freezing of water in a dispersed medium. J Colloid Interface Sci 49:165–172

Vité JP, Rudinsky JA 1959 The water-conducting system in conifers and their importance to the distribution of trunk-injected chemicals. Contrib Boyce Thompson Inst 20:27–38

von Faber FC 1915 Physiologische Fragmente aus einem tropischen Urwald. Jahrb Wiss Bot 56:197–220

von Minden M 1899 Beiträge zur anatomischen und physiologischen Kenntnis Wasser-sezernierender Organe. Bibl Bot 9(46):1–76

Waisel Y, Agami M, Shapira Z 1982 Uptake and transport of ^{86}Rb, ^{32}P, [36]CI, and [22]Na by four submerged hydrophytes. Aquatic Bot 13:179–186

Wei C, Tyree MT, Steudle E 1999 Direct measurement of xylem pressure in leaves of intact maize plants. A test of the cohesion-tension theory taking hydraulic architecture into consideration. Plant Physiol 121:1191–1205

Wei C, Tyree MT, Bennink JP 2000 The transmission of gas pressure to xylem fluid pressure when plants are inside a pressure bomb. J Exp Bot 51:309–316

Weiner G 1992 Zur stammanatomic der Rattanpalmen. PhD dissertation, University of Hamburg, Germany

Weiner G, Liese W 1992 Zellarten und Faserlängen innerhalb des Stammes verschiedenen Rattansgattungen. Holz Roh Werkst 50:457–464

Weiner G, Liese W 1993 Generic identification key to rattan palms based on stem anatomical characters. IAWA J 14:55–61

West DW, Gaff DF 1976 Xylem cavitation in excised leaves of *Malus sylvestris* Mill and measurement of leaf water status with the pressure chamber. Planta 129:15–18

Wiegand KM 1906 Pressure and flow of sap in the maple. Am Nat 40:409–453

Wieler A 1888 Über den Anteil des sekundären Holzes der dikotylen Gewächse an der Saftleitung und über die Bedeutung der Anastomosen für die Wasserversorgung der transpirierenden Flächen. Jahrb Wiss Bot 19:82–137

Wieler A 1893 Das Bluten der Pflanzen. Beitr Biol Pflanz 6:1–211

Wilson K 1947 Water movements in submerged aquatic plants, with special reference to cut shoots of *Ranunculus fluitans*. Ann Bot 11:91–122

Woodhouse RM, Nobel PS 1982 Stipe anatomy, water potentials, and xylem conductances in seven species of ferns (Filicopsida). Am J Bot 69:135–140

Worrall JJ, Parmeter JR Jr 1982 Formation and properties of wetwood in white fir. Phytopathology 72:1209–1212

Yang S, Tyree MT 1992 A theoretical model of hydraulic conductivity recovery from embolism with comparison to experimental data on *Acer saccharum*. Plant Cell Environ 15:633–643

Yang Y, Tyree MT 1993 Hydraulic resistance in the shoots of *Acer saccharum* and its influence on leaf water potential and transpiration. Tree Physiol 12:231–242

Yang S, Tyree MT 1994 Hydraulic architecture of *Acer saccharum* and *A. rubrum*: comparison of branches to whole trees and the contribution of leaves to hydraulic resistance. J Exp Bot 45:179–186

Zajaczkowski S, Wodzicki T, Romberger JA 1983 Auxin waves and plant morphogenesis. In: Scott TK (ed) Functions of hormones in growth and development at levels of organization from the cell up to the whole plant. Encyclopedia of plant physiology, new series, vol 10. Springer, Berlin Heidelberg New York

Ziegenspeck H 1928 Zur Theorie der Wachstums- und Bewegungserscheinungen bei Pflanzen. Bot Arch 21:449–647

Ziegler H 1967 Biologische Aspekte der Kernholzbildung. Proc 14th IUFRO Congr Munich 9:93–116

Ziegler H 1968 Biologische Aspekte der Kernholzbildung. Holz Roh Werkst 26:61–68

Zimmermann MH 1960 Longitudinal and tangential movement within the sieve-tube system of white ash (*Fraxinus americana* L.). Beih Z Schweiz Forstver 30:289–300

Zimmermann MH 1964 Effect of low temperature on ascent of sap in trees. Plant Physiol 39:568–572

Zimmermann MH 1965 Water movement in stems of tall plants. In: 19th Symposium of the Soc for Exp Biol. The state and movement of water in living organisms. Fogg GE (ed) Cambridge University Press, Cambridge, 151–155

Zimmermann MH 1971 Dicotyledonous wood structure made apparent by sequential sections. Film E 1735 (Film data and summary available as a reprint) Inst wiss Film, Nonnenstieg 72, Göttingen, Germany

Zimmermann MH 1976 The study of vascular patterns in higher plants. In: Wardlaw IF, Passioura JB (eds) Transport and transfer processes in plants. Academic Press, New York, 221–235

Zimmermann MH 1978 Hydraulic architecture of some diffuse-porous trees. Can J Bot 56:2286–2295

Zimmermann MH 1979 The discovery of tylose formation by a Viennese lady in 1845. IAWA Bull 2-3:51–56

Zimmermann MH 1983 Xylem structure and the ascent of sap, 1st edn. Springer, Berlin Heidelberg New York

Zimmermann MH, Brown CL 1971 Trees. Structure and function. Springer, Berlin Heidelberg New York, 336 pp

Zimmermann U, Hüsken D 1979 Theoretical and experimental exclusion of errors in the determination of the elasticity and water transport parameters of plant cells by the pressure probe technique. Plant Physiol 64:18–24

Zimmermann MH, Jeje AA 1981 Vessel-length distribution in stems of some American woody plants. Can J Bot 59:1882–1892

Zimmermann MH, Mattmuller MR 1982a The vascular pattern in the stem of the palm *Rhapis excelsa*. I. The mature stem. Film C 1404. (Film data and summary available as a reprint) Inst wiss Film, Nonnenstieg 72, Göttingen, Germany

Zimmermann MH, Mattmuller MR 1982b The vascular pattern in the stem of the palm *Rhapis excelsa*. II. The growing tip. Film D 1418. (Film data and summary available as a reprint) Inst wiss Film, Nonnenstieg 72, Göttingen, Germany

Zimmermann MH, McDonough J 1978 Dysfunction in the flow of food. In: Horsfall JG, Cowling EB (eds) Plant disease. An advanced treatise, vol 3. Academic Press, New York, 117–140

Zimmermann MH, Potter D 1982 Vessel-length distribution in branches, stem, and roots of *Acer rubrum*. IAWA Bull 3:103–109

Zimmermann MH, Sperry JS 1983 Anatomy of the palm *Rhapis excelsa*. IX. Xylem structure of the leaf insertion. J Arnold Arbor 64:599–609

Zimmermann U, Steudle E 1974 The pressure dependence of the hydraulic conductivity, the membrane resistance and membrane potential during turgor pressure regulation. J Membr Biol 16:331–352

Zimmermann MH, Tomlinson PB 1965 Anatomy of the palm *Rhapis excelsa*. I. Mature vegetative axis. J Arnold Arbor 46:160–180

Zimmermann MH, Tomlinson PB 1966 Analysis of complex vascular systems in plants: optical shuttle method. Science 152:72–73

Zimmermann MH, Tomlinson PB 1967 Anatomy of the palm *Rhapis excelsa*. IV. Vascular development in apex of vegetative aerial axis and rhizome. J Arnold Arbor 48:122–142

Zimmermann MH, Tomlinson PB 1968 Vascular construction and development in the aerial stem of *Prionium* (Juncaceae). Am J Bot 55:1100–1109

Zimmermann MH, Tomlinson PB 1969 The vascular system of *Dracaena fragrans* (Agavaceae). I. Distribution and development of primary strands. J Arnold Arbor 50:370–383

Zimmermann MH, Tomlinson PB 1970 The vascular system of *Dracaena fragrans* (Agavaceae). II. Distribution and development of secondary vascular tissue. J Arnold Arbor 51:478–491

Zimmermann MH, Tomlinson PB 1974 Vascular patterns in palm stems: variations of the Rhapis principle. J Arnold Arbor 55:402–424

Zimmermann U, Räde G, Steudle E 1969 Kontinuierliche Druckmessung in Pflanzenzellen. Naturwissenschaften 56:634

Zimmermann MH, Tomlinson PB, LeClaire J 1974 Vascular construction and development in the stems of certain Pandanaceae. Bot J Linn Soc 68:21–41

Zimmermann MH, McCue KF, Sperry JS 1982 Anatomy of the palm *Rhapis excelsa*. VIII. Vessel network and vessel-length distribution in the stem. J Arnold Arbor 63:83–95

Zotz G, Tyree MT, Patiño S, Carlton MR 1998 Hydraulic architecture and water use of selected species from a lower montane forest in Panama. Trees 12:302–309

Zweypfenning RCVJ 1978 A hypothesis on the function of vestured pits. IAWA Bull 1:13–15

Zwieniecki MA, Melcher PJ, Holbrook NM 2001 Hydrogel control of xylem hydraulic resistance in plants. Science 291:1059–1062

Subject Index

Printing (Computer to Film): Saladruck Berlin
Binding: Stürtz AG, Würzburg